CAMBRIDGE TRACTS IN MATHEMATICS

General Editors

B. BOLLOBAS, P. SARNAK, C. T. C. WALL

103   Designs and their codes

E. F. ASSMUS, JR.
Department of Mathematics, Lehigh University

J. D. KEY
Department of Mathematical Sciences, Clemson University

Designs and their codes

CAMBRIDGE
UNIVERSITY PRESS

Published by the Press Syndicate of the University of Cambridge
The Pitt Building, Trumpington Street, Cambridge CB2 1RP
40 West 20th Street, New York, NY 10311-4211, USA
10 Stamford Road, Oakleigh, Victoria 3166, Australia

First published 1992

First published in paperback (with corrections) 1993

Library of Congress cataloguing in publication data available
British Library cataloguing in publication data available

ISBN 0 521 41361 3   hardback
ISBN 0 521 45839 0   paperback

Transferred to digital printing 2004

Preface

The aim of this book is to study applications of algebraic coding theory to the analysis and classification of designs. Designs are usually classified by their parameter sets, by their inclusion in infinite families, or according to the type of automorphism group they admit — or in other ways related to their geometric properties. Here we have tried to confront the following questions. To what extent can algebraic coding theory address classification questions concerning designs? Can it assist in their construction? Does it provide insight into the structure and nature of classes of designs? We have tried to outline ways in which these questions have been answered and to point to areas where coding theory can make a valuable contribution to design theory.

These matters occupy the last three chapters of the book; the first five chapters are of a background nature and form an introduction to both the theory of designs and algebraic coding theory. Although parts of these chapters are elementary we hope they will be of interest even to the expert. We have rethought much of the material and some of the development is new. We have not always given complete proofs and in several instances we have given no proof at all. On the other hand there are occasions where we have given more than one proof for the same result. Our guiding principle has been to use the proofs as didactic aids rather than verifications of the assertions; in particular, where the proof in the literature is clear and easily available but would add little to the exposition, we have omitted it.

We have included a glossary of terms and symbols that we hope will aid the expert who wishes to jump into one of the last three chapters without consulting the first five. Thus, for example, someone interested only in what coding theory might have to say about Hadamard matrices will be able to begin by reading Chapter 7 — using the glossary and index in the event that a concept or the notation is unfamiliar. Similarly, someone interested in projective planes — and their coding-theoretic habitat — ought to be able to go directly to Chapter 6 and the Steiner-system expert directly to Chapter 8.

The numbering is standard and is consecutive within sections: thus the third exercise of the second section of the first chapter is Exercise 1.2.3. The theorems, propositions, corollaries, lemmas and examples are similarly listed. In the **Bibliography** the abbreviations for the titles of journals are those used by *Mathematical Reviews*; the full journal title is given in the event *Mathematical Reviews* does not list an abbreviation or the abbreviation might prove confusing. Volume numbers are in Arabic numerals in keeping with current library custom. We have tried to be accurate in reporting the historical developments and assigning credit; moreover, we have cited all the works that we have consulted. We apologize in advance for any omissions and slights that perhaps remain.

The book was begun on 11 February 1990; the construction of the "manuscript" and the collaborative effort was almost exclusively electronic, using the EMACS editor in conjuction with LaTeX and relying on NSFNET to transfer files between us. We have, for this second printing, updated the bibliography and corrected all the typographical errors that have come to our attention since publication, but otherwise the book is essentially the same as the original.

Contents

List of Figures

List of Tables

Chapter 1

Designs

1.1 Introduction

We begin by discussing those basic concepts from design theory needed in our development of the use of linear codes as an aid in organizing and classifying designs. The more specific properties of those designs that we use as examples of this development will follow in the chapters devoted to the class of designs in question. Much of the material in this chapter is now quite standard and treatments can be found in related books on designs and geometries, in particular the books of Beth, Jungnickel and Lenz [37], Dembowski [85], Hall [122], and Hughes and Piper [144]. The books of Cameron [63], Cameron and van Lint [64], and Tonchev [281] are also useful for related material, and Batten [31] has an elementary account that highlights the geometry. We restrict our attention to *finite* structures, however, and whenever a set, group, or other mathematical structure is mentioned the reader should assume it to be finite. For background material the books by Wielandt [296] and Passman [232] can be consulted for permutation groups, and that by Lidl and Niederreiter [185] for Galois theory and finite fields.

1.2 Basic definitions

The most basic structure of the theory is a **finite incidence structure** which we denote by $\mathcal{S} = (\mathcal{P}, \mathcal{B}, \mathcal{I})$, and which consists of two disjoint finite sets $\mathcal{P}$ and $\mathcal{B}$, and a subset $\mathcal{I}$ of $\mathcal{P} \times \mathcal{B}$. The members of $\mathcal{P}$ are called **points** and are generally denoted by lower-case Roman letters; the members of $\mathcal{B}$ are called **blocks** and are generally denoted by upper-case Roman letters. If the ordered pair (p, B) is in $\mathcal{I}$ we say that p is **incident** with B, or that

1

B contains the point p, or that p is on B, using the general phraseology of geometry, and viewing the incidence structure geometrically. The pair (p, B) is called a **flag** if it is in $\mathcal{I}$, an **anti-flag** if not.

Example 1.2.1 Let $\mathcal{P}$ be any set and take $\mathcal{B}$ to be any subset of the power set $2^{\mathcal{P}}$, the set of all subsets of $\mathcal{P}$. Define incidence by $(p, B) \in \mathcal{I}$ if and only if $p \in B$.

As this example shows, it is sometimes convenient to identify a block of an incidence structure $\mathcal{S}$ with the set of points incident with it, and we may write $p \in B$ or $B \ni p$ instead of $(p, B) \in \mathcal{I}$. In the event that a block of $\mathcal{S}$ becomes a block of a new structure it may then be incident with a different set of points, and hence the abstract definition is needed. Sometimes we will not mention the set $\mathcal{I}$ at all, particularly when we have an instance of the above example with $\mathcal{B}$ a collection of subsets of $\mathcal{P}$ and $\mathcal{I}$ the membership relation.

We will be concerned almost exclusively with those structures that have a particular degree of regularity and that are traditionally called *block designs*, a term that arose in the statistical literature. We will simply call them **designs**, or **t-designs** when the degree of regularity is to be emphasized. Here is the formal definition:

Definition 1.2.1 *An incidence structure $\mathcal{D} = (\mathcal{P}, \mathcal{B}, \mathcal{I})$ is a t-(v, k, λ)* **design***, or simply a t-**design***, where t, v, k and λ are non-negative integers, if*

(1) $|\mathcal{P}| = v$;

(2) every block $B \in \mathcal{B}$ is incident with precisely k points;

(3) every t distinct points are together incident with precisely λ blocks.

Example 1.2.1 above is a t-design provided that $|B|$, the cardinality of the subset B of $\mathcal{P}$, is k for every subset $B \in \mathcal{B}$ and that, for every subset T of $\mathcal{P}$ of cardinality t, $|\{B \in \mathcal{B} | B \supseteq T\}| = \lambda$. Thus the blocks all have the same cardinality and every t-subset (i.e. a subset of cardinality t) is contained in the same number of blocks.

The non-negative integers t, v, k and λ are referred to as the **parameters** of the design and we will sometimes refer to a t-(v, k, λ) design as a "design with parameters t-(v, k, λ)". It is possible that distinct blocks could be incident with the same set of points, but for the vast majority of designs appearing in this book we shall not have such so-called **repeated blocks**. In the literature, since structures with repeated blocks can be important, a *t*-design without repeated blocks is distinguished by calling it a **simple** *t*-design, and the further property that if two blocks are incident with the same set of *k* points, they are equal, is often included in the definition. Here the designs we treat are *always simple* and we will omit this qualifying adjective. When $t = 2$ the restriction that all blocks have the same cardinality is sometimes relaxed and such designs are called **pairwise-balanced** designs; although they are quite important, especially in recursive construction techniques, we shall make no use of the notion.

Although the definition does not demand it, we shall almost always have $0 \leq t \leq k$. As is customary, we set $b = |\mathcal{B}|$. A design is called **trivial** if every set of *k* points is incident with a block,[1] in which case $b = \binom{v}{k}$. A 1-design is called a **tactical configuration**; a non-trivial 2-design is known as a **balanced incomplete block design** (BIBD) in the statistical literature (and in much of the combinatorial literature as well). For 2-designs another notation that is frequently used, but which we will not employ at all, includes more of the parameters, thus (b, v, r, k, λ), where *r* is the **replication number** and denotes the number of blocks incident with a point. This number is independent of the point chosen, for, as we will see below, any *t*-design is an *s*-design for any $s \leq t$. The parameters shown in this notation are not independent, and the most crucial parameter has not yet appeared. If $t = \lambda = 1$, a design is simply a partition of the point set into subsets of cardinality *k*. It follows therefore that in this case the parameters are constrained by the condition that *k* must divide *v*; we shall soon see the generalization of this simple fact to arbitrary *t*-designs.

If $k = 2$, a t-(v, k, λ) design is a **graph** (undirected, no loops), and points are called **vertices** and blocks are called **edges**. Two vertices on the same edge are referred to as **adjacent**. A graph is **regular** with **valency** *r* if $t \geq 1$, and **complete** if $t = 2$ (i.e. all possible edges are present); similarly it is called **null** if it has no edges. A **bipartite** graph is one in which the vertex set is the disjoint union of two sets such that no two vertices in any one of these sets are adjacent, i.e. any edge has exactly one vertex from each of the sets. Of particular relevance to design theory are **strongly regular graphs** : a regular graph which is neither complete nor null is strongly regular if the number of vertices adjacent to both of two distinct vertices *p* and *q* depends only on whether *p* and *q* are adjacent or not. Its parameters are sometimes specified by the ordered 4-tuple (n, a, λ, μ) where *n* is the number of vertices, *a* the valency, λ the number of vertices adjacent to both of two adjacent vertices, and μ the number of vertices adjacent to both of two non-adjacent vertices.

---

[1] When there are repeated blocks a trivial design is one for which every set of *k* points is incident with the same number of blocks.

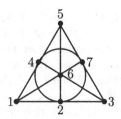

Figure 1.1: The Fano plane

Exercise 1.2.1 If Γ is a strongly regular graph with parameter set (v, k, λ, λ), show that a 2-(v, k, λ) design $\mathcal{D}$ can be defined in the following way: for the points of $\mathcal{D}$ take the vertices of Γ; for each point p define a block B_p to consist of all the points (vertices) adjacent to p. Then this set of points and blocks defines a 2-design with equally many points and blocks, a so-called symmetric design.

A t-(v, k, λ) design is also sometimes described as an $S_\lambda(t, k, v)$ design in the literature: see, for example, Beth *et al.* [37]. If $t \geq 2$ and $\lambda = 1$, then a t-design is called a **Steiner system** or Steiner t-design, and $S_1(t, k, v)$ becomes $S(t, k, v)$. We shall make no further use of the $S(t, k, v)$ notation. For Steiner systems with $t = 2$, blocks are often called **lines**, since any two points determine a unique block. More generally, and for the same reason, an incidence structure is called a **linear space** if every two points are on a unique block, or a **partial linear space** if any two points are on at most one block: see Doyen [98] for a survey article on linear spaces and Steiner systems. For a 2-(v, k, λ) design in general, a **line** is defined to be any set of points that is the intersection of all blocks that contain both of a pair of distinct points of the design, where here we are regarding blocks as sets of points. Thus for Steiner 2-designs the lines are essentially the blocks, and the terminology is consistent.

Example 1.2.2   (1) Let $\mathcal{P} = \{1, 2, 3, 4, 5, 6, 7\}$ and set

$$\mathcal{B} = \{\{1, 2, 3\}, \{1, 4, 5\}, \{1, 6, 7\}, \{2, 4, 7\}, \{2, 5, 6\}, \{3, 5, 7\}, \{3, 4, 6\}\}.$$

Then $(\mathcal{P}, \mathcal{B})$ with the natural incidence forms a 2-(7,3,1) design. This is the well-known Fano plane, a Steiner system, and is the smallest design arising from a projective geometry. It is an example of many classes of designs and will appear frequently throughout the book. Its usual representation is by the diagram shown in Figure 1.1, where the labelling coincides with that given above.

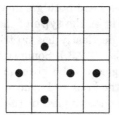

Figure 1.2: The block $B_{3,2}$ of the biplane

(2) Consider a 4×4 array and let the point set $\mathcal{P}$ be the 16 positions (i, j) of this array. For each (i, j) define a block B_{ij} to be incident with the six points (i, k) for $k \neq j$ and (k, j) for $k \neq i$. Figure 1.2 indicates a typical block. Then $|\mathcal{B}| = |\mathcal{P}| = 16$, every two points are incident with two blocks, and $(\mathcal{P}, \mathcal{B})$ is a 2-(16,6,2) design. It is known as a **biplane of order 4**: see page 120 for a general definition. (Notice that an analogous construction with an $n \times n$ array where $n \neq 4$, will only give a 1-design.)

(3) Consider the complete graph on six vertices. Let $\mathcal{P}$ be the set of 15 edges together with another point which we denote by ∞. The blocks containing the point ∞ will consist of ∞ and the five edges of a 5-claw, i.e. the five edges containing a fixed vertex of the graph. The blocks not containing ∞ will consist of the edges of those subgraphs of the complete graph which are two disjoint triangles. It is not difficult to see that once again we have a 2-$(16, 6, 2)$ design; it is not as easy to see that it is essentially the same as the one above.

(4) Let G be a t-transitive permutation group on a set $\mathcal{P}$. With $\mathcal{B}$ the union of any collection of orbits of G on k-subsets, where $k > t$, a t-$(|\mathcal{P}|, k, \lambda)$ design is obtained. If $k = t$, the design is trivial.

(5) A 2-$(n^2 + n + 1, n + 1, 1)$ design, for $n \geq 2$, is called a **projective plane of order n**. (The Fano plane is a projective plane of order 2.) Not only do any two points lie on a unique line, but it is also a consequence of the nature of the parameters that any two lines meet in a unique point.

(6) A 1-(v, k, λ) design with the property that any two points are together incident with *at most* one block, and that for x a point and B a block not containing x there is precisely one point on B that is on a block together with x, is called a **generalized quadrangle** provided that

$k = s + 1$ and $\lambda = t + 1$ are both greater than 1. Then the pair (s, t) is generally referred to as the **type** of the generalized quadrangle. A generalized quadrangle clearly has no triangles, but plenty of quadrangles. (See Exercise 1.2.3.) Some standard classical examples of generalized quadrangles will be given in Chapter 3 (Example 3.7.2, Exercise 3.7.1): see the book of Payne and Thas [233] for more properties of generalized quadrangles, or the survey article of Thas [272] for more examples and a general bibliography.

Example 1.2.2 (3) above illustrates an important theme of finite geometry in its group-theoretical aspect: the transitive extension of a permutation group. The doubly-transitive group which is the automorphism group of the biplane of order 4 that has been constructed is a transitive extension of the symmetric group on six letters viewed as a permutation group on 15 letters, the $\binom{6}{2}$ 2-subsets of the six letters. The following exercise reveals an even simpler example of such a transitive extension, in this case of the symmetric group on four letters viewed as a permutation group on six letters, once again a set of 2-subsets, but now of the four letters. The resulting transitive extension is the automorphism group of the Fano plane, a group of order $168 = 7 \times 24$.

Exercise 1.2.2 Consider the complete graph on four vertices. Let $\mathcal{P}$ be the set of six edges together with another point which we denote by ∞. Construct a biplane of order 2 as follows: the blocks containing ∞ consist of ∞ together with the three edges of a 3-claw of the complete graph, one for each vertex. The blocks not containing ∞ consist of the four edges of the squares of the complete graph, one for each square. Show that the resulting incidence structure is a 2-$(7, 4, 2)$ design, i.e. a biplane of order 2. How is this structure related to the Fano plane?

We next explore the most elementary consequences of Definition 1.2.1 and add to the array of parameters that are naturally associated with a design.

Theorem 1.2.1 *Let* $\mathcal{D} = (\mathcal{P}, \mathcal{B})$ *be a* t-(v, k, λ) *design. Then for every integer* s *such that* $0 \leq s < t$, *the number* λ_s *of blocks incident with* s *distinct points is independent of the* s *points, and is given by*

$$\lambda_s = \lambda \frac{(v - s)(v - s - 1) \cdots (v - t + 1)}{(k - s)(k - s - 1) \cdots (k - t + 1)}.$$

In particular $\mathcal{D}$ *is an* s-(v, k, λ_s) *design for every* s *with* $1 \leq s \leq t$.

Proof: Let S be a set of s points, where $0 \le s < t$. Let m be the number of blocks that contain every point of S. Let $\mathcal{T} = \{(T, B) | S \subset T \subseteq B, |T| = t, B \in \mathcal{B}\}$. Count $|\mathcal{T}|$ in two ways to get:

$$\lambda \binom{v - s}{t - s} = m \binom{k - s}{t - s},$$

which shows that m is independent of S and given by the formula in the theorem. $\square$

Setting $\lambda_t = \lambda$ the theorem shows that the integers λ_s are most easily calculated from the recursion

$$\lambda_s = \frac{(v - s)}{(k - s)} \lambda_{s+1}.$$

For example, we shall see later (Corollary 7.4.3) that there is a design with parameters 5-$(12, 6, 1)$ and hence, not only for that design but for any design with these parameters, it follows that $\lambda_4 = 4, \lambda_3 = 12, \lambda_2 = 30, \lambda_1 = 66$ and the number of blocks is $b = \lambda_0 = 132$.

Note that for $s = 0$ we get

$$\lambda_0 = b = \lambda \frac{v(v - 1) \dots (v - t + 1)}{k(k - 1) \dots (k - t + 1)} \tag{1.1}$$

and, since $\lambda_1 = r$, we have from the recursion the well-known identities

$$bk = vr \tag{1.2}$$

and, for $t = 2$,

$$r(k - 1) = \lambda(v - 1), \tag{1.3}$$

indicating that the parameters are not independent. Moreover, the fact that λ_s is always an integer means that there are divisibility constraints on the parameters; for our purposes these constraints are much too crude to be helpful and the reader will not encounter in this book a set of parameters that is not at least feasible.

Exercise 1.2.3 (1) Show that a generalized quadrangle $\mathcal{D} = (\mathcal{P}, \mathcal{B})$ with parameters 1-$(v, s+1, t+1)$ (i.e. of type (s, t)) has $v = (st+1)(s+1)$ and $b = (st+1)(t+1)$. Further, if x and y are distinct points that are not together on a block (i.e. non-collinear), then for any two blocks B and C through x there is a unique pair of blocks B' and C' through y such that B meets B' and C meets C'. Show also that the graph $\Gamma = (\mathcal{P}, \mathcal{E})$ defined to have vertex set $\mathcal{P}$, and edge set $\mathcal{E}$ defined by two distinct points x and y being adjacent if they are together on a block, is strongly regular with parameter set $((st + 1)(s + 1), s(t + 1), s - 1, t + 1)$.

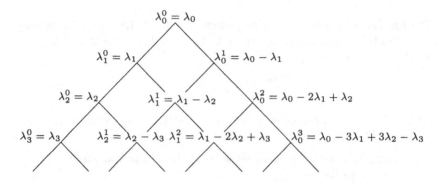

$$\lambda_0^0 = \lambda_0$$

$$\lambda_1^0 = \lambda_1 \qquad\qquad \lambda_0^1 = \lambda_0 - \lambda_1$$

$$\lambda_2^0 = \lambda_2 \qquad \lambda_1^1 = \lambda_1 - \lambda_2 \qquad \lambda_0^2 = \lambda_0 - 2\lambda_1 + \lambda_2$$

$$\lambda_3^0 = \lambda_3 \quad \lambda_2^1 = \lambda_2 - \lambda_3 \ \ \lambda_1^2 = \lambda_1 - 2\lambda_2 + \lambda_3 \quad \lambda_0^3 = \lambda_0 - 3\lambda_1 + 3\lambda_2 - \lambda_3$$

Figure 1.3: Pascal triangle for designs

(2) Let $\mathcal{D}$ be a 2-(v, k, λ) design. Show that there is a unique line through any two distinct points, that the number of points on any line is at most $(b - \lambda)/(r - \lambda)$ (where b and r are as above), and that equality holds for some line L if and only if every block meets L. (See the definition of a line in a 2-design on page 4.)

A kind of **Pascal triangle** can be used to compute certain more refined lambdas first explicitly discussed by R. M. Wilson. Let $\mathcal{D}$ be a design with parameters t-(v, k, λ), and suppose that i and j are non-negative integers. Then for disjoint point sets I and J with $|I| = i$ and $|J| = j$ the number of blocks containing I and disjoint from J is independent of these point sets provided $i + j \le t$. We denote this number by λ_i^j; the fact that it is independent of the point sets follows easily from Theorem 1.2.1 and inclusion-exclusion, but, as we shall soon see, the independence follows from a simple recursion and the fact that $\lambda_i^0 = \lambda_i$.

In fact these refined lambdas fit into the "Pascal triangle[2]" shown in Figure 1.3. With λ_i^j in the $(i, j)^{\text{th}}$ position, and $\lambda_i^0 = \lambda_i$ for $0 \le i \le t$, the Pascal property, *viz.*

$$\lambda_i^j = \lambda_{i+1}^j + \lambda_i^{j+1}, \qquad (1.4)$$

follows immediately from the definition of the λ_i^j's in precisely the same way that the similar recursion for the binomial coefficients does: take an additional point not in $I \cup J$. Thus (1.4) yields the independence of the λ_i^j for $i + j \le t$.

Moreover, if $\mathcal{D}$ is a Steiner system, then these integers remain independent of the point sets for $i + j \le k$ *provided* $I \cup J \subseteq B$ for some block B and we set $\lambda_i^0 = 1$ for $t \le i$. We have, of course, $\lambda_t = \lambda = 1$. This is easily

---

[2]Our Pascal triangle is a mirror image of the triangle most usually depicted.

established by induction. Then the triangle can be extended to the $(k+1)^{\text{st}}$ row, corresponding to $i = k$.[3] The reader may wish to look at Gross [111] where these integers are called "block intersection numbers".

Ray-Chaudhuri and Wilson [245] have given a formula that is more compact than the one obtained from inclusion-exclusion and we record it in the following:

Proposition 1.2.1 *For i and j non-negative integers with $i + j \leq t$ the number of blocks of a t-(v, k, λ) design containing an i-set and disjoint from a j-set is*

$$\lambda_i^j = \lambda \frac{\binom{v-i-j}{k-i}}{\binom{v-t}{k-t}}.$$

Proof: It is only necessary to check that the numbers given in the proposition satisfy the initial conditions, namely that $\lambda_i^0 = \lambda_i$, and the recursion $\lambda_i^j = \lambda_{i+1}^j + \lambda_i^{j+1}$. The initial conditions are almost obvious and the recursion is satisfied because of the recursion for the binomial coefficients of the numerator. $\square$

It is a simple matter, for a particular design, to start at the lower left vertex of the triangle working up to the top vertex via the recursion $\lambda_s = \lambda_{s+1}(v - s)/(k - s)$ and then to fill in the triangle using the recursion $\lambda_i^j = \lambda_{i+1}^j + \lambda_i^{j+1}$. For example, for the Fano plane of Figure 1.1, the triangle is as follows:

$$7$$
$$3 \quad 4$$
$$1 \quad 2 \quad 2$$
$$1 \quad 0 \quad 2 \quad 0$$

with the last line existing because the design is a Steiner system.

The difference of a particular pair of lambdas has an especially important role to play in the theory, so we introduce a special notation and name for it.

Definition 1.2.2 *If $\mathcal{D}$ is a t-design, where $t \geq 2$, then the **order** of $\mathcal{D}$ is $n = \lambda_1 - \lambda_2 = \lambda_1^1$, i.e. $r - \lambda$ if $t = 2$.*

The order of a design is crucial in determining those primes ℓ for which the linear codes — taken over the prime field $\mathbf{F}_\ell$ — associated with the

---

[3]It does not seem to have been observed that the triangle can also be extended to the $(k + 1)^{\text{st}}$ row when the design is a biplane; this fact does not, however, appear to be useful.

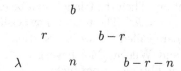

Figure 1.4: Triangle for a 2-design

design may be of significance. The Fano plane has order 2 and the biplanes of Example 1.2.2, as we have already indicated, have order 4. Observe that for a 2-(v, k, λ) design the Pascal triangle is as shown in Figure 1.4, and that the order n is centrally located; for symmetric designs especially, it is the most important parameter.

Even for Steiner systems the order can be brought into a prominent position by observing that the divisibility constraints amount to demanding that k divide $n(n+1)$ when $t = 2$. Fisher's inequality (see Corollary 1.4.1) is expressed in this case by $k \leq n + 1$, and the parameters of a 2-$(v, k, 1)$ design are given in terms of n and k by 2-$(k + n(k - 1), k, 1)$. For any particular n there are only finitely many feasible parameter sets that might yield such a Steiner system. The more usual approach has been to fix k and let n vary so as to get, possibly, an infinite class. Historically, the starting point was $k = 3$, and these designs are known as **Steiner triple systems.** Here the divisibility constraint is that the order be congruent to 0 or 2 modulo 3, and the parameters are 2-$(3 + 2n, 3, 1)$; the first possible order is 2, and this is a design we have already seen, the Fano plane. For order 3 we have the affine plane of that order, the residual (in a sense that will be made clear in Section 1.3) of the projective plane of order 3. It has been shown that for all admissible orders Steiner triple systems exist; for order 5 there are two distinct systems, in a sense that we are about to make clear, and, as the order increases, the number of distinct systems of that order tends rapidly to infinity. There is an enormous literature devoted to this topic and it is much too extensive to be dealt with here, but we include a short section on Steiner triple systems in Chapter 8.

The question of the "essential" identity of designs, and of incidence structures in general, through **isomorphism**, must arise: it is defined in the natural way. Since conceptually we should not distinguish between the role of points and blocks, mappings that interchange points and blocks will also have an important part to play.

Definition 1.2.3 *Let $\mathcal{S} = (\mathcal{P}, \mathcal{B}, \mathcal{I})$ and $\mathcal{T} = (\mathcal{Q}, \mathcal{C}, \mathcal{J})$ be incidence structures, and let φ be a bijection from $\mathcal{P} \cup \mathcal{B}$ to $\mathcal{Q} \cup \mathcal{C}$. Then:*

(1) *If $\varphi(\mathcal{P}) = \mathcal{Q}$ and $\varphi(\mathcal{B}) = \mathcal{C}$ with $p \in \mathcal{P}$ incident with $B \in \mathcal{B}$ if and only if $\varphi(p) \in \mathcal{Q}$ is incident with $\varphi(B) \in \mathcal{C}$, then φ is an **isomorphism** from $\mathcal{S}$ to $\mathcal{T}$ and we write $\mathcal{S} \approx \mathcal{T}$. If $\mathcal{S} = \mathcal{T}$, then φ is an **automorphism** or a **collineation**.*

(2) *If $\varphi(\mathcal{P}) = \mathcal{C}$ and $\varphi(\mathcal{B}) = \mathcal{Q}$ with $p \in \mathcal{P}$ incident with $B \in \mathcal{B}$ if and only if $\varphi(B) \in \mathcal{Q}$ is incident with $\varphi(p) \in \mathcal{C}$, then φ is an **anti-isomorphism** from $\mathcal{S}$ to $\mathcal{T}$. If $\mathcal{S} = \mathcal{T}$, then φ is an **anti-automorphism** or, more commonly, a **correlation**; it is called a **polarity** when $\varphi \circ \varphi$, the composition of φ with itself, is the identity map. A point p (respectively, block B) is **absolute** if p is incident with $\varphi(p)$ (respectively, B incident with $\varphi(B)$). If every point is absolute, then φ is a **null polarity**.*

We have written the bijections in the definition above on the left but we will also write maps on the right, and when we do they will sometimes be written as superscripts; for example the image of the block B under a polarity will frequently be written B^φ. The **automorphism group** of $\mathcal{S}$ will be denoted by $\mathrm{Aut}(\mathcal{S})$.

Example 1.2.3 Any element of the direct product, $G = \mathrm{Sym}(4) \times \mathrm{Sym}(4)$ (where $\mathrm{Sym}(n)$ denotes the symmetric group on n letters), with the first factor operating on the rows and the second on the columns, is an automorphism of the biplane $\mathcal{D}$ of Example 1.2.2 (2); thus G is a subgroup of $\mathrm{Aut}(\mathcal{D})$. It follows that $\mathrm{Aut}(\mathcal{D})$ acts transitively on the points of this design (and transitively on the blocks as well). In fact, taking any transitive subgroup H of $\mathrm{Sym}(4)$ in the first factor and any transitive subgroup K in the second, $H \times K \subseteq G$ will act transitively on the points and on the blocks; taking H to be the Klein 4-group or a cyclic group of order 4, with a similar choice for K, will yield a group acting regularly on both points and blocks and hence a *difference set* (see Chapters 4 and 7). It is very easy to verify that the correspondence $(i, j) \leftrightarrow B_{i,j}$ is an anti-isomorphism, and in fact a polarity, of $\mathcal{D}$.

For the biplane $\mathcal{E}$ given by the graph of Example 1.2.2 (3), $\mathrm{Sym}(6)$ acting on the complete graph is a subgroup of $\mathrm{Aut}(\mathcal{E})$, and as such it fixes ∞ and acts transitively on the other 15 points (the edges of the complete graph). Since it is a fact (not difficult to verify) that $\mathcal{D} \approx \mathcal{E}$ it follows that its automorphism group acts doubly-transitively on points (and doubly-transitively on blocks). We thus have here a specific instance of Example 1.2.2 (4).

Exercise 1.2.4 Let $\mathcal{D}$ be the design defined by a strongly regular graph as in Exercise 1.2.1. Show that the map $p \mapsto B_p$ and $B_p \mapsto p$ is a polarity of the design which has no absolute points.

Isomorphic designs clearly have the same parameters, t, v, k, λ; it is a major problem in design theory when enumerating designs to determine whether or not those with the same parameters are isomorphic. Properties of the automorphism group, and various geometrical configurations, and other intersection properties, may distinguish amongst them. Following the classification theorem of finite simple groups, and hence of all doubly-transitive groups of finite degree, Steiner systems with flag-transitive automorphism groups were classified: see Buekenhout *et al.* [59]. But most designs have little symmetry—in fact the vast majority have trivial automorphism groups; here the geometry must play a bigger role, and it is here also that the properties of the associated codes (see Section 2.4) can come more forcefully into play.

We need also the concept of an **incidence matrix** of a structure; it is a complete description of the structure in much the same way that a matrix is a complete description of a linear transformation and it can be viewed as the matrix of an incidence transformation.

Definition 1.2.4 *Let* $\mathcal{S} = (\mathcal{P}, \mathcal{B}, \mathcal{I})$, *with* $|\mathcal{P}| = v$ *and* $|\mathcal{B}| = b$. *Let the points be labelled* $\{p_1, p_2, \ldots, p_v\}$ *and the blocks be labelled* $\{B_1, B_2, \ldots, B_b\}$. *An* **incidence matrix** *for* $\mathcal{S}$ *is a* $b \times v$ *matrix* $A = (a_{ij})$ *of* 0*'s and* 1*'s such that*

$$a_{ij} = \begin{cases} 1 & \text{if } (p_j, B_i) \in \mathcal{I} \\ 0 & \text{if } (p_j, B_i) \notin \mathcal{I} \end{cases}$$

When $\mathcal{S}$ is a t-design with $t \neq 0$, any incidence matrix for $\mathcal{S}$ will have precisely k entries equal to 1 in each row, and r entries equal to 1 in each column. In fact, any matrix of 0's and 1's with constant row sums and constant column sums is the incidence matrix of a 1-design. It is easy to see that two structures are isomorphic if and only if they have incidence matrices, A and B respectively, which are related by $PAQ = B$, where P and Q are permutation matrices.

Note that many books on design theory, for example Dembowski [85] and Hughes and Piper [144], define the incidence matrix by labelling the rows by points and columns by blocks, thus using the transpose of our incidence matrix. This is, initially, more natural since $\mathcal{I}$ is a subset of $\mathcal{P} \times \mathcal{B}$ and hence $\mathcal{I}$ can be identified with the matrix. Our convention stems from our requirements for the codes of the design where we prefer to use the row space of our incidence matrix for one of the codes. In fact the correct way to define an incidence "matrix" is really independent of the actual matrix: we first make a definition in order to set our terminology.

Definition 1.2.5 *For any field* F *and any set* Ω, *denote by* F^Ω *the vector space over* F *of functions from* Ω *to* F, *the addition and scalar multipli-*

*cation being pointwise. For any subset Y of Ω, denote the **characteristic function** on Y by the vector v^Y, i.e.*

$$v^Y(\omega) = \begin{cases} 1 & \textit{if } \omega \in Y \\ 0 & \textit{if } \omega \notin Y \end{cases}$$

Here we are writing our functions on the left, i.e. $f(x)$ is the image of x under f. The standard basis for F^Ω is $\{v^{\{\omega\}} \mid \omega \in \Omega\}$; when circumstances permit we shall drop the braces on a singleton set and therefore write v^ω instead of $v^{\{\omega\}}$.

Now, with this terminology and $\Omega = \mathcal{P} \times \mathcal{B}$, the incidence corresponds precisely to the characteristic function of the subset $\mathcal{I}$ of $\mathcal{P} \times \mathcal{B}$. Moreover we obtain a natural **incidence transformation** from $F^{\mathcal{P}}$ to $F^{\mathcal{B}}$ given by $v \mapsto w$ where $w(B) = \sum_{p \in \mathcal{P}} v^{\mathcal{I}}(p, B)v(p)$. If an actual matrix is needed an ordering of the points and blocks must be chosen, and clearly different orderings will usually give different matrices. What does follow easily from the definition is that if A and B are incidence matrices from two different orderings of the points and blocks, then A and B are related by $PAQ = B$, where P is a $b \times b$ permutation matrix, and Q is a $v \times v$ permutation matrix.

1.3   Related structures

Given any design $\mathcal{D}$, there are some natural, related structures which in some cases yield new designs. The principal ideas are those of internal and external structures (see Dembowski [85]), involving also derived and extended designs, and also the concepts of complementation and duality. We will see that it is very often instructive to consider these related structures along with the given structure, particularly when we are considering the related linear codes (see Chapter 2). We need some definitions.

Definition 1.3.1 *If $\mathcal{S} = (\mathcal{P}, \mathcal{B}, \mathcal{I})$, then the structure $\mathcal{S}^t = (\mathcal{P}^t, \mathcal{B}^t, \mathcal{I}^t)$, where $\mathcal{P}^t = \mathcal{B}$, $\mathcal{B}^t = \mathcal{P}$, and $(B, p) \in \mathcal{I}^t$ if and only if $(p, B) \in I$, is called the **dual** of $\mathcal{S}$.*

Exercise 1.3.1 Verify that the designs of Example 1.2.2 except (4) and (6) have duals that are designs with the same parameters as the original. Show that the dual design of a generalized quadrangle of type (s, t) is a generalized quadrangle of type (t, s).

The four examples (1), (2), (3) and (5) of Example 1.2.2 exhibit a general property of a class of designs that we will spend some time on in Chapter 4, *viz.* the *symmetric* designs—those for which $b = v$. We will prove that the property exhibited in the four examples holds for all symmetric designs, i.e. that the dual of a symmetric design is a design with the same parameters. An obvious essential property a t-design $\mathcal{D}$ must have in order that its dual $\mathcal{D}^t$ be a t-design is that any t blocks meet in the same number of points. Thus the symmetric designs will be characterized as those 2-designs with the property that any two distinct blocks meet in λ points.

The dual of a 1-design is a 1-design, but it may have repeated blocks; we will not be concerned with these designs. It is clear that if A is an incidence matrix for $\mathcal{S}$, then A^t, the transpose of A, is an incidence matrix for $\mathcal{S}^t$, which is why we have chosen the notation as we have.

Definition 1.3.2 *A structure $\mathcal{S}$ is **self-dual** if it is isomorphic to its dual $\mathcal{S}^t$, i.e. if there exists an isomorphism $\varphi : \mathcal{S} \to \mathcal{S}^t$.*

Clearly an incidence structure is self-dual if and only if it has a correlation. Further, given a correlation φ, any other correlation can be written as $\varphi \circ \sigma$ where σ is an automorphism.

Examples 1.2.2 (1) and (2) give instances of self-dual symmetric designs; in fact, as we remarked in Example 1.2.3, it is easy to see that the design in Example 1.2.2 (2) is self-dual, the correspondence $(i,j) \leftrightarrow B_{ij}$ being a polarity. Example 1.2.2 (1) gives a desarguesian projective plane which is self-dual as we shall see in Chapter 3; in fact, we will encounter many classes of self-dual symmetric designs in that chapter.

Another natural structure to consider is the *complementary* structure:

Definition 1.3.3 *Let $\mathcal{S} = (\mathcal{P}, \mathcal{B}, \mathcal{I})$. Then the **complement** of $\mathcal{S}$ is the structure $\bar{\mathcal{S}} = (\bar{\mathcal{P}}, \bar{\mathcal{B}}, \bar{\mathcal{I}})$, where $\bar{\mathcal{P}} = \mathcal{P}$, $\bar{\mathcal{B}} = \mathcal{B}$, and $\bar{\mathcal{I}} = \mathcal{P} \times \mathcal{B} - \mathcal{I}$.*

Thus for $p \in \mathcal{P}$ and $B \in \mathcal{B}$, $(p, B) \in \bar{\mathcal{I}}$ if and only if $(p, B) \notin \mathcal{I}$. If A is an incidence matrix for $\mathcal{S}$, then $J - A$ is an incidence matrix for $\bar{\mathcal{S}}$ (where J is the matrix all of whose entries are 1).[4]

Theorem 1.3.1 *If $\mathcal{D}$ is a t-(v, k, λ) design with $v - k \geq t$, then $\bar{\mathcal{D}}$ is a t-$(v, v - k, \bar{\lambda})$ design, where*

$$\bar{\lambda} = \lambda \frac{(v - k)(v - k - 1) \cdots (v - k - t + 1)}{k(k - 1) \cdots (k - t + 1)}.$$

Proof: This is essentially immediate from Proposition 1.2.1 since the number of blocks of $\bar{\mathcal{D}}$ containing a t-subset of points is $\lambda_0^t = \lambda \binom{v-t}{k} / \binom{v-t}{k-t} = \lambda(v - k)(v - k - 1) \cdots (v - k - t + 1)/k(k - 1) \cdots (k - t + 1)$. $\square$

---

[4]As is customary, J, when it is a matrix, will always denote a matrix of the appropriate size all of whose entries are 1.

The complement of the Fano plane is a 2-$(7, 4, 2)$ design. Since it has seven blocks, it is a symmetric design; it is a biplane of order 2. The complement of a 5-$(12, 6, 1)$ design has the same parameters. In fact there is only one design with these parameters and it is self-complementary, i.e. the complement of any block is again a block. This remarkable design is closely related to a class of self-complementary designs called Hadamard 3-designs, a class of designs we treat extensively in Chapter 7.

We will see that the complementary structures have an important role to play when considering codes associated with designs.

We turn next to *derived* and *residual* structures, also known as *internal* and *external* structures. We are interested mainly in the following type of structures: the *contraction of a design at a point*, which is the design obtained by deleting the point and retaining only those blocks incident with it; the *residual of a design at a block*, which is the design obtained by deleting the block and retaining those points *not* incident with the block. Each of these constructions apply also to the dual, of course, and Dembowski [85] examines all four of them in detail. Here is the formal definition:

Definition 1.3.4 *Let* $S = (\mathcal{P}, \mathcal{B}, \mathcal{I})$, $p \in \mathcal{P}$, *and* $B \in \mathcal{B}$. *Define sets as follows:* $\mathcal{P}_p = \mathcal{P} - \{p\}$; $\mathcal{P}^B = \mathcal{P} - \{p | p \in \mathcal{P}, p \in B\}$; $\mathcal{B}_B = \mathcal{B} - \{B\}$; $\mathcal{B}_p = \{B | B \in \mathcal{B}, B \ni p\}$; $\mathcal{B}^p = \mathcal{B} - \mathcal{B}_p$. *Then the* **contraction** S_p *of* S *at the point* p *is given by*

$$S_p = (\mathcal{P}_p, \mathcal{B}_p)$$

and the **residual** S^B *of* S *at the block* B *by*

$$S^B = (\mathcal{P}^B, \mathcal{B}_B),$$

where incidence is defined through $\mathcal{I}$ *by restriction.*

The contraction at p is sometimes called the **restriction**, but then, more appropriately, the restriction is to $\mathcal{P} - \{p\}$; a similar construction in coding theory is referred to as "puncturing". Dually we have the following derived or internal structure, S_B, whose point set is $\mathcal{P}_B = \{p | p \in \mathcal{P}, p \in B\}$ and whose block set is $\mathcal{B}_B$, and the external structure, S^p, whose point set is $\mathcal{P}_p$ and whose block set is $\mathcal{B}^p = \mathcal{B} - \{B | B \ni p\}$.

Definition 1.3.5 *A structure* $S = (\mathcal{P}^*, \mathcal{B}^*, \mathcal{I}^*)$ *is an* **extension** *of the structure* $S = (\mathcal{P}, \mathcal{B}, \mathcal{I})$ *if there is a point* $p \in \mathcal{P}^*$ *such that the contraction* S_p^* *is isomorphic to* S.

Since our particular interest is in structures that are t-designs, we examine the immediate implications for these; Hughes and Piper [144] have much more concerning general structures.

Theorem 1.3.2 *If $\mathcal{D} = (\mathcal{P}, \mathcal{B})$ is a t-(v, k, λ) design, where $t \geq 2$, then for any $p \in \mathcal{P}$, $\mathcal{D}_p$ is a $(t-1)$-$(v-1, k-1, \lambda)$ design.*

Proof: Clearly, $|\mathcal{P}_p| = v - 1$, and for $B \in \mathcal{B}_P$, $|B| = k - 1$. Any set T of $t - 1$ points of $\mathcal{D}_p$ does not include p, so can be extended to a set of t points of $\mathcal{D}$. There are λ blocks of $\mathcal{D}$ that contain these t points, and hence they are blocks of $\mathcal{D}_p$ containing T. $\square$

Remark Similar reasoning shows that that $\mathcal{D}^p$ is a $(t-1)$-$(v-1, k, \lambda_{t-1} - \lambda)$ design. This design consists of all the points of $\mathcal{D}$ except p and those blocks not containing p. Although this is an important construction, we shall make no use of it.

The structures $\mathcal{D}_B$ and $\mathcal{D}^B$ require a certain uniformity of block intersections, which will occur for certain classes of designs, in particular for the symmetric designs. We leave this discussion until Chapter 4.

The following necessary condition for a design to extend is a simple divisibility criterion, but it is useful:

Proposition 1.3.1 *If a t-(v, k, λ) design $\mathcal{D}$ has an extension, then $(k + 1)$ divides $b(v + 1)$.*

Proof: Suppose $\mathcal{D}$ has an extension $\mathcal{E}$. Then $\mathcal{E}$ is a $(t + 1)$-$(v + 1, k + 1, \lambda)$ design, and, by definition, the value of λ_1 for $\mathcal{E}$ is b, the number of blocks of $\mathcal{D}$. Writing b^* for the number of blocks for $\mathcal{E}$, we have, by the recursion of Section 1.2, $b^* = (v + 1)\lambda_1/(k + 1)$, which proves the Proposition. $\square$

Example 1.3.1 (1) The Fano plane, Example 1.2.2 (1), is a 2-(7,3,1) design, $\mathcal{D}$. If p is any point of $\mathcal{D}$ and L any line, then $\mathcal{D}_p$ is a 1-(6,2,1) design, which is a partition of the 6-set into 2-subsets, and $\mathcal{D}^L$ is a 2-(4,2,1) design, which is a complete graph and the affine plane of order 2. It is somewhat more interesting to observe that $\mathcal{D}$ extends to a 3-(8,4,1) design, which is obtained by including the complements of the lines as the blocks not containing the extra point. This design does not extend, since 5 does not divide 63.

(2) The 5-(12,6,1) design of Corollary 7.4.3 already mentioned contracts three times to a 2-(9,3,1) design which is the affine plane of order 3. Equivalently, the affine plane of order 3 extends three times; this example has been heavily studied: see, among other references, Lüneburg [196]. Once again divisibility prevents the possibility of a further extension.

The reader may have observed that we have not yet produced an example of a t-design with $t > 5$, and, in fact, none are known for $t > 5$ and $\lambda = 1$, although Teirlinck [271] has shown that there do exist non-trivial t-designs for all values of t. However, it does not seem that the methods of this book are of very much help in either the construction or understanding of these highly-regular designs; they seem to be very much the product of set-theoretical and recursive methods and are, for the moment at least, untouched by either group-theoretical or coding-theoretical methods.[5] Even the 5-(12,6,1) design mentioned above in Example 1.3.1 was barely caught in the net we are casting; it is one of two remarkable 5-designs examined very closely by Witt [300]; the other design, with parameters 5-$(24, 8, 1)$, might very well be described as the centrepiece of the relationship of design theory to algebraic coding theory.

1.4 Ranks of incidence matrices

If $\mathcal{D}$ is a t-design, where $t \geq 2$, then it is also a 2-design by Theorem 1.2.1; thus we prove the following theorem for 2-designs, but it holds for t-designs, $t \geq 2$.

Theorem 1.4.1 *Let $\mathcal{D}$ be a 2-(v, k, λ) design of order n and let A be any incidence matrix for $\mathcal{D}$. Set $r = \lambda_1$, and let I_v denote the $v \times v$ identity matrix, and J_v the $v \times v$ matrix with all entries equal to 1. Then $A^t A = (r - \lambda)I_v + \lambda J_v$, and $\det(A^t A) = rkn^{v-1}$. Further, if $\mathcal{D}$ is non-trivial, then the rank of the matrix A over the rational field $\mathbf{Q}$ is v.*

Proof: The $(i, j)^{\text{th}}$ entry of $A^t A$ is the inner product of the ith column of A with the jth column of A, and is thus equal to r if $i = j$, and to λ if $i \neq j$. Thus

$$A^t A = \begin{pmatrix} r & \lambda & \dots & \lambda \\ \lambda & r & \dots & \lambda \\ \vdots & \vdots & \ddots & \vdots \\ \lambda & \lambda & \dots & r \end{pmatrix} = (r - \lambda)I_v + \lambda J_v.$$

That $\det(A^t A) = rkn^{v-1}$ follows easily using elementary row and column operations, but it also can easily be seen by noting that the all-one vector is an eigenvector with eigenvalue $r + (v - 1)\lambda = rk$ and that there are $v - 1$ eigenvectors $((1, -1, 0, \dots, 0), (0, 1, -1, 0, \dots, 0)$, etc.) with eigenvalue $r - \lambda = n$.

---

[5]Except for $t = 6$ where designs were first obtained by orbiting under a group: see Magliveras and Leavitt [205].

Now suppose that $\mathcal{D}$ is non-trivial. Since $\mathrm{rank}_{\mathbf{Q}}(A) \geq \mathrm{rank}_{\mathbf{Q}}(A^t A)$ and the latter matrix is a square $v \times v$ matrix, its rank is v if and only if its determinant is non-zero. Now $\det(A^t A) = rkn^{v-1}$ and clearly r and k are non-zero. If $n = 0$, then $r = (v-1)\lambda/(k-1) = \lambda$ implies that $v = k$, and the design is trivial, contrary to assumption. $\square$

Corollary 1.4.1 (Fisher) *If $\mathcal{D}$ is a non-trivial t-design where $t \geq 2$, then there are at least as many blocks as points, i.e. $b \geq v$.*

Proof: If $\mathcal{D}$ is a non-trivial t-design, then it is a non-trivial 2-design, so that $\mathrm{rank}_{\mathbf{Q}}(A) = v$ which must be at most the number of rows or columns, i.e. $v \leq b$. $\square$

Remark Fisher's original proof of this is quite different: we give a version of this direct proof due to Wilson [299]; it is part of a much more general set of inequalities for designs. See also Ray-Chaudhuri and Wilson [245] for a generalization of Fisher's inequality to t-designs.

Proof: (Fisher's inequality) Let $\mathcal{D} = (\mathcal{P}, \mathcal{B})$ be a 2-(v, k, λ) design, let B be a fixed block of $\mathcal{D}$, and let $B_1, B_2, \ldots, B_{b-1}$ be the other blocks. Let $\mu_i = |\{p | p \in \mathcal{P}, p \in B \cap B_i\}|$, for $i = 1, \ldots, b-1$. Counting the number of elements in the set

$$\{(p, C) | p \in \mathcal{P}, p \in B \cap C, C \in \mathcal{B}, C \neq B\}$$

gives $k(r-1) = \sum_{i=1}^{b-1} \mu_i$. Counting the number of elements in the set

$$\{(p, q, C) | p, q \in \mathcal{P}, p \neq q, C \in \mathcal{B}, C \neq B, p \in B \cap C, q \in B \cap C\}$$

gives $k(k-1)(\lambda-1) = \sum_{i=1}^{b-1} \mu_i(\mu_i - 1)$. Setting $\bar{\mu} = \sum_i \mu_i/(b-1)$, we then have that $\bar{\mu} = k(r-1)/(b-1)$. Hence

$$
\begin{aligned}
\sum_i (\mu_i - \bar{\mu})^2 &= \sum \mu_i^2 - 2\bar{\mu} \sum \mu_i + \bar{\mu}^2 (b-1) \\
&= \sum \mu_i(\mu_i - 1) - (2\bar{\mu} - 1)\bar{\mu}(b-1) + \bar{\mu}^2(b-1) \\
&= k(k-1)(\lambda-1) - \left(\frac{2k(r-1)}{b-1} - 1 \right) + \frac{k^2(r-1)^2}{b-1} \\
&= \frac{k}{b-1} \left(kb\lambda - k\lambda - b\lambda + \lambda - kb - kr^2 + 2rk + rb - r \right).
\end{aligned}
$$

Substituting on the right for $\lambda = r(k-1)/(v-1)$ and $b = vr/k$, and after some substantial rearranging, we get

$$\sum_i (\mu_i - \bar\mu)^2 = \frac{r^2(b-v)(v-k)^2}{b(b-1)(v-1)} \geq 0,$$

giving $b \geq v$. $\square$

We will soon be returning to the rank of the incidence matrix of a design when we consider linear codes associated with the design: then we will take the rank over a finite field, and show that some finite fields will lead to ranks considerably less than v. These will be the codes that will be of assistance in distinguishing designs with the same parameters.

1.5 Arcs and ovals

The configurations we consider now for 2-designs not only have an interesting and informative role to play in the geometry of the design, but also show themselves in the related codes.

Definition 1.5.1 *Let $\mathcal{D} = (\mathcal{P}, \mathcal{B})$ be a 2-(v,k,λ) design. A set of points $S \subseteq \mathcal{P}$ such that no three points of S are together on a block of $\mathcal{B}$ is called an **arc** of $\mathcal{D}$, or, more precisely, an **s-arc** if $|S| = s$. A block $B \in \mathcal{B}$ is*

- **tangent** *to S if B meets S in a single point;*

- **secant** *to S if B meets S in two points;*

- **exterior** *to S if B does not meet S.*

Example 1.5.1 (1) In the Fano plane, Example 1.2.2 (1), $S = \{1,2,4,6\}$ is a 4-arc, the line $\{1,2,3\}$ is a secant, the line $\{3,5,7\}$ is exterior, and there are no tangents.

(2) For the biplane given in Example 1.2.2 (2) the set of vertices of any subrectangle of the array is easily seen to be a 4-arc; the four positions of the 1's of any 4×4 permutation matrix superimposed on the array are also easily seen to form a 4-arc. These 60 arcs are the only 4-arcs of this design; none of them have tangents.

We examine now the maximum size an arc can have. The following theorem is from Andriamanalimanana [4].

Theorem 1.5.1 *Let $\mathcal{D} = (\mathcal{P}, \mathcal{B})$ be a 2-(v,k,λ) design of order n, where $k \geq 3$, and let S be an s-arc of $\mathcal{D}$. Then*

(1) if n is odd, or n is even and λ does not divide r, $s \leq (r + \lambda - 1)/\lambda$;

(2) if n is even and λ divides r, $s \leq (r + \lambda)/\lambda$.

Proof: Let S be an s-arc, and without loss of generality take $s \geq 1$. Take $p \in S$, and count the number of elements in the set $\{(q, B)|q \in S - \{p\}, B \ni p, q \in B \in \mathcal{B}\}$, in two ways:

$$(s - 1)\lambda = \sum_{B \ni p} |B \cap (S - \{p\})| \leq r,$$

since a block can meet S in at most one point other than p and there are r blocks through p. Thus $s \leq (r + \lambda)/\lambda$, whatever the order of $\mathcal{D}$.

Now suppose that S has a tangent at some point p. Then there is a block $B \ni p$, for which $|B \cap (S - \{p\})| = 0$, so that $\lambda(s - 1) \leq r - 1$, i.e. $s \leq (r - 1 + \lambda)/\lambda$.

If S has no tangents, then $|B \cap (S - \{p\})| = 1$ for all $B \ni p$, and so $s = (r + \lambda)/\lambda$, whence λ must divide r. Since $k \geq 3$, there is a point p with $p \notin S$; suppose there are m secants to S through p. Counting $\{(q, B)|q \in S, \{p, q\} \subset B\}$ in two ways gives $2m = s\lambda = \lambda(r + \lambda)/\lambda$, i.e. $r + \lambda = 2m$, so that $n = r - \lambda = 2m - 2\lambda$, and n is even. $\square$

This leads to the following definition:

Definition 1.5.2 *Let $\mathcal{D}$ be a 2-(v, k, λ) design of order n. Then an **oval** in $\mathcal{D}$ is an arc of maximum size m, where*

(1) $m = (r + \lambda - 1)/\lambda$ if n is odd, or if n is even and λ does not divide r (i.e. every point on the arc is on a unique tangent);

(2) $m = (r + \lambda)/\lambda$ if n is even and λ divides r (i.e. there are no tangents).

We will not have occasion to speak about **complete** arcs—those to which no point can be adjoined without their losing the property of being an arc. By definition an oval is a complete arc, but the converse certainly need not be true: see Hirschfeld [136].

1.6   Block's theorem

A theorem of Block [43] concerning decompositions of an incidence matrix of a structure has an important consequence for the orbit structure of any subgroup of the automorphism group of a design. We give his theorem and its proof in a somewhat more general setting due to Siemons [264]. The theorem has its origins in work of Brauer related to the representation theory of finite groups (see Section 4.3); it involves the concept of a *tactical decomposition*, which is extensively treated by Dembowski [85], but the ideas go back at least as far as Moore [217] who built on ideas of Cayley.

Definition 1.6.1 *Let* $\mathcal{S} = (\mathcal{P}, \mathcal{B}, \mathcal{I})$ *be an incidence structure. Let* $X_\mathcal{P}$ *denote a partition of the point set* $\mathcal{P}$ *into non-empty point classes and let* $X_\mathcal{B}$ *denote a partition of block set* $\mathcal{B}$ *into non-empty block classes. Then* $(X_\mathcal{P}, X_\mathcal{B})$ *is a* **tactical decomposition** *of* $\mathcal{S}$ *if the following dual conditions are satisfied:*

Tac($\mathcal{P}$): *for any point* p *and any block class* $\mathcal{B}'$ *the number of blocks in* $\mathcal{B}'$ *incident with* p *depends only on the point class containing* p;

Tac($\mathcal{B}$): *for any block* B *and any point class* $\mathcal{P}'$ *the number of points in* $\mathcal{P}'$ *incident with* B *depends only on the block class containing* B.

A decomposition is called **point-tactical** *if* Tac($\mathcal{P}$) *is satisfied and* **block-tactical** *if* Tac($\mathcal{B}$) *is satisfied.*

Any incidence structure $\mathcal{S}$ has the *trivial* tactical decomposition where each class consists of a single element. A more important tactical decomposition is defined by any subgroup G of Aut($\mathcal{S}$); here the orbits of G on the point set are the point classes and the orbits of G on the block set are the block classes. Should $\mathcal{S}$ have a decomposition with exactly one point class, $\mathcal{S}$ is called **resolvable** with $X_\mathcal{B}$ the **resolution**. Then each point of $\mathcal{P}$ is incident with the same number of blocks within each class; if each point in $\mathcal{P}$ is incident with precisely one block in each block class, then the resolution is a **parallelism**, and each block class is a **parallel class**. Examples of designs with parallelisms are affine planes (see Chapter 6) and oval designs (see Chapters 7 and 8). For 1-designs there is the example of a **net**:

Exercise 1.6.1 A 1-(v, k, r) design $\mathcal{N}$ with a parallelism and the additional properties that $\mathcal{N}$ is a partial linear space (i.e. any two points are on at most one block) and that blocks either meet or are in the same parallel class, is what is called a **net**. Show that $v = k^2$ and that each parallel class contains exactly k blocks.

For the proof of Block's theorem we follow Siemons, but we use the notation of Definition 1.2.5 of Section 1.2. For a more general treatment the reader should consult Bridges [49].

Recall that if Y is a subset of Ω then v^Y in F^Ω denotes the characteristic function of Y and that the set $\{v^{\{\omega\}} \,|\, \omega \in \Omega\}$ is the standard basis of the vector space F^Ω; in particular, $v^Y = \sum_{\omega \in Y} v^{\{\omega\}}$. Next, for any field F, consider the vector spaces $F^{\mathcal{P}}$ and $F^{\mathcal{B}}$ of dimensions $|\mathcal{P}|$ and $|\mathcal{B}|$ respectively.

Now there is a natural linear transformation $\Delta : F^{\mathcal{P}} \to F^{\mathcal{B}}$ that arises from the incidence relation; it is most easily described on the standard basis:

$$\Delta : \begin{cases} F^{\mathcal{P}} & \to & F^{\mathcal{B}} \\ v^{\{p\}} & \mapsto & \sum_{X \ni p, X \in \mathcal{B}} v^{\{X\}}. \end{cases} \tag{1.5}$$

Thus the image of $v^{\{p\}}$ under Δ simply "records" those blocks containing the point p. Similarly define

$$\nabla : \begin{cases} F^{\mathcal{B}} & \to & F^{\mathcal{P}} \\ v^{\{B\}} & \mapsto & \sum_{p \in B, p \in \mathcal{P}} v^{\{p\}}, \end{cases} \tag{1.6}$$

so that the image of $v^{\{B\}}$, a standard basis element of $F^{\mathcal{B}}$, simply records the points incident with the block B. After ordering the points and blocks the corresponding incidence matrix A of $\mathcal{S}$ is the matrix of Δ in the standard bases while A^t is the matrix of ∇.

Now take F to be any field of characteristic 0 (for present purposes F can be taken to be the rational numbers $\mathbf{Q}$). Let $(X_{\mathcal{P}}, X_{\mathcal{B}})$ be an arbitrary partition of the points and blocks of $\mathcal{S}$ into non-empty classes. The characteristic functions of the point classes clearly generate a subspace of $F^{\mathcal{P}}$ of dimension $|X_{\mathcal{P}}|$; we denote this subspace by $F(\mathcal{P})$; similarly the subspace of $F^{\mathcal{B}}$ generated by the characteristic functions of the block classes is of dimension $|X_{\mathcal{B}}|$ and is denoted by $F(\mathcal{B})$.

Lemma 1.6.1 *With $\mathcal{S}$ and $(X_{\mathcal{P}}, X_{\mathcal{B}})$ as above, $(X_{\mathcal{P}}, X_{\mathcal{B}})$ satisfies* Tac$(\mathcal{P})$ *if and only if $\Delta(F(\mathcal{P})) \subseteq F(\mathcal{B})$. Dually, $(X_{\mathcal{P}}, X_{\mathcal{B}})$ satisfies* Tac$(\mathcal{B})$ *if and only if $F(\mathcal{P}) \supseteq \nabla(F(\mathcal{B}))$.*

Proof: We have for any point class $\mathcal{P}'$

$$\Delta(v^{\mathcal{P}'}) = \sum_{p \in \mathcal{P}'} \Delta(v^{\{p\}}) = \sum_{p \in \mathcal{P}'} \sum_{B \ni p} v^{\{B\}} = \sum_{B} N_B v^{\{B\}},$$

where N_B is the number of points of $\mathcal{P}'$ incident with B. Now Tac$(\mathcal{P})$ holds if and only if N_B is constant over each block class and hence if and only if $\sum_B N_B v^{\{B\}} \in F(\mathcal{B})$. $\square$

The proof of Block's theorem now follows easily:

Theorem 1.6.1 (Block) *Let $\mathcal{S} = (\mathcal{P}, \mathcal{B}, \mathcal{I})$ be an incidence structure with a partition $(X_{\mathcal{P}}, X_{\mathcal{B}})$ of points and blocks, and let A be an incidence matrix for $\mathcal{S}$. Then, if $(X_{\mathcal{P}}, X_{\mathcal{B}})$ satisfies* Tac$(\mathcal{P})$,

$$|X_{\mathcal{P}}| \leq |X_{\mathcal{B}}| + |\mathcal{P}| - \mathrm{rank}_F(A).$$

Proof: From Lemma 1.6.1 we have that $\Delta(F(\mathcal{P})) \subseteq F(\mathcal{B})$, and hence that

$$
\begin{aligned}
\dim(F(\mathcal{P})) &= \dim(\Delta(F(\mathcal{P})) + \dim(\ker(\Delta) \cap F(\mathcal{P})) \\
&\leq |X_\mathcal{B}| + \dim(\ker(\Delta) \cap F(\mathcal{P})).
\end{aligned}
$$

Now $\dim(\ker(\Delta) \cap F(\mathcal{P})) \leq \dim(\ker(\Delta)) = |\mathcal{P}| - \mathrm{rank}_F(A)$, and thus the result follows. $\square$

Note that there is, obviously, a dual theorem involving $\mathrm{Tac}(\mathcal{B})$ that we have not stated. This would be used when applying Block's theorem to symmetric designs, but in fact we will give an even simpler proof in that case: see Section 4.3. The application of Block's theorem is to the following result concerning orbits of an automorphism group of a structure with incidence matrix of rank equal to v, the number of points.

Corollary 1.6.1 *If $\mathcal{S} = (\mathcal{P}, \mathcal{B}, \mathcal{I})$ is an incidence structure with incidence matrix A of rank equal to $|\mathcal{P}|$ over $\mathbf{Q}$ and if G is a subgroup of $\mathrm{Aut}(\mathcal{S})$, then the number of orbits of G on blocks is greater than or equal to the number of orbits of G on points.*

Proof: It is quite clear that the point and block orbits under G form a tactical decomposition of points and blocks, and thus satisfy both $\mathrm{Tac}(\mathcal{P})$ and $\mathrm{Tac}(\mathcal{B})$; we invoke only $\mathrm{Tac}(\mathcal{P})$ of course. $\square$

Remark Block's theorem applies in particular if $\mathcal{S}$ is a non-trivial t-(v, k, λ) design with $t \geq 2$, since a non-trivial 2-design always has A of rank v over $\mathbf{Q}$, by Theorem 1.4.1. This is the situation for which Block's theorem is normally used and could be thought of simply as saying that there are at least as many block orbits as point orbits. In particular, for any 2-design, if a group of automorphism of the design is transitive on blocks, i.e. is *block-transitive*, then it is also *point-transitive*.

Chapter 2

Codes

2.1 Introduction

Error-correcting codes were introduced in the 1940s in order to implement a theorem of Shannon's [260] which guaranteed that virtually error-free communication could be obtained even over a noisy channel. The message to be communicated is first "encoded", i.e. turned into a codeword, by adding "redundancy". The codeword is then sent through the channel and the received message is "decoded" by the receiver into a message resembling, as closely as possible, the original message. The degree of resemblance will depend on how good the code is in relation to the channel. The usual pictorial representation of this communication link is shown in the schematic diagram Figure 2.1.

Here a message is first given by the *source* to the *encoder* that turns the message into a *codeword*, i.e. a string of letters from some alphabet, chosen according to the code used; the encoded message is then sent through the channel, where it may be subjected to *noise* and hence altered; this possibly-altered message, when it arrives at the *decoder* belonging to the receiver, is, first of all, equated with the most likely codeword, i.e. the one (should that exist) that, in a probabilistic sense depending on the channel, was probably sent; finally this "most likely" codeword is decoded and the message is passed on to the receiver.

Example 2.1.1 Suppose we use an alphabet of just two symbols, 0 and 1, and we have only two messages, for example "no" corresponding to 0, and "yes" corresponding to 1. We wish to send the message "no", and we add redundancy by simply repeating the message five times. Thus we encode the message as the codeword (00000). The channel interferes, perhaps, with

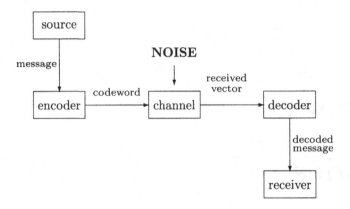

Figure 2.1: A noisy communications channel

the message and could, for example, change it to (01100). The decoder assesses the message and decides that of the two possible codewords, i.e. (00000) and (11111), the former is the more likely, and hence the message is decoded, correctly, as "no".

Notice that we have made several assumptions here: for example we have assumed that the probability of an error at any position in the word is less than $\frac{1}{2}$, that each codeword is equally likely to be sent, and that the receiver is aware of the code used. This example is not at all representative of the complexity of the subject and at every stage in Figure 2.1 there are deep and interesting mathematical questions to be answered. There are many excellent books that deal with this sort of question—both at the information-theoretical level and at the coding-theoretical; the reader can consult Peterson [235] for an early lay article from the point of view of the engineer, Hill [134] for an elementary introduction to coding theory, and MacWilliams and Sloane [203] for further information and a massive bibliography.

We will make no use here of information theory, but we shall rely heavily on the coding theory that has been developed over the past 40 years by both engineers and mathematicians. We collect together in this chapter the basic tools and results that we shall need; later we will discuss some of the deeper results from coding theory that are crucial to its application to the theory of designs and, in particular, to the classification of finite projective planes.

Our main interest will be in *linear codes* associated with designs, but we will first give the more general definition and, in fact, endeavour to make the chapter a self-contained introduction to coding theory; the reader interested *only* in such an introduction may want to omit Section 2.4.

Definition 2.1.1 *Let F be a finite set, or* **alphabet,** *of q elements. A* **q-ary code** *C is a set of finite sequences of symbols of F, called* **codewords** *and written $x_1x_2 \ldots x_n$, or $(x_1, x_2, \ldots, x_n)$, where $x_i \in F$ for $i = 1, \ldots, n$. If all the sequences have the same length n, then C is a* **block code** *of* **block length** *n.*

The code used in the example above is called the **repetition code** of length 5: it can be generalized to length n and to any alphabet of size q, and hence will have q codewords of the form $xx \cdots x$, where $x \in F$.

Given an alphabet F, it will be convenient, and also consistent with terminology for cartesian products of sets and for vector spaces when F is a field, to denote the set of all sequences of length n of elements of F by F^n and to call these sequences *vectors*, referring to the member of F in the ith position as the coordinate at i. We use either notation, $x_1x_2 \cdots x_n$ or $(x_1, x_2, \ldots, x_n)$, for the vectors — the former being customary for binary codes where the x_i take only two values, 0 or 1 (the *bits*) and confusion cannot arise. A code over F of block length n is thus any subset of F^n. The following example appears in Hill [134, Chapter 1].

Example 2.1.2 Headquarters needs to transmit a safe route through enemy territory to an agent in the field. Both headquarters and the agent have the same map and grid and the agent knows what code headquarters will use. Using the alphabet $F = \{0, 1\}$ denote the four compass directions N, S, E, W by the four pairs, 00,11,01,10 respectively. If only these are used to denote the route, the agent will not be aware of any errors in transmission. Redundancy is added as follows: $N = 000, S = 110, E = 011, W = 101$. Now if the route, say $NNWNESWWN$, is transmitted and *at most* single errors, per block of three, occur in the transmitted message,

$$000000101000011110101101000,$$

the agent will know whether or not there have been errors and can ask for retransmission if there are. If further redundancy is added, the agent will be able to *correct* a single error in any codeword: $N = 00000, S = 11011, E = 01101, W = 10110$. Now if only *detection* is required, then up to *two* errors can be detected; if *correction* is being sought, then *one* per codeword can be corrected. This is based on the assumption we stated above: correct transmission of any alphabet letter at any time is more probable than incorrect transmission.

The formal process in the reasoning of the two simple examples above uses the concept of the *distance* between codewords.

Definition 2.1.2 *Let $v = (v_1, v_2, \ldots, v_n)$ and $w = (w_1, w_2, \ldots, w_n)$ be two vectors in F^n. The* **Hamming distance***, $d(v,w)$, between v and w is the number of coordinate places in which they differ:*

$$d(v,w) = |\{i|v_i \neq w_i\}|.$$

We will usually refer to the Hamming distance as simply the **distance** between two vectors.

Proposition 2.1.1 *The Hamming distance defines a metric on F^n, i.e.*

(1) $d(v,w) = 0$ if and only if $v = w$;

(2) $d(v,w) = d(w,v)$ for all $v, w \in F^n$;

(3) $d(u,w) \leq d(u,v) + d(v,w)$ for all $u, v, w \in F^n$.

Proof: Only the third property is not immediate. To prove it observe that $d(u,w)$ is the smallest number of changes required to change u to w. But u can be changed to w by first changing it to v — in $d(u,v)$ steps — and then changing v to w — in $d(v,w)$ steps. This gives the third property. □

Nearest-neighbour decoding picks the codeword v' nearest (in terms of Hamming distance) to the received vector, should such a vector be uniquely determined. This method maximizes the decoder's likelihood of correcting errors — provided that each symbol has the same probability (less than $\frac{1}{2}$) of being received in error and each symbol of the alphabet is equally likely to occur. A channel with these two properties is called a **symmetric q-ary channel**.

Definition 2.1.3 *The* **minimum distance** *$d(C)$ of a code C is the smallest of the distances between distinct codewords; i.e.*

$$d(C) = \min\{d(v,w)|v, w \in C, v \neq w\}.$$

The following simple result shows the vital importance of this concept for codes used in symmetric channels.

Theorem 2.1.1 *Let C be a code of minimum distance d. If $d \geq s+1 > 1$, then C can be used to detect up to s errors in any codeword or, if $d \geq 2t+1$, C can be used to correct up to t errors.*

Proof: If $d \geq s+1$ and v is transmitted, suppose s or fewer errors occur. Then the received vector w is at distance at most s from v and hence cannot be a codeword (unless no errors have occurred and $w = v$). Thus, were there errors, they would be detected. On the other hand suppose v is transmitted, w is received, and t or fewer errors, where $d \geq 2t+1$, have occurred. Then $d(v, w) \leq t$. If u is any codeword other than v, then $d(u, w) + d(w, v) \geq d(u, v) \geq 2t+1$, so that $d(u, w) \geq 2t+1-d(v, w) \geq t+1$. Thus v is the nearest codeword to w, and thus nearest-neighbour decoding corrects the errors. □

Corollary 2.1.1 *If* $d(C) = d$ *then* C *can detect up to* $d-1$ *errors or correct up to* $\lfloor (d-1)/2 \rfloor$ *errors.*

Exercise 2.1.1 Suppose we want to simultaneously correct t or fewer errors and detect whether or not s or fewer errors were made where $1 \leq t < s$. How large must the minimum distance be?

Notation: If C is a code of block length n having M codewords and minimum distance d, then we say that C is an (n, M, d) **q-ary code,** where $|F| = q$. We will frequently refer to n as the **length** of the code rather than the block length.

From Corollary 2.1.1, we see that for a good (n, M, d) code C, i.e. one that detects or corrects many errors, we need d to be large. However we also require n to be small (for fast transmission) and M to be large (for a large number of messages). These are clearly conflicting aims, since for a q-ary code, $M \leq q^n$. In fact there are many bounds connecting these three parameters, the simplest of which is the **Singleton bound:**

Theorem 2.1.2 *If* C *is an* (n, M, d) *q-ary code, then* $M \leq q^{n-d+1}$.

Proof: Consider the M codewords and remove their last $d-1$ coordinates. The pruned vectors, of length $n-d+1$, are still all different (since no two codewords differed in less than d places), and are all inside F^{n-d+1}. Thus $M \leq q^{n-d+1}$. □

Definition 2.1.4 *Let* C *and* C' *be two q-ary codes of the same length* n *and having the same number of codewords. Then they are* **equivalent,** *written* $C \sim C'$, *if each can be obtained from the other by a combination of operations of the following types:*

(1) any permutation on the n *coordinate positions;*

(2) any permutation on the letters of the alphabet in any fixed coordinate position.

Clearly equivalent codes will preserve the important aspects of having the same length, the same number of codewords, and the same minimum distance. The converse of this statement does not follow, and we will give counter-examples later. The notion of equivalence of codes will change slightly when we introduce linear codes, and it will be this finer equivalence that we will mean when we say in this book that codes are equivalent.

Another bound, usually better than the Singleton bound, is the so-called **sphere-packing bound**. We first describe what coding theorists have traditionally called a sphere; a more accurate name might have been *closed ball*.

Definition 2.1.5 *Let F be any alphabet and suppose $u \in F^n$. Then for any integer $r \geq 0$, the **sphere of radius r with centre u** is the set of vectors $S(u, r) = \{v | v \in F^n, d(u, v) \leq r\}$.*

Let C be an (n, M, d) code. Then the spheres of radius $\rho = \lfloor (d-1)/2 \rfloor$ with centre in C do not overlap, i.e. they form M pairwise disjoint subsets of F^n. The integer ρ is called the **packing radius** of C.

Theorem 2.1.3 (Sphere-packing bound) *If C is an (n, M, d) q-ary code of packing radius ρ, then*

$$M \left(1 + (q-1)n + (q-1)^2 \binom{n}{2} + \cdots + (q-1)^\rho \binom{n}{\rho} \right) \leq q^n.$$

Proof: Consider the M spheres of radius ρ with centre at each of the codewords. The number of vectors in each sphere is

$$1 + (q-1)n + (q-1)^2 \binom{n}{2} + \cdots + (q-1)^\rho \binom{n}{\rho}.$$

Since the M spheres are pairwise disjoint and $|F^n| = q^n$, we have the inequality. $\square$

The **covering radius**[1] of a code is defined to be the smallest integer R such that spheres of radius R with their centres at the codewords cover all the words of F^n. As we have just seen, the packing radius is immediately determined by the minimum distance; the covering radius is a much more elusive integer and is, in general, much harder to determine. If the covering

---

[1]Both the packing radius and covering radius were introduced by Prange [241], where he called them the "packing integer" and "covering integer" respectively. This short technical report introduced most of the major ideas involving cyclic codes and also shows that the binary Golay code is cyclic — although Prange was not then certain that his cyclic $(23, 2^{12}, 7)$ code was equivalent to Golay's.

radius is equal to the packing radius—and equal to t say—the code is called a **perfect t-error-correcting code**. Thus perfect codes are those that meet the sphere-packing bound and hence, for perfect codes, q^n must be divisible by the cardinality (given above for $t = \rho$) of a sphere of radius t. It was Golay [107] who, having observed the divisibility relation for $q = 2, n = 23, t = 3$ and for $q = 3, n = 11, t = 2$, first produced the two perfect codes that now bear his name.

The binary (i.e. $|F| = 2$) repetition codes of odd length n are *trivial* perfect $(n-1)/2$-error-correcting codes; the infinite class of binary perfect 1-error-correcting codes with $n = 2^m - 1$ and hence $M = 2^{n-m}$ was discovered by Hamming [128] and generalized to the p-ary case by Golay. These perfect 1-error-correcting codes are now almost universally called Hamming codes. All perfect codes which are linear (see the next section) will be constructed in this book.

2.2 Linear codes

We turn next to *linear codes* and from now on devote our attention solely to them, since they will be the codes that arise in our treatment of designs. The alphabet will be a finite field $F = \mathbf{F}_q$ of order q and the ambient domain for the codewords will be in the vector space $V = F^n$ of n-tuples, $(x_1, x_2, \ldots, x_n)$, over F — again sometimes written as simply $x_1 x_2 \cdots x_n$ when there is no possibility of confusion.

Definition 2.2.1 *A code C over $F = \mathbf{F}_q$ of length n is **linear** if C is a subspace of $V = F^n$. If $\dim(C) = k$ and $d(C) = d$, then we write $[n, k, d]$ for the q-ary code C — instead of (n, q^k, d); if the minimum distance is not specified we simply write $[n, k]$. The **information rate** is k/n and the **redundancy** is $n - k$.*

Note that the size of the alphabet, i.e. the order of the field, is once again omitted from the notation and again for the historical reason that most codes were, at the outset, binary.

In fact a q-ary linear code is any subspace of a finite-dimensional vector space over a finite field $\mathbf{F}_q$, *but with reference to a particular basis*. The standard basis for F^n, as the space of n-tuples, has a natural ordering through the numbers 1 to n, and this coincides with the spatial layout of a codeword as a sequence of alphabet letters sent over a channel. To avoid ordering the basis we may take $V = F^X$, where X is any set of size n. Then a linear code is any subspace of V. In this case the basis is specified through the coordinate functions v^i, for $i \in X$ (see the notation introduced in Definition 1.2.5), i.e. $v^i(i) = 1$ and $v^i(j) = 0$ for $j \neq i$, and this basis is not yet ordered through the numbers $1, 2, \ldots, |X| = n$. This approach will be particularly convenient when we look at codes defined by designs.

In Assmus and Mattson [19] linear codes are looked at through their coordinate functionals and this approach illustrates very clearly the way redundancy is added: if C is a subspace of $V = F^n$ (with the standard basis), then the *coordinate functionals*, $\varphi_i : C \to F$ for $i = 1, \ldots, n$, are defined by $c\varphi_i = c_i$ where c_i is the coordinate of $c \in C$ at i, i.e. $c = (c_1, c_2, \ldots, c_n)$. Thus $\varphi_i \in C^t$, the dual space (of all linear functionals: see page 94) of C. Now we can redefine a linear code of length n and dimension k over $F = \mathbf{F}_q$ (i.e. having q^k codewords) as follows:

Proposition 2.2.1 *Let U be a vector space of dimension k over the field $F = \mathbf{F}_q$ and let S be a sequence, $\{f_1, \ldots, f_n\}$, of functionals in U^t such that S spans U^t. With $V = F^n$, the set*

$$C = \{(uf_1, uf_2, \ldots, uf_n) | u \in U\}$$

is a linear code of length n and dimension k over F.

Proof: Clearly we have defined a linear transformation of U into F^n. The kernel of this transformation is $\{0\}$ because the functionals span U^t; thus C has dimension k as required. $\square$

The process of encoding and the way redundancy is added are made very clear by this approach: since the functionals span U^t there must be at least k functionals and some k-subset of S is a basis for U^t with the remaining $n - k$ functionals playing the "redundant" role. The codewords *are* the vectors in U and the functionals inject U into the ambient space V. The codes that have recently been derived from algebraic geometry are usually constructed in precisely this manner: see van Lint [189] or van Lint and van der Geer [191] for applications of algebraic geometry to coding theory.

The minimum distance, $d(C)$, for a linear code is simply the *minimum weight* of the non-zero codewords. We make the appropriate formal definition:

Definition 2.2.2 *Let $V = F^n$. For any vector $v = (v_1, v_2, \ldots, v_n) \in V$ set $S = \{i \mid v_i \neq 0\}$. Then S is called the **support** of v and the **weight** of v is $|S|$. The **minimum weight** of a code is the minimum of the weights of the non-zero codewords.*

We denote the support of a vector v by $\mathrm{Supp}(v)$ and its weight by $\mathrm{wt}(v)$. As we shall see, it is the supports of the minimum-weight codewords that give the most immediate connection of codes to designs.

Proposition 2.2.2 *Let C be a $[n, k, d]$ code. Then the minimum distance $d = d(C)$, is the minimum weight of C.*

Proof: Notice that in F^n, $d(v, w) = \text{wt}(v - w)$. Since C is a subspace, $v - w \in C$ for any $v, w \in C$, and the result follows. $\square$

For linear $[n, k, d]$ q-ary codes the Singleton bound (Theorem 2.1.2) and the Sphere-packing bound (Theorem 2.1.3) become the following:

Singleton bound: $d \leq n - k + 1$;

sphere-packing bound: $\sum_{i=0}^{\rho}(q - 1)^i \binom{n}{i} \leq q^{n-k}$.

Code equivalence changes a little for linear codes, because of the added structure of linearity.

Definition 2.2.3 *Two linear codes in F^n are* **equivalent** *if each can be obtained from the other by permuting the coordinate positions in F^n and multiplying each coordinate by a non-zero field element. The codes will be said to be* **isomorphic** *if a permutation of the coordinate positions suffices to take one to the other.*

In terms of the distinguished basis that is present when discussing codes, code equivalence is obtained by reordering the basis and multiplying each of the basis elements by a non-zero scalar. Thus the codes C and C' are equivalent if there is a linear transformation of the ambient space F^n, given by a monomial matrix in the standard basis, that carries C onto C'. When the codes are isomorphic, a permutation matrix can be found with this property. When $q = 2$, the two concepts are identical. Once again, equivalent linear codes must have the same parameters $[n, k, d]$.

Equivalence and isomorphism of codes can be looked at in a more general setting if we consider the codes as subspaces of F^X for some set X. Isomorphism is obtained by allowing only the symmetric group, $\text{Sym}(X)$, to act on F^X in the natural way: for $\sigma \in \text{Sym}(X)$, and $i \in X$, define $(f\sigma)(i) = f(\sigma(i))$ for $f \in F^X$, where we are viewing the action of $\text{Sym}(X)$ on X on the left and on F^X on the right.[2] Equivalence is obtained by introducing monomial actions, and we can be more general than in Definition 2.2.3.

Thus let G be an arbitrary subgroup of $F^\times$, the multiplicative group of the field. Set $A = G^X$, the group of all functions from X to G with the binary operation of point-wise multiplication: $ab(i) = a(i)b(i)$ for $a, b \in A$ and $i \in X$. Let $\mathcal{G} = \text{Sym}(X) \times A$ be the semi-direct product of A by

---

[2]If both actions are to be on the left then one must have $(\sigma f)(i) = f(\sigma^{-1}(i))$; we will sometimes use this notation.

Sym(X). Thus, for $\sigma, \tau \in$ Sym(X) and $a, b \in A$, $(\sigma, a)(\tau, b) = (\sigma\tau, (a\tau)b)$, where the action of Sym(X) on A is defined as it is on F^X, i.e. $(a\tau)(i) = a(\tau(i))$. Then $\mathcal{G}$ acts naturally on F^X by $(v(\sigma, a))(i) = v(\sigma(i))a(i)$. It is easy to check that $v((\sigma, a)(\tau, b)) = (v(\sigma, a))(\tau, b)$ and it is clear that $\mathcal{G}$ acts as a group of linear transformations of the vector space F^X. Further, in the standard basis for F^X, the action is monomial. Taking G to be the trivial subgroup of $F^\times$ gives the finest equivalence — *code isomorphism* — and taking $G = F^\times$ gives the coarsest — *code equivalence*. There is also a distinguished role to be played by the subgroup $G = \{\pm 1\}$ of $F^\times$ as we shall see in Chapter 7; it is the natural choice when discussing self-orthogonal codes: see Definition 2.3.2.

Of course, for $C = C'$ and any subgroup $G \subseteq F^\times$ the G-equivalences form a group under composition. There is then, for any code C, a collection of groups that might be called the "automorphism group" of the code. The smallest of these groups will play the dominant role in most of the book and we therefore make it our automorphism group.

Definition 2.2.4 *If C is a linear code of length n over a field F, then any isomorphism of C onto itself is called an **automorphism** of C. The set of all automorphisms of C is the **automorphism group** of C, denoted by* Aut(C).

The automorphism group of C is thus a subgroup of Sym(n), or of Sym(X) if $C \subseteq F^X$. The existence of automorphisms for C can provide a richer structure for the code and allow one to make use of deeper results from algebra. This is particularly the case when C has a *regular* automorphism group $G \subseteq$ Aut(C); this means that G is transitive on X and that $|G| = |X| = n$, the block length of C. In this case we can define another binary operation on F^X, making it an algebra. To do this we introduce the *group algebra*, $F[G]$; the code C then becomes a left ideal in $F[G]$.

Recall how the group algebra $F[G]$ (or the *group ring*, $R[G]$ where R is a ring) is defined: form the vector space over F with basis the elements of G, so that the vectors are simply formal sums $\sum_{g \in G} a_g g$, where $a_g \in F$. A multiplication is defined through the group's binary operation so that the distributive law holds:

$$\left(\sum_{g \in G} a_g g \right) \left(\sum_{h \in G} b_h h \right) = \sum_{k \in G} \left(\sum_{gh=k} a_g b_h \right) k.$$

Now we put C in $F[G]$ by a correspondence between the sets X and G: pick any element $e \in X$ and let e correspond to the identity element of G. We identify $eg \in X$ with the element $g \in G$. Since G acts regularly, this is a bijective correspondence. We view the ambient space F^X as F^G via this correspondence. Now place C in $F[G]$ by writing, for $c \in C$,

$$c = \sum_{g \in G} c(g)g.$$

Then for any $h \in G$, hc, as a product of elements in the group algebra, becomes

$$hc = \sum_{g \in G} c(g)hg = \sum_{g \in G} c(h^{-1}g)g = \sum_{g \in G} (ch^{-1})(g)g,$$

which is just the image of the codeword c under the code automorphism h^{-1} — and thus is in C. So C, since it is a subspace, is a left ideal in $F[G]$. Furthermore, any left ideal in $F[G]$ is a code over F of block length $|G|$, with G in its automorphism group through the left regular representation.

Later we will show how this property can be exploited in some of the most important classes of codes.

2.3 Parity-check matrices

Since we are considering subspaces of finite dimension, it is frequently convenient to express the code, which is a subspace, in terms of a basis for that subspace — given as an array of vectors in F^n. This can, of course, be done in many ways.

Definition 2.3.1 *If C is a q-ary $[n, k]$ code, a **generator matrix** for C is a $k \times n$ array obtained from any k linearly independent vectors of C.*

Via elementary row operations, a generator matrix G for C can be brought into reduced row echelon form and still generate C, and then, by permuting columns, it can be brought into a so-called *standard form* $G' = [I_k | A]$, where this is now a generator matrix for an equivalent (in fact, isomorphic) code. Here A is a $k \times (n - k)$ matrix over F.

Now we come to another important way of describing a linear code, *viz.* through its *orthogonal*. For this we need an *inner product* defined on our space; it is the standard inner product: for $v, w \in F^n$, $v = (v_1, v_2, \ldots, v_n)$, $w = (w_1, w_2, \ldots, w_n)$, we write the inner product of v and w as (v, w) where

$$(v, w) = \sum_{i=1}^{n} v_i w_i. \tag{2.1}$$

Definition 2.3.2 *Let C be a q-ary $[n, k]$ code. The* **orthogonal code** *is denoted by $C^\perp$ and is given by*

$$C^\perp = \{v \in F^n | (v, c) = 0 \text{ for all } c \in C\}.$$

We call C **self-orthogonal** *if $C \subseteq C^\perp$ and* **self-dual** *if $C = C^\perp$.*

Remark $C^\perp$ is often referred to as the *dual code* of C, but we prefer to use the term "orthogonal" and reserve the word "dual" for the dual space, i.e. the space of linear functionals. We will, however, use the term "self-dual" as required.

Proposition 2.3.1 *For any linear code C of length n,*

$$\dim(C) + \dim(C^\perp) = n.$$

Proof: If G is a generator matrix for C, where C has dimension k, then $C^\perp$ is simply given by the kernel of the linear transformation defined by G^t from F^n onto F^k. The result follows. $\square$

The vector in V all of whose entries are 1 is called the **all-one** vector and is denoted by $\jmath$. If $C = F\jmath$, then $C^\perp$ is the subspace of V of dimension $n - 1$ spanned by the vectors with precisely two non-zero entries, 1 and -1, in any two positions.

Exercise 2.3.1 Show that a standard form for $C^\perp$, where $C = F\jmath$, is an $(n - 1) \times (n - 1)$ identity matrix followed by a column of -1's. What is the standard form for C?

Taking G to be a generator matrix for C, a generator matrix H for $C^\perp$ satisfies $GH^t = 0$, i.e. $c \in C$ if and only if $cH^t = 0$, or, equivalently, $Hc^t = 0$. Any generator matrix H for $C^\perp$ is called a **parity-check** matrix for C. If G is written in the standard form $[I_k | A]$, then $H = [-A^t | I_{n-k}]$ is a parity-check matrix for the code with generator matrix G. A generator matrix in standard form simplifies the encoding: suppose data consisting of q^k messages are to be encoded by adding redundancy using the code C with generator matrix G. First identify the data with the vectors in F^k, where $F = \mathbf{F}_q$. Then for $u \in F^k$, encode u by forming uG. If $u = (u_1, u_2, \ldots, u_k)$ and G has rows $R_1, R_2, \ldots, R_k$, where each R_i is in F^n, then $uG = \sum_i u_i R_i = (x_1, x_2, \ldots, x_k, x_{k+1}, \ldots, x_n) \in F^n$, which is now encoded. But when G is in standard form, the encoding takes the simpler form

$$u \mapsto (u_1, u_2, \ldots, u_k, x_{k+1}, \ldots, x_n),$$

and here the $u_1, \ldots, u_k$ are the *message* or *information* symbols, and the last $n - k$ entries are the *check* symbols, and represent the redundancy.

In general it is not possible to say anything about the minimum weight of $C^\perp$ knowing only the minimum weight of C but, of course, either a generator matrix or a parity-check matrix gives complete information about both C and $C^\perp$. In particular, a parity-check matrix for C determines the minimum weight of C in a useful way:

Theorem 2.3.1 *Let H be a parity-check matrix for an $[n, k, d]$ code C. Then every choice of $d - 1$ or fewer columns of H forms a linearly independent set. Moreover if every $d - 1$ or fewer columns of a parity-check matrix for a code C are linearly independent, then the code has minimum weight at least d.*

Proof: Suppose a relationship

$$\sum_{j=1}^{r} a_j C_{i_j} = 0$$

holds amongst some r columns, C_{i_j} of H. Then the vector a with i_j^{th} coordinate a_j, for $j = 1, \ldots, r$, and the other coordinates 0, is in C, so the number of non-zero a_j's is at least d. Clearly any codeword of weight m yields a set of m columns that are linearly dependent. These observations complete the proof. $\square$

This theorem gives another derivation of the Singleton bound and we record the proof here:

Corollary 2.3.1 *For a linear $[n, k, d]$ code, $d \leq n - k + 1$.*

Proof: A parity-check matrix has only $n - k$ rows and hence some $n - k + 1$ columns must be linearly dependent. $\square$

Notice that in terms of generator matrices, two codes C and C' with generator matrices G and G' are equivalent if and only if there exist a non-singular matrix M and a monomial matrix N such that $MGN = G'$, with isomorphism if N is a permutation matrix and equality if $N = I_n$, n being the block length of the codes.

Example 2.3.1 The smallest Hamming code is a [7,4,3] binary code, which is a perfect single-error-correcting code. It can be given by the generator

Figure 2.2: Hamming code

matrix G in standard form $[I_4|A]$ where

$$A = \begin{pmatrix} 1 & 1 & 1 \\ 0 & 1 & 1 \\ 1 & 0 & 1 \\ 1 & 1 & 0 \end{pmatrix}.$$

Thus a parity-check matrix will be

$$H = \begin{pmatrix} 1 & 0 & 1 & 1 & 1 & 0 & 0 \\ 1 & 1 & 0 & 1 & 0 & 1 & 0 \\ 1 & 1 & 1 & 0 & 0 & 0 & 1 \end{pmatrix}.$$

Taking $\{a, b, c, d\}$ as the information symbols, and $\{b', c', d'\}$ as the check symbols, the diagram shown in Figure 2.2 (due to McEliece — see, for example, [199]) can be used to correct a single error, for any vector received, if at most a single error has occurred. The rule is that the sum of the coordinates in any of the three circles must be 0, which constitute the "parity checks" as seen from the matrix H above. Thus, for example, if the vector 1011111 is received, checking the parity in the three circles shows that an error occurred at the information symbol b, so that the error is corrected, yielding 1111111.

This example also illustrates the connection between designs and codes: the seven vectors of weight 3 in this code are easily discovered using Figure 2.2. For example the vector with 1's only at a, b and b' is such a vector and similarly, by symmetry, a, c, and c' and a, d and d'. The reader will have no difficulty in seeing that these seven vectors are precisely the lines of the Fano plane of Figure 1.1(relabelled). It follows immediately that the row space of any incidence matrix of the Fano plane is isomorphic to the [7, 4, 3] Hamming code.

Exercise 2.3.2 Using Figure 2.2, explicitly determine the weight-3 vectors of this Hamming code and verify that their supports are the lines of the

Fano plane. Show that the supports of the weight-4 vectors yield the biplane of order 2, i.e. a $(7, 4, 2)$ design. Finally, verify that the subspace generated by the weight-4 vectors is the orthogonal of this Hamming code.

A general method of decoding for linear codes is a method, due to Slepian [266], that uses nearest-neighbour decoding and is called **standard-array decoding**. The **error vector** is defined to be $e = w - v$, where v is the codeword sent and w is the received vector. Given the received vector we wish to determine the error vector. We look for that coset of the subgroup C in F^n that contains w and observe that the possible error vectors are just the members of this coset. The strategy is thus to look for a vector e of minimum weight in the coset $w + C$, and decode as $v = w - e$. A vector of minimum weight in a coset is called a *coset leader*; of course it might not be unique, but it will be in the event that its weight is at most ρ, where ρ is the packing radius, and this will always happen when at most ρ errors occurred during transmission. It should be noted that there may be a unique coset leader even when the weight of that leader is greater than ρ and thus a complete analysis of the probability of success of nearest-neighbour decoding will involve analysing the weight distribution of all the cosets of C; in the engineering literature this is known as "decoding beyond the minimum distance". For examples of such explicit information on the weight distribution of cosets see Berlekamp and Welch [35] and Maiorana [206]. Use of a parity-check matrix H for C to calculate the *syndrome*, viz. wH^t, of the received vector w assists this decoding method, the syndrome being constant over a coset and equal to the zero vector when a codeword has been received. Note that the covering radius of C is simply the maximum of the weights of the coset leaders; in a perfect t-error-correcting code all the coset leaders have weight at most t. Decoding is an important and heavily studied aspect of coding theory, but it will not play a role in our application of this theory to the investigation of designs.

There are various natural constructions for building codes; a full discussion of these constructions giving "new codes from old" can be found in MacWilliams and Sloane [203]. Our main use will be for codes extended by adding an overall parity check:

Definition 2.3.3 *Let C be a q-ary $[n, k, d]$ code. Define the* **extended code** $\widehat{C}$ *to be the code of length $n + 1$ in F^{n+1} of all vectors $\hat{c}$ for $c \in C$ where, if $c = (c_1, \ldots, c_n)$, then*

$$\hat{c} = \left(c_1, \ldots, c_n, -\sum_{i=1}^{n} c_i \right).$$

This is called **adding an overall parity check** , for we see that if
$v = (v_1, \ldots, v_{n+1})$ then $v \in \widehat{C}$ satisfies $\sum v_i = 0$. If C has generator
matrix G and parity-check matrix H, then $\widehat{C}$ has generator matrix $\widehat{G}$ and
parity-check matrix $\widehat{H}$, where $\widehat{G}$ is obtained from G by adding an $(n+1)^{\text{th}}$
column such that the sum of the columns of $\widehat{G}$ is the zero column, and $\widehat{H}$
is obtained from H by attaching an $(n-k+1)^{\text{th}}$ row and $(n+1)^{\text{th}}$ column
onto H, the row being all 1's and the column being $(0, 0, \ldots, 0, 1)^t$. If C is
binary with d odd, then $\widehat{C}$ will be a $[n+1, k, d+1]$ code.

The inverse process to extending an $[n, k, d]$ code is that of **puncturing**:
this is achieved by simply deleting a coordinate, thus producing a linear
code of length $n - 1$. The dimension will be k or $k - 1$, clearly, but in the
great majority of cases the dimension will remain k; the minimum weight
may change in either way, but unless the minimum weight of the original
code is 1, the minimum weight of the punctured code will be either $d - 1$
or d and in the great majority of cases $d - 1$.

Example 2.3.2 Extending the $[7, 4, 3]$ binary Hamming code gives an
$[8, 4, 4]$ binary code. From the description of the Hamming code as the
row space of the incidence matrix of the Fano plane, it follows easily that
this extended code is self-dual. (For an elementary account of another use
of error-correcting codes in data storage and retrieval and a full account of
this classic code, see McEliece [199])

Before proceeding to the linear codes associated with designs, we give
another useful construction.

Proposition 2.3.2 *Let C and C' be $[n, k, d]$ and $[n, k', d']$ q-ary codes,
respectively. For any $u, v \in F^n$ let $(u|v)$ denote the vector of length $2n$
formed by writing u followed by v. Then*

$$\{(u|u + v) : u \in C, v \in C'\}$$

is a $[2n, k + k', \min\{2d, d'\}]$ q-ary code.

Proof: The code is spanned by vectors of the form $(u|u)$ and $(0|v)$ where u
and v range over spanning sets for C and C', respectively; moreover, these
vectors will give a basis for the code if the u's and v's give bases for C and
C'. The code has vectors of weight $2d$ and d', so the minimum weight is at
most the minimum of these. Let $(u|u + v)$ be any vector in the code with
$v \neq 0$. Then

$$\begin{aligned} \text{wt}((u|u + v)) &= \text{wt}(u) + \text{wt}(u + v) \\ &\geq \text{wt}(u) + \text{wt}(v) - \text{wt}(u) \\ &\geq \text{wt}(v) \geq d' \end{aligned}$$

which gives the stated result. □

2.4 Codes from designs

We now examine the most immediate way in which a linear code can be associated with a design — or, indeed, with any incidence structure. In good cases the p-ary code of the structure, as we define it below, will provide a rough but handy characterization of the structure, but there are some very notable exceptions, the Witt 5-(12,6,1) design's connection with the ternary $[12, 6, 6]$ Golay code being one. We will, of course, examine this more subtle connection, and some other constructions, later (see, for example, Section 7.4).

Let $\mathcal{S} = (\mathcal{P}, \mathcal{B}, \mathcal{I})$ be an incidence structure. For any field F the vector space $F^{\mathcal{P}}$ of functions from $\mathcal{P}$ to F has a distinguished basis given by the characteristic functions of the singleton subsets of $\mathcal{P}$ and thus a setting for linear codes is naturally present.

Definition 2.4.1 *The* **code** *of* $\mathcal{S} = (\mathcal{P}, \mathcal{B}, \mathcal{I})$ *over the field F is the subspace $C_F(\mathcal{S})$ of $F^{\mathcal{P}}$ spanned by the vectors corresponding to the characteristic functions of the blocks of $\mathcal{S}$. Thus*

$$C_F(\mathcal{S}) = \langle v^B | B \in \mathcal{B} \rangle.$$

The field F will in practice be a prime field $\mathbf{F}_p$, in which case we may also write $C_p(\mathcal{S})$ for the code. In the definition we have used the notation introduced in Definition 1.2.5 and Section 1.6, writing

$$v^B = \sum_{x\mathcal{I}B} v^{\{x\}} \tag{2.2}$$

for the block $B \in \mathcal{B}$. Technically this notation involves an abuse of language when the block is not a set of points and, in this case, in order to define v^B we must identify B with the set of points incident with it. We will, nevertheless, go ahead and speak of "characteristic functions of blocks" in all that follows. We will also speak of v^B as the **incidence vector** of the block B.

Notice that our definition of the code over F of $\mathcal{S}$ does not specify an *ordering* of the distinguished basis. In some situations it will be convenient to label the points (from 1 to v) and consider the code to be, effectively, the row space of an incidence matrix of $\mathcal{S}$. Whatever the labelling, these codes are all isomorphic, so we may speak of *the* code of $\mathcal{S}$ over F. However, thinking of an incidence matrix naturally raises the question of the *column* space of the matrix. This is the code of the dual structure, S^t, and is a subspace of $F^{\mathcal{B}}$.

Example 2.4.1 The Fano plane $\Pi = (\mathcal{P}, \mathcal{L}, \mathcal{I})$ has $C_F(\Pi) = F^{\mathcal{P}}$ when F is of characteristic $p \neq 2, 3$ (since the determinant of an incidence matrix is 24); $C_3(\Pi) = (\mathbf{F_3 \jmath})^{\perp}$ and $C_2(\Pi)$ is the smallest Hamming code: see Exercise 2.3.2 and Section 2.5.

Incidence structures are far too general for us to be able to say much about their codes; in fact, we will focus on the special case of t-designs, where $t \geq 2$. We can then immediately dismiss all but a finite number of possibilities for the characteristic p of the finite field if the code is to be of any assistance in classifying the design. Notice that we might as well take only prime fields, since our generating vectors have all entries equal to 1 or 0.

Definition 2.4.2 *If $S = (\mathcal{P}, \mathcal{B}, \mathcal{I})$ is any incidence structure and p is any prime, the **p-rank** of S is the dimension of the code $C_F(S)$, where $F = \mathbf{F}_p$, and is written*

$$\mathrm{rank}_F(S) = \mathrm{rank}_p(S) = \dim(C_F(S)).$$

Our first result is one that is immediately discovered by anyone who begins to think about the subject and it has appeared in almost every elementary discussion of codes and designs. It even gives another proof of Fisher's inequality, as we remark below, but, as far as we know, this proof was not discovered until the connection between codes and designs was beginning to be understood.

Theorem 2.4.1 *Let $\mathcal{D} = (\mathcal{P}, \mathcal{B}, \mathcal{I})$ be a non-trivial 2-(v, k, λ) design of order n. Let p be a prime and let F be a field of characteristic p where p does* not *divide n. Then*

$$\mathrm{rank}_p(\mathcal{D}) \geq (v - 1)$$

with equality if and only if p divides k; in the case of equality we have that $C_F(\mathcal{D}) = (F\jmath)^{\perp}$ and otherwise $C_F(\mathcal{D}) = F^{\mathcal{P}}$.

Proof: Fix a prime p not dividing n and let $C = C_F(\mathcal{D})$. Set $w = \sum_{B \in \mathcal{B}} v^B = r\jmath$. For $x \in \mathcal{P}$, set $w_x = \sum_{xIB} v^B$, so that $w_x(x) = r$, and $w_x(y) = \lambda$ for $y \neq x$. Then $w - w_x = n(\jmath - v^{\{x\}}) \in C$, so $\jmath - v^{\{x\}} \in C$ for any $x \in \mathcal{P}$, and hence $v^{\{x\}} - v^{\{y\}} \in C$ for any $x, y \in \mathcal{P}$. Thus $C \supseteq (F\jmath)^{\perp}$, and so $\mathrm{rank}_p(\mathcal{D}) \geq v - 1$. We have equality if and only if $C = (F\jmath)^{\perp}$ and this occurs if and only if $v^B \in (F\jmath)^{\perp}$ for $B \in \mathcal{B}$, that is if and only if p divides k since $(v^B, \jmath) = k$. $\square$

Remark Observe that this theorem gives another proof of Fisher's inequality, Corollary 1.4.1, since, there being infinitely many primes, there is at least one that does not divide nk and hence $b \geq \mathrm{rank}_p(\mathcal{D}) \geq v$ for any such prime p.

Exercise 2.4.1 Prove that if p is a prime dividing $n + 1$ then $C_p(\mathcal{D}) = (\mathbf{F}_p \jmath)^{\perp}$ when $\mathcal{D}$ is a $(n^2 + n + 1, n + 1, 1)$-design and thus verify the assertion we made about the ternary code of the Fano plane in Example 2.4.1.

It follows that for p-ary codes to assist in a classification of designs, we must certainly have p dividing n. This condition is necessary, but not sufficient: there are designs whose associated codes always have rank v or $v - 1$ and hence are of no assistance in this way. We shall soon see (Theorem 2.4.2) that, for a symmetric design $\mathcal{D}$, the p-rank for p dividing n satisfies $\mathrm{rank}_p(\mathcal{D}) \leq \frac{1}{2}(v+1)$ and in this case the codes play an important role in classification. These matters will be treated fully in Chapter 4.

In studying an incidence structure $\mathcal{S}$ through the codes $C_p(\mathcal{S})$, the orthogonal code $C_p(\mathcal{S})^{\perp}$, and its relationship to $C_p(\mathcal{S})$, need to be investigated. The intersection of these two codes has, as we will see, a leading role to play. Thus we make the following definition:

Definition 2.4.3 *If $\mathcal{S} = (\mathcal{P}, \mathcal{B}, \mathcal{I})$ is any incidence structure and F any field, then the **hull** of $\mathcal{S}$ at F is the code*

$$\mathrm{Hull}_F(\mathcal{S}) = C_F(\mathcal{S}) \cap C_F(\mathcal{S})^{\perp}.$$

We also write $H_F(\mathcal{S})$ or $H_p(\mathcal{S})$ if $F = \mathbf{F}_p$; moreover, if the field is understood we write simply $H(\mathcal{S})$. Further, we define

$$B_F(\mathcal{S}) = \mathrm{Hull}_F(\mathcal{S})^{\perp} = C_F(\mathcal{S}) + C_F(\mathcal{S})^{\perp},$$

and write $B_p(\mathcal{S})$ or $B(\mathcal{S})$ similarly.

*Two structures are **linearly equivalent** (over F) if their hulls (over F) are code-isomorphic.*

Structures may be linearly equivalent over some field but have non-isomorphic codes over that field: see, for example, Section 8.3, Figure 8.1, where the codes from the designs actually have different dimension. Thus linear equivalence is a coarser equivalence relation than design isomorphism.

Another subcode of $C_F(\mathcal{S})$ that will in certain cases be closely related to the hull is the following:

Definition 2.4.4 *If $\mathcal{S} = (\mathcal{P}, \mathcal{B}, \mathcal{I})$ is any incidence structure and F any field, define $E_F(\mathcal{S})$ to be the code generated by the differences of the incidence vectors of the structure:*

$$E_F(\mathcal{S}) = \langle v^B - v^C | B, C \in \mathcal{B} \rangle = \langle v^B - v^{B_0} | B \in \mathcal{B} \rangle,$$

where $B_0 \in \mathcal{B}$.

For a given incidence structure $\mathcal{S}$ and a given field F, the three codes $C(\mathcal{S}), H(\mathcal{S})$ and $E(\mathcal{S})$, along with their orthogonals, need to be studied in conjunction. Further, the codes associated with the *dual* structure, $\mathcal{S}^t$, and the *complementary* structure, $\bar{\mathcal{S}}$, and its dual, necessarily also play a part. Now the blocks of $\mathcal{S}^t$ are the points of $\mathcal{S}$, and although the notation introduced in Section 1.6 is explicit — v^B denoting the incidence vector of a block of $\mathcal{S}$ and $v^{\{B\}}$ a basis vector of $F^{\mathcal{B}}$ — we are, for clarity's sake, going to introduce redundancy into the notation: the characteristic function of a subset X of $\mathcal{B}$ will be denoted by w^X rather than v^X. Since the blocks of $\mathcal{S}^t$ are the points of $\mathcal{S}$ and $C(\mathcal{S}^t)$ is the subspace of $F^{\mathcal{B}}$ spanned by the characteristic functions of its blocks, we have

$$C(\mathcal{S}^t) = \langle w^x | x \in \mathcal{P} \rangle,$$

where, as before,

$$w^x = \sum_{B : x \mathcal{I} B} w^{\{B\}}. \tag{2.3}$$

Just as the block B is identified with the points incident with it, so the point x is identified with the blocks incident with it. This leads on rather naturally to the flag space $F^{\mathcal{I}}$, which has as distinguished basis the characteristic functions of the flags $(x, B) \in \mathcal{I}$. For the same reason we mentioned above, we write the characteristic functions in this space differently again, as u^X, where $X \subseteq \mathcal{I}$. Thus the standard basis for $F^{\mathcal{I}}$ is

$$\{u^{\{(x,B)\}} | (x, B) \in \mathcal{I}\},$$

and

$$F^{\mathcal{I}} = \langle u^{\{(x,B)\}} | (x, B) \in \mathcal{I} \rangle. \tag{2.4}$$

Corresponding to $v^B \in F^{\mathcal{P}}$ and $w^x \in F^{\mathcal{B}}$, we have[3]

$$u^B = \sum_{x : x \mathcal{I} B} u^{\{(x,B)\}} \quad \text{and} \quad u^x = \sum_{B : x \mathcal{I} B} u^{\{(x,B)\}}. \tag{2.5}$$

---

[3]With this notation the cumbersome braces used for the characteristic functions of the singleton sets, i.e. $v^{\{x\}}$, $w^{\{B\}}$ and $u^{\{(x,B)\}}$, can be omitted, since the distinction between v^B and w^B, for example, is clear. We include the braces in what follows for clarity.

These concepts fit more generally into *geometries of rank n* (where ours have rank 2): see Ott [227, 228] and also Ghinelli-Smit [103], where the flag space $F^{\mathcal{I}}$ is called the *standard module*. Here we only need introduce the *boundary map* δ from $F^{\mathcal{I}}$ to $F^{\mathcal{P}} \oplus F^{\mathcal{B}}$, to obtain the *Steinberg module* of the structure; later in this section we will see that this module provides a lower bound for the rank over F of a certain type of structure by Theorem 2.4.3 of Hillebrandt [135]. Note that since we have chosen bases for both $F^{\mathcal{P}}$ and $F^{\mathcal{B}}$, the basis for $F^{\mathcal{P}} \oplus F^{\mathcal{B}}$ is forced upon us. It is the set

$$\{(v^{\{x\}}, 0)|x \in \mathcal{P}\} \cup \{(0, w^{\{B\}})|B \in \mathcal{B}\}.$$

We could, alternatively, think of the functions $v^{\{x\}}$ and $w^{\{B\}}$ as being defined on the union $\mathcal{P} \cup \mathcal{B}$, taking, for example, $v^{\{x\}}(B) = 0$ for all $B \in \mathcal{B}$. Having done this we can identify $F^{\mathcal{P}} \oplus F^{\mathcal{B}}$ with $F^{\mathcal{P} \cup \mathcal{B}}$ and then have that

$$(v^{\{x\}}, 0) = v^{\{x\}} \quad \text{and} \quad (0, w^{\{B\}}) = w^{\{B\}}.$$

Now the meaning of $v^X + w^Y$, where $X \subseteq \mathcal{P}$ and $Y \subseteq \mathcal{B}$, should be clear to the reader; for example, $v^{\mathcal{P}} + w^{\mathcal{B}}$ is the all-one vector.

Definition 2.4.5 *For an incidence structure* $\mathcal{S} = (\mathcal{P}, \mathcal{B}, \mathcal{I})$ *and any field* F, *the* **boundary map** $\delta : F^{\mathcal{I}} \to F^{\mathcal{P}} \oplus F^{\mathcal{B}} = F^{\mathcal{P} \cup \mathcal{B}}$ *is defined on the basis elements by*

$$\delta : u^{\{(x,B)\}} \mapsto v^{\{x\}} - w^{\{B\}}$$

and extended linearly. The **Steinberg module,** $St_F(\mathcal{S})$, *is the kernel of the map* δ.

Since $\delta(F^{\mathcal{I}}) \subseteq (F(v^{\mathcal{P}} + w^{\mathcal{B}}))^{\perp}$, it follows immediately that the dimension of the Steinberg module is at least $|\mathcal{I}| - |\mathcal{P}| - |\mathcal{B}| + 1$. Later in this section we will see that this is precisely the dimension for a certain type of structure that we will be concerned with.

We have already seen that the presence of the all-one vector $\jmath$ in the code or its orthogonal leads to certain conclusions. If we look at the dual structure, $\mathcal{S}^t$, we can obtain more. For the sake of clarity we emphasise which all-one vector we are referring to by letting $\jmath_{\mathcal{P}}$ be the all-one vector in $F^{\mathcal{P}}$ and $\jmath_{\mathcal{B}}$ the all-one vector in $F^{\mathcal{B}}$. Thus

$$\jmath_{\mathcal{P}} = v^{\mathcal{P}} \quad \text{and} \quad \jmath_{\mathcal{B}} = w^{\mathcal{B}}.$$

Now recall (see (1.6)) the map

$$\nabla : F^{\mathcal{B}} \to F^{\mathcal{P}}$$

which, on the standard basis of $F^\mathcal{B}$ and in our new notation, is given by $w^{\{B\}} \mapsto v^B$ and then extended linearly. Thus

$$\nabla(f) = \sum_{B \in \mathcal{B}} f(B) v^B$$

for each $f \in F^\mathcal{B}$. Then $\nabla(F^\mathcal{B}) = C_F(\mathcal{S}) = \langle v^B | B \in \mathcal{B} \rangle$ and its kernel is $\ker(\nabla) = C_F(\mathcal{S}^t)^\perp$.

Next we define

$$\sigma : F^\mathcal{B} \to F$$

by $\sigma(f) = \sum_{B \in \mathcal{B}} f(B)$. Clearly σ is onto with $\ker(\sigma) = (F\jmath_\mathcal{B})^\perp$ and we have immediately that $\jmath_\mathcal{B} \in C_F(\mathcal{S}^t)$ if and only if $C_F(\mathcal{S}^t)^\perp \subseteq (F\jmath_\mathcal{B})^\perp$, i.e. if and only if $\ker(\nabla) \subseteq \ker(\sigma)$. Finally, observe that $\nabla(\ker \sigma) = E_F(\mathcal{S})$.

Proposition 2.4.1 *Let $\mathcal{S} = (\mathcal{P}, \mathcal{B}, \mathcal{I})$ be an incidence structure and F a field. Then $E_F(\mathcal{S})$ has codimension at most 1 in $C_F(\mathcal{S})$ and $E_F(\mathcal{S}) = C_F(\mathcal{S})$ if and only if $\jmath_\mathcal{B} \notin C_F(\mathcal{S}^t)$. Dually, $E_F(\mathcal{S}^t)$ has codimension at most 1 in $C_F(\mathcal{S}^t)$ and $E_F(\mathcal{S}^t) = C_F(\mathcal{S}^t)$ if and only if $\jmath_\mathcal{P} \notin C_F(\mathcal{S})$.*

Proof: That $E_F(\mathcal{S})$ has codimension at most 1 is obvious from the fact that $E_F(\mathcal{S}) = \langle v^B - v^{B_0} | B \in \mathcal{B} \rangle$. Observe first that if $\jmath_\mathcal{B} \notin C_F(\mathcal{S}^t)$, then $F^\mathcal{B} = \ker(\nabla) + \ker(\sigma)$. For $v \in C_F(\mathcal{S})$ and $w \in F^\mathcal{B}$ with $\nabla(w) = v$, express w as $w' + w''$ with $w' \in \ker(\nabla)$ and $w'' \in \ker(\sigma)$. Then $v = \nabla(w'') \in E_F(\mathcal{S})$ and therefore $E_F(\mathcal{S}) = C_F(\mathcal{S})$. If $\jmath_\mathcal{B} \in C_F(\mathcal{S}^t)$, then $\ker(\nabla) \subset \ker(\sigma)$ and since $\ker(\sigma)$ is of codimension 1 in $F^\mathcal{B}$ its image under ∇, namely $E_F(\mathcal{S})$, is of codimension 1 in $\nabla(F^\mathcal{B}) = C_F(\mathcal{S})$. $\square$

Corollary 2.4.1 *For any incidence structure $\mathcal{S}$ and any field F we have that $E_F(\mathcal{S})$ is of codimension 1 in $C_F(\mathcal{S})$ if and only if the all-one vector is in $C_F(\mathcal{S}^t)$.*

Now consider the complementary structure $\bar{\mathcal{S}}$. If B is a block of $\mathcal{S}$ and $\bar{B}$, the complementary block, then $v^{\bar{B}} = \jmath_\mathcal{P} - v^B$ and it follows immediately that

$$E_F(\mathcal{S}) = E_F(\bar{\mathcal{S}}).$$

Moreover, it is clear that

$$C_F(\mathcal{S}) + F\jmath_\mathcal{P} = C_F(\bar{\mathcal{S}}) + F\jmath_\mathcal{P}, \tag{2.6}$$

with $C_F(\mathcal{S}) = C_F(\bar{\mathcal{S}})$ if and only if $\jmath_\mathcal{P} \in C_F(\mathcal{S}) \cap C_F(\bar{\mathcal{S}})$. In any event we have

$$\dim(C_F(\mathcal{S})) - 1 \leq \dim(C_F(\bar{\mathcal{S}})) \leq \dim(C_F(\mathcal{S})) + 1. \tag{2.7}$$

We use these simple facts to give conditions sufficient to ensure that the all-one vector is in various codes given by a design.

Proposition 2.4.2 *Let $S = (\mathcal{P}, \mathcal{B}, \mathcal{I})$ be an incidence structure and set $F = \mathbf{F}_2$. Then $C_F(S) = C_F(\bar{S})$ if and only if $\jmath_{\mathcal{P}} \in E_F(S)$.*

Proof: If $\jmath_{\mathcal{P}} \in E_F(S)$, then clearly $\jmath_{\mathcal{P}} \in C_F(S) \cap C_F(\bar{S})$ and by (2.6) above $C_F(S) = C_F(\bar{S})$. On the other hand suppose $C_F(S) = C_F(\bar{S})$. Then, if $E_F(S) = C_F(S)$, clearly $\jmath_{\mathcal{P}} \in E_F(S)$. Suppose $E_F(S) \neq C_F(S)$ and $\jmath_{\mathcal{P}} \notin E_F(S)$. Then, since $E_F(S)$ is of codimension 1 in $C_F(S)$, $C_F(S) = F\jmath_{\mathcal{P}} + E_F(S)$ and for any block B of S we have that $v^B = \jmath_{\mathcal{P}} + e$ for some $e \in E_F(S)$. But then $v^{\bar{B}} = \jmath_{\mathcal{P}} + v^B = e \in E_F(S)$ and $E_F(S) = C_F(\bar{S}) = C_F(S)$, a contradiction. $\square$

We next prove a result due to Klemm [172] that generalizes an earlier result concerning the codes of symmetric designs. Klemm's result applies to *any* 2-design.

Theorem 2.4.2 (Klemm) *Let $\mathcal{D} = (\mathcal{P}, \mathcal{B})$ be a 2-(v, k, λ) design of order n and let p be a prime dividing n. Then*

$$\mathrm{rank}_p(\mathcal{D}) \leq \frac{|\mathcal{B}| + 1}{2};$$

moreover, if p does not divide λ and p^2 does not divide n, then

$$C_p(\mathcal{D})^\perp \subseteq C_p(\mathcal{D})$$

and $\mathrm{rank}_p(\mathcal{D}) \geq v/2$.

Proof: Let A be an incidence matrix for $\mathcal{D}$ and set $F = \mathbf{F}_p$. Denote by C the row space of A over F and by D the column space of A over F. Thus C is isomorphic to $C_p(\mathcal{D})$ and D is isomorphic to the code of the dual structure, $C_p(\mathcal{D}^t)$. Recall that $A^t A = nI + \lambda J$ and $JA = rJ$. Also, $C = \{wA | w \in F^{\mathcal{B}}\}$ and $C^\perp = \{u | u \in F^{\mathcal{P}}, uA^t = 0\}$, where we are making the obvious identifications since we are dealing with the incidence matrix of $\mathcal{D}$. We equip both $F^{\mathcal{P}}$ and $F^{\mathcal{B}}$ with the standard inner products given by the standard bases.

Now $D = C_p(\mathcal{D}^t) = \langle w^x | x \in \mathcal{P} \rangle$. Then, with the standard inner product, $(w^x, w^y) = \lambda$ if $x \neq y$, and $(w^x, w^x) = r$. Thus

$$(w^x - w^y, w^z) \equiv 0 \pmod{p}$$

for any points x, y, z, so that

$$E_F(\mathcal{D}^t) \subseteq D \cap D^\perp = H_F(\mathcal{D}^t).$$

Now the codimension of $E_F(\mathcal{D}^t)$ in $C_F(D)$ is at most 1 and hence

$$\dim(D^\perp) \geq \dim(D \cap D^\perp) \geq \dim(E_F(\mathcal{D}^t)) \geq \dim(D) - 1,$$

and thus $|\mathcal{B}| - \dim(D) \geq \dim(D) - 1$, and since $\dim(D) = \dim(C)$, we get that $\dim(C) \leq (|\mathcal{B}| + 1)/2$ or, in other words, that $\mathrm{rank}_p(\mathcal{D}) \leq (|\mathcal{B}| + 1)/2$.

Working next over the rational field $\mathbf{Q}$ we obtain, from $A^t A = nI + \lambda J$, that $I = \frac{1}{n}(A^t A - \lambda J)$. Since $JA = rJ = A^t J$,

$$I = \frac{1}{n}(A^t A - \frac{\lambda}{r} JA) = \frac{1}{n}(A^t - \frac{\lambda}{r} J)A. \qquad (2.8)$$

Let $u = (u_1, u_2, \ldots, u_v)$ be a vector with integral components and suppose that $u \in C^{\perp}$ when u is read modulo p. Then $uA^t \equiv 0 \pmod{p}$. Now set

$$w = \frac{1}{n} u (A^t - \frac{\lambda}{r} J).$$

Then, from (2.8),

$$wA = \frac{1}{n} u (A^t - \frac{\lambda}{r} J) A = uI = u, \qquad (2.9)$$

where w is a vector with components in $\mathbf{Q}$. In order to show that u is in C when read modulo p we need to show that the components of w are in the ring

$$\mathbf{Q}_p = \{ \frac{x}{y} | x, y \in \mathbf{Z}, p \text{ relatively prime to } y \},$$

the ring of fractions with respect to the prime ideal (p). Since we are assuming that p^2 does not divide n and that all the components of uA^t are divisible by p, the components of $\frac{1}{n} uA^t$ are indeed in $\mathbf{Q}_p$ and the components of

$$\frac{r}{n} uJ = \frac{1}{n} urJ = \frac{1}{n} uA^t J$$

are also. Since we are assuming that p does *not* divide λ and that it *does* divide n, it cannot divide $r = n + \lambda$ and we have that the components of

$$\frac{r}{n} \frac{\lambda}{r^2} uJ = \frac{\lambda}{rn} uJ$$

are in $\mathbf{Q}_p$. Thus

$$w = \frac{u}{n}(A^t - \frac{\lambda}{r} J) = \frac{1}{n} uA^t - \frac{\lambda}{rn} uJ$$

has the same property and hence $wA = u \in C$. We have thus shown that $C^{\perp} \subseteq C$ and hence $\mathrm{rank}_p(\mathcal{D}) \geq |\mathcal{P}|/2 = v/2$. $\square$

Remark The bound $(|\mathcal{B}|+1)/2$ is best possible when $|\mathcal{B}| = |\mathcal{P}|$, i.e. when the design is symmetric, but usually $|\mathcal{P}|$ is small compared to $|\mathcal{B}|$ and the upper bound is of no use whatsoever: for example, for the unitary design 2-$(28,4,1)$, $|\mathcal{B}| = 63$, and the bound tells us nothing. The lower bound is also best possible for p sharply dividing n but here the important part of Klemm's result is that the code contains its orthogonal in this case.

Exercise 2.4.2 (1) Use Klemm's theorem to show that the dimension of the binary code of the Fano plane is 4 and that it contains its orthogonal. Why does Klemm's theorem not give the same result for the binary code of the biplane of order 2?

(2) Prove that the p-ary code of a projective plane of order p, i.e. a 2-$(p^2+p+1, p+1, 1)$ design, contains its orthogonal and is of dimension $(p^2+p+2)/2$. Show that the minimum weight is $p+1$ and that there are $(p-1)(p^2+p+1)$ minimum-weight vectors. (It is believed that the minimum weight of the orthogonal is $2p$ and this is known for the classical projective planes. See Chapter 6.)

Most of the characterizations we will obtain for designs using the associated codes will involve the divisibility properties of the prime p with respect to the design parameters, as in Theorems 2.4.1 and 2.4.2. There is no guarantee that if p divides n the code of the design will be of any use in the characterization: inversive planes (see Chapter 8) and the Witt 5-$(12,6,1)$ design are examples. Klemm's theorem gives a lower bound on the dimension when p^2 does not divide n and, as we remarked above, is best possible in this case. More general results, due to Bruen and Ott [56] and, more recently, to Hillebrandt [135], give lower bounds for the rank over any field F, but only for *linear* or *partial linear* spaces — defined below (or see Section 1.2). In particular the results yield lower bounds on the ranks of Steiner 2-designs. Thus neither approach will yield information when p^2 divides the order of a symmetric design with λ greater than 1 and we know of no such lower bound; moreover, we shall see in Chapter 6 that the lower bound obtained here is far from satisfactory for projective planes.

We develop next a modification of Hillebrandt's result since it gives a better bound than that obtained in Bruen and Ott [56] and uses similar, although more general, arguments. We start with a definition:

Definition 2.4.6 *An incidence structure $\mathcal{S} = (\mathcal{P}, \mathcal{B}, \mathcal{I})$ is said to be connected if, for any two elements $X, Y \in \mathcal{P} \cup \mathcal{B}$, there exists a sequence $X_1, \ldots, X_n$ of elements of $\mathcal{P} \cup \mathcal{B}$ such that X is incident with X_1, X_1 is incident with $X_2, \ldots,$ and X_n is incident with Y.*

Notice that any 2-design is connected. Observe also that, for a connected incidence structure, the Steinberg module $St_F(S)$ has dimension precisely $|\mathcal{I}| - |\mathcal{P}| - |\mathcal{B}| + 1$ since, in this case, the image of δ is of codimension 1: for example, $\langle \delta(F^{\mathcal{I}}), v^{\{x\}} \rangle = F^{\mathcal{P}} \oplus F^{\mathcal{B}}$ for any point x.

The method of Hillebrandt involves the construction of spanning elements of $St_F(S)$ of a particular geometric nature. Since we do not need the full generality of his results, we will modify them a little for our purposes. The following results on bounds are thus all due to Hillebrandt and can be found in a more general, and possibly slightly different, form in Hillebrandt [135].

Definition 2.4.7 *Let $x_1, x_2, \ldots, x_n$ be a sequence of not necessarily distinct points and let $B_1, B_2, \ldots, B_n$ be a sequence of blocks, also not necessarily distinct, of an incidence structure S. Set $x_{n+1} = x_1$. If $f_i = (x_i, B_i)$ and $g_i = (x_{i+1}, B_i)$, for $i = 1, 2, \ldots, n$, are flags of S, then the vector*

$$c = \sum_{i=1}^{n} (u^{\{f_i\}} - u^{\{g_i\}})$$

of $F^{\mathcal{I}}$ is called a **circuit** *of S.*

Lemma 2.4.1 *If S is an incidence structure and F is any field, then the Steinberg module $St_F(S)$ is spanned by circuits of S.*

Proof: Let c be a circuit as in Definition 2.4.7. Then, if δ is the boundary map as defined in Definition 2.4.5,

$$\delta(c) = \delta \left(\sum_{i=1}^{n} (u^{\{f_i\}} - u^{\{g_i\}}) \right) = \sum_{i=1}^{n} (v^{\{x_i\}} - w^{\{B_i\}} - v^{\{x_{i+1}\}} + w^{\{B_i\}}) = 0,$$

showing that any circuit c is in $St_F(S)$.

Now suppose that $St_F(S)$ is not spanned by circuits. Let c be an element of minimum weight in $St_F(S)$ that is not a linear combination of circuits and suppose the flag $f = (x, B)$ is in the support of c. Since our structure is finite, and since $c \in St_F(S)$, there must be a sequence of flags in the support of c that defines a circuit c' that includes the flag f, and which is, by the above, in $St_F(S)$. Then $c - c(f)c' \in St_F(S)$, and has weight less than c, a contradiction. $\square$

We will next define an *incidence map* from the tensor product over F of $F^{\mathcal{P}}$ and $F^{\mathcal{B}}$ — which, as is customary, we denote by $F^{\mathcal{P}} \otimes F^{\mathcal{B}}$ — into $F^{\mathcal{I}}$. Once again the standard bases for the components of the tensor product force a standard basis for the tensor product: $\{v^{\{x\}} \otimes w^{\{B\}} | x \in \mathcal{P}, B \in \mathcal{B}\}$

Definition 2.4.8 *Let* $\mathcal{S} = (\mathcal{P}, \mathcal{B}, \mathcal{I})$ *and let* F *be any field. Define the map* $\varphi : F^{\mathcal{P}} \otimes F^{\mathcal{B}} \to F^{\mathcal{I}}$ *on the basis of the tensor product as follows:*

$$\varphi(v^{\{x\}} \otimes w^{\{B\}}) = \begin{cases} u^{\{(x,B)\}} & \textit{if } x\mathcal{I}B \\ 0 & \textit{otherwise} \end{cases},$$

extending linearly. The map φ *is called the* **incidence map.**

Thus, for any $B \in \mathcal{B}$ and $x \in \mathcal{P}$, we have that

$$\begin{aligned} \varphi(v^B \otimes w^x) &= \sum_{y:y\mathcal{I}B} \sum_{C:x\mathcal{I}C} \varphi(v^{\{y\}} \otimes w^{\{C\}}) \\ &= \sum_{(y,C):y\mathcal{I}B, x\mathcal{I}C, y\mathcal{I}C} u^{\{(y,C)\}}, \end{aligned}$$

and, as we shall presently see, the right-hand-side of this expression can be simplified for linear and partial linear spaces — which we now define (see also Section 1.2, page 4):

Definition 2.4.9 *An incidence structure* $\mathcal{S} = (\mathcal{P}, \mathcal{B}, \mathcal{I})$ *is a* **linear space** *if for any two points* x *and* y *in* $\mathcal{P}$ *there is* precisely one *block* $B \in \mathcal{B}$ *that is incident with both* x *and* y. *The structure is a* **partial linear space** *if for any two points* x *and* y *in* $\mathcal{P}$ *there is* at most *one block incident with both.*

Since Steiner 2-designs are precisely those 2-designs that have $\lambda = 1$, they are the linear spaces that have constant block size. Notice also that a linear space must be connected, provided that we exclude the trivial cases that still satisfy the definition, i.e. we take all our parameters, including $|B|$, the number of points incident with the block B, and $|x|$, the number of blocks incident with the point x, for any $B \in \mathcal{B}$ and $x \in \mathcal{P}$, to be non-zero.

From the definition of the incidence map φ we have, whenever $x\mathcal{I}B$, in a partial linear space, the following:

$$\begin{aligned} \varphi(v^B \otimes w^x) &= \sum_{y:y\mathcal{I}B} u^{\{(y,B)\}} + \sum_{C:x\mathcal{I}C} u^{\{(x,C)\}} - u^{\{(x,B)\}} \\ &= u^B + u^x - u^{\{(x,B)\}}. \end{aligned}$$

Proposition 2.4.3 *Let* $\mathcal{S} = (\mathcal{P}, \mathcal{B}, \mathcal{I})$ *be a partial linear space,* F *any field, and* φ *the incidence map. If* c *is a circuit, then* $c \in \varphi(E_F(\mathcal{S}) \otimes E_F(\mathcal{S}^t))$.

Proof: Let $c = \sum_{i=1}^{n}(u^{\{f_i\}} - u^{\{g_i\}})$, where f_i and g_i are as given in Definition 2.4.7. Then it can be verified directly that, for any point $x \in \mathcal{P}$,

$$c = \varphi\left(\sum_{i=1}^{n}(v^{B_i} - v^{B_{i+1}}) \otimes (w^{x_{i+1}} - w^x)\right),$$

which gives the result. $\square$

Corollary 2.4.2 *If S is a partial linear space then*

$$St_F(S) \subseteq \varphi(E_F(S) \otimes E_F(S^t)).$$

If, further, S is connected, then

$$(\mathrm{rank}_F(S))^2 \geq \dim(E_F(S)) \dim(E_F(S^t)) \geq |\mathcal{I}| - |\mathcal{B}| - |\mathcal{P}| + 1.$$

Adapted to our proof and requirements, Hillebrandt's theorem ([135, Theorem II]) becomes:

Theorem 2.4.3 *Suppose S is a non-trivial linear space, F any field, and B a block of S with $|B| \geq 2$. Let $\{B_j | j \in J\}$ index the set of blocks of S that do not meet B. Then*

$$\dim(E_F(S)) \dim(E_F(S^t)) \geq (|\mathcal{P}| - |B|)(|B| - 1) + \sum_{j \in J}(|B_j| - 1).$$

Moreover, if S is a Steiner 2-$(v, k, 1)$ design and $F = \mathbf{F}_p$ with p dividing n, then

$$\mathrm{rank}_p(S)[\mathrm{rank}_p(S) - 1] \geq \frac{(v-1)(v-k)}{k},$$

and if, further, p does not divide k, then

$$\mathrm{rank}_p(S) \geq 1 + \sqrt{(v-1)(v-k)/k}.$$

Proof: We first construct a set of circuits that are linearly independent from triangles and quadrilaterals involving a fixed flag (x, B). Here the terminology is obvious, i.e. a triangle in S will be a set of three distinct points $\{x, y, z\}$, three distinct blocks $\{A, B, C\}$, and six distinct flags arising from the requirement that the blocks are the "sides" of the triangle: for example, A is the side opposite the point x and hence both (y, A) and (z, A) must be flags. Similarly a quadrilateral will consist of four distinct points, four distinct blocks and eight flags. (These are particular cases of n-gons, which give the simple circuits spanning the Steinberg module, as is shown in one of the main theorems of [135].)

Consider first the blocks B_j that do not meet the block B. Fix another point y on B and, for each $j \in J$, fix a point b_j on B_j. For each point b'_j of the remaining $|B_j| - 1$ points of B_j construct a quadrilateral on the points $\{x, y, b_j, b'_j\}$ and blocks $\{B, yb_j, B_j, b'_j x\}$, obtaining $|B_j| - 1$ quadrilaterals. We get a linearly independent set of $\sum_{j \in J}(|B_j| - 1)$ quadrilaterals by the nature of the construction.

Now consider triangles. There are $|\mathcal{P}| - |B|$ points off the block B, and thus $(|\mathcal{P}| - |B|)(|B| - 1)$ triangles can be formed with (x, B) as an included flag. The triangular circuits so produced together with those from the quadrilaterals are easily seen to be linearly independent, thus giving the result by Proposition 2.4.3.

If $\mathcal{S}$ is a 2-$(v, k, 1)$ design and p divides n, then, since $n = r - 1$, p does not divide r and hence $\jmath_{\mathcal{P}} \in C_p(\mathcal{S})$. By Proposition 2.4.1, $E_p(\mathcal{S}^t) \neq C_p(\mathcal{S}^t)$ and thus $\dim(E_p(\mathcal{S}^t)) = \mathrm{rank}_p(\mathcal{S}) - 1$. The inequality follows. If, further, p does not divide k, then also $\jmath_B \in C_p(\mathcal{S}^t)$ and thus, by the dual result, $\dim(E_p(\mathcal{S})) = \mathrm{rank}_p(\mathcal{S}) - 1$ yielding the final inequality. $\square$

Remark   (1) The right-hand side of the first inequality of the Theorem gives the dimension of the Steinberg module; this is, in fact, Hillebrandt's result.

(2) The construction of Bruen and Ott [56] uses triangular circuits only, so their bound is exactly the same as that in Theorem 2.4.3 in the case of a projective plane, since then all blocks meet. If $\mathcal{S}$ is a Steiner 2-$(v, k, 1)$ design with $\mathrm{rank}_p(\mathcal{S}) = \rho$, then the bounds in [56] for p dividing n are

$$\rho(\rho - 1) \geq (k - 1)(v - k)$$

and, if also p does not divide k,

$$\rho \geq 1 + \sqrt{(k - 1)(v - k)}.$$

These are not as good as the bounds given above, unless the design is symmetric, i.e. a projective plane, since $(k-1) = (v-1)/r \leq (v-1)/k$, by Fisher's inequality, Corollary 1.4.1. In the case of a projective plane of order n, the bounds are the same: $\rho \geq n\sqrt{n} + 1$.

Example 2.4.2   (1) For a projective plane Π of order n the bound is $\rho \geq 4$ for $n = 2$, which is the 2-rank of the unique Fano plane. For $n = 3$, the bound is $\rho \geq 7$, which is the 3-rank of the plane of order 3. For $n \geq 4$, however, the lower bound is less than the p-rank of the desarguesian plane (where n is a power of p), and in this case the conjecture of Chapter 6 puts the lower bound at the p-rank of the desarguesian plane.

(2) If $\mathcal{D}$ is a 2-$(28, 4, 1)$ design, then $p = 2$ is the only prime of interest, and the bound gives $\rho(\rho - 1) \geq 162$, so $\rho \geq 13$. In fact the smallest 2-rank known for a design of these parameters is 19.

(3) If $\mathcal{D}$ is a 2-$(v, 3, 1)$ design, i.e. a Steiner triple system, then only $p = 2, 3$ give low rank (see Chapter 8). In case $p = 2$, we get

$$\rho \geq 1 + \sqrt{(v-1)(v-3)/3},$$

where elementary considerations show that $v \equiv 1$ or 3 (mod 6). For $v = 7$ and $p = 2$, again we have the Fano plane, and the bound is precise. For $v = 9$ and $p = 2$ the bound is 4, and the actual 2-rank is 9, and for $p = 3$ the bound is 5, and the 3-rank is 6. In fact, a specific formula for the lower bound of the 2-rank and 3-rank of Steiner triple systems is given in Chapter 8 by Theorem 8.2.1 — a result of Doyen, Hubaut and Vandensavel [99].

To get the full power of coding theory to apply to designs, we must, of course, look not only at the rank of the codes (or incidence matrix), but also at the nature of the codewords. In particular, the minimum weight of the code is immediately bounded above by the block size. Asking the same question of the orthogonal code, $C^\perp$, we see that the ovals for the design (should they exist) provide some information concerning the minimum weight of $C^\perp$. From Definition 1.5.2, we have oval size for a 2-(v, k, λ) either $(r + \lambda - 1)/\lambda$ or $(r + \lambda)/\lambda$.

Lemma 2.4.2 *Let $\mathcal{D}$ be a 2-(v, k, λ) design, and let C be its code over some field F. Then if $C \neq F^P$ the minimum weight of $C^\perp$ is at least $(r + \lambda)/\lambda$. Further, if F has characteristic 2 and $\mathcal{D}$ has even order, then a word of this weight in $C^\perp$ will have support that is an oval for $\mathcal{D}$.*

Proof: Let w be any vector in $C^\perp$ and let $\mathcal{A}$ be its support. Assuming that $w \neq 0$ (taking $C \neq F^P$), let x be a point in $\mathcal{A}$. Then each block of $\mathcal{D}$ through x must meet $\mathcal{A}$ again, since $w \in C^\perp$, and so counting incidences gives $(|\mathcal{A}| - 1)\lambda \geq r$, and so $|\mathcal{A}| \geq r/\lambda + 1$, as asserted. If $F = \mathbf{F}_2$, then a word of this weight in $C^\perp$ will have support that is an oval, by our Definition 1.5.2. $\square$

Now we prove a result (see Assmus [6]) that gives information on the minimum weight of $C_2(\mathcal{D})$ and its orthogonal whenever the design has a sufficiently regular collection of ovals.

Theorem 2.4.4 *Let $\mathcal{D}$ be a 2-(v, k, λ) design that has a set of ovals of size $k' = (r + \lambda)/\lambda$ that forms the block set of a 2-(v, k', λ') design $\mathcal{D}'$. Then the blocks of $\mathcal{D}$ are ovals for $\mathcal{D}'$ and the minimum weight of $C_2(\mathcal{D})$ is k with the supports of the minimum-weight vectors of $C_2(\mathcal{D})$ forming a set of ovals of $\mathcal{D}'$. Moreover, the minimum weight of $C(\mathcal{D}')$ is k' and the supports of its minimum-weight vectors are ovals of $\mathcal{D}$.*

Proof: Every block of $\mathcal{D}$ meets a block of $\mathcal{D}'$ in zero or two points, by Theorem 1.5.1 (since the oval size is $(r + \lambda)/\lambda$), so blocks of $\mathcal{D}$ are certainly arcs of $\mathcal{D}'$. Now, denoting the parameters of $\mathcal{D}'$ by the usual symbol primed, then $r'/\lambda' = (v-1)/(k'-1) = (v-1)/(r/\lambda) = k-1$, so that any block of $\mathcal{D}$ is an oval for $\mathcal{D}'$. The oval size then shows that both $\mathcal{D}$ and $\mathcal{D}'$ are of even order (by Theorem 1.5.1). Setting $C = C_2(\mathcal{D})$ and $C' = C_2(\mathcal{D}')$, we have $C \subseteq C'^\perp$, $C' \subseteq C^\perp$, and, by Lemma 2.4.2, the minimum weight of $C'^\perp$ is at least the oval size for $\mathcal{D}'$, i.e. k, and hence the minimum weight of $C'^\perp$ and C is precisely k, since vectors of this weight are present. Similarly for C'. $\square$

Example 2.4.3 (1) The Fano plane and its orthogonal provide the simplest example of the theorem. Here the minimum-weight vectors of the orthogonal also form a symmetric design, the biplane of order 2.

(2) The projective plane of order 4 furnishes a more interesting example. Here the ovals are certain subsets of cardinality 6, and it is possible to choose 56 that form the blocks of a 2-(21,6,4) design. There are, however, 168 ovals in this plane and this is an instance in which further ovals may appear; i.e. starting with the design of the 56 ovals (rather than with the plane itself), the code would provide a further 112 ovals.

(3) The Ree unital on 28 points has ovals of cardinality 10 and they form a design since the automorphism group of the unital is doubly-transitive. See Chapter 8 for the details.

2.5 Hamming codes

These codes were first fully described by Golay [107] and Hamming [128, 129] although the [7,4] binary code had already appeared in Shannon's fundamental paper.[4] They provide an infinite class of perfect codes. We look first at the binary case, the one considered by Hamming; our treatment is traditional and essentially that given by Hamming. Consider a binary code C with parity-check matrix H: the transposed syndrome, Hy^t, of a received vector y is, in the binary case, simply the sum of those columns of H where the errors occurred. To design a single-error-correcting code, we want H not to have any zero columns, for errors in that position would not be detected; similarly, we want H not to have any equal columns, for then errors in those positions would be indistinguishable. If such an H has

---

[4]For an interesting account of the question of priority see Golay [108].

r rows, n columns and is of rank r, then it will be a parity-check matrix for a single-error-correcting $[n, n-r]$ code; in order, therefore, to maximize the dimension, n should be chosen as large as possible. The number of distinct non-zero r-tuples available for columns is $2^r - 1$. We take for H the $r \times (2^r - 1)$ binary matrix whose columns are all the distinct non-zero r-tuples.

Definition 2.5.1 *The binary Hamming code of length* $2^r - 1$ *is the code* $\mathcal{H}_r$ *that has for parity-check matrix the* $r \times (2^r - 1)$ *matrix H of all non-zero r-tuples over* $\mathbf{F}_2$.

Theorem 2.5.1 *The binary code* $\mathcal{H}_r$ *is a* $[2^r - 1, 2^r - 1 - r, 3]$ *perfect single-error-correcting code for all* $r \geq 2$.

Proof: Clearly H has rank r and hence $\mathcal{H}_r$ has dimension $2^r - 1 - r$. That the minimum distance is 3 follows from the form of the matrix H: any two columns of H are linearly independent — and hence by Theorem 2.3.1 the minimum weight is at least 3 — but there are three columns that are dependent since if c and c' are any two distinct columns, then c, c' and $c+c'$ are all present and form a linearly dependent set.

   To show that $\mathcal{H}_r$ is perfect, set $n = 2^r - 1$ and observe that the number of codewords is 2^{n-r}. Each codeword is the centre of a sphere of radius 1 that does not contain another codeword. Counting all the words in all the spheres of radius 1 about the codewords gives

$$2^{n-r}(1 + n) = 2^{n-r}(1 + 2^r - 1) = 2^n,$$

so that all the vectors in $\mathbf{F}_2^n$ are covered, and the code is perfect. $\square$

Example 2.5.1 (1) If $r = 2$ then

$$H = \begin{pmatrix} 1 & 0 & 1 \\ 0 & 1 & 1 \end{pmatrix}$$

and $\mathcal{H}_2$ is a $[3, 1, 3]$ code, which is simply the binary repetition code of length 3.

(2) If $r = 3$, $\mathcal{H}_3$ is a $[7, 4, 3]$ binary code; it is the code we described in Example 2.3.1

   Decoding a binary Hamming code is very easy. We first arrange the columns of H, a parity-check matrix for $\mathcal{H}_r$, so that the jth column represents the binary representation (transposed) of the integer j. Now we decode as follows. Suppose the vector y is received. We first find the syndrome, $\text{synd}(y) = yH^t$. If $\text{synd}(y) = 0$, then decode as y, since then $y \in \mathcal{H}_r$. If $\text{synd}(y) \neq 0$, then, assuming one error has occurred, it must have occurred at the jth position where the vector $(\text{synd}(y))^t$ is the binary representation of the integer j. Thus decode y to $y + e_j$, where e_j is the vector of length $n = 2^r - 1$ with 0 in every position except the jth, where it has a 1.

If we form the extended binary Hamming code $\widehat{\mathcal{H}_r}$, we obtain a $[2^r, 2^r - 1 - r, 4]$ code which is still single-error-correcting, but which is capable of simultaneously detecting two errors — see Exercise 2.1.1. This code is useful for *incomplete decoding*: see Hill [134].

Exercise 2.5.1 Prove that the supports of the vectors of weight 3 in $\mathcal{H}_r$ form a Steiner triple system, and prove that the supports of the vectors of weight 4 in $\widehat{\mathcal{H}_r}$ form a Steiner quadruple system, i.e. a design with parameters 3-$(2^r, 4, 1)$.

Remark Steiner [268] surely had the supports of the weight-3 vectors of $\mathcal{H}_r$ in mind when he posed, in 1852, his "combinatorial problems" but, being a geometer, he was thinking of the lines of the projective space over $\mathbf{F}_2$. One of the questions[5] Steiner asked, "Does every Steiner triple system extend to a Steiner quadruple system?", is still open. As the exercise above shows, the Steiner triple systems coming from the binary Hamming codes do extend. For a fuller account of this matter the reader may wish to consult Assmus and Mattson [17]. We shall define all the projective and affine designs in Chapter 3, and Delsarte's treatment of the coding-theoretical aspects can be found in Chapter 5.

Definition 2.5.2 *The orthogonal code $\mathcal{H}_r^\perp$ of $\mathcal{H}_r$ is called the binary* **simplex** *code. It is of length $2^r - 1$ and is denoted by $\mathcal{S}_r$.*

Remark The simplex code $\mathcal{S}_r$ clearly has length $2^r - 1$ and dimension r. The generator matrix is H. Viewing this code via the functional approach described in Proposition 2.2.1, every non-zero vector of $\mathbf{F}_2^r$ can be viewed itself as a functional on its dual space, the space of r-tuples which are, except for the zero vector, precisely the columns of H. It follows that $\mathcal{S}_r$ consists of the zero vector and $2^r - 1$ vectors of weight 2^{r-1}, so that it is a $[2^r - 1, r, 2^{r-1}]$ binary code. Any two codewords are at Hamming distance 2^{r-1} and, if the codewords are placed at the vertices of a unit cube in $2^r - 1$ dimensions, they form a simplex. The simplex codes are closely related to

---

[5] We are taking some liberties in rephrasing this question, but we are conveying something quite close to the truth. In fact, the main question Steiner asked in [268] had already been answered by Kirkman [169] but the subsequent questions have never been fully treated.

the first-order Reed-Muller codes which we discuss in Chapter 5 and exploit in Chapter 7.

Exercise 2.5.2 Prove that for $r \geq 2$ the supports of the non-zero vectors in $\mathcal{S}_r$ form a symmetric design (see Definition 4.1.1) with parameters 2-$(2^r - 1, 2^{r-1}, 2^{r-2})$. Does this design extend to a 3-design?

We turn next to the q-ary analogues of the Hamming codes that were discovered by Golay; they are called, however, q-ary Hamming codes. Here q is a prime power and what is really involved is the finite field $\mathbf{F}_q$. Let F be a field and set $U = F^r$. We still wish to build perfect, single-error-correcting codes using the alphabet F; i.e. we are still requesting minimum distance 3. Thus we require that our parity-check matrix does not have two columns that are linearly dependent. This will happen if we take a non-zero representative from each 1-dimensional subspace of U and use their transposes for the columns of the parity-check matrix H.

Definition 2.5.3 *The q-ary Hamming code $\mathcal{H}_r(\mathbf{F}_q)$ is the q-ary code having as parity-check matrix a matrix H whose columns consist of non-zero representatives of each of the 1-dimensional subspaces of $\mathbf{F}_q^r$.*

Theorem 2.5.2 *The q-ary code $\mathcal{H}_r(\mathbf{F}_q)$ is a*

$$\left[\frac{q^r - 1}{q - 1}, \frac{q^r - 1}{q - 1} - r, 3 \right]$$

perfect single-error-correcting code for all $r \geq 2$.

Proof: The length of the code is $n = (q^r - 1)/(q - 1)$ since this is the number of 1-dimensional subspaces of $U = \mathbf{F}_q^r$. The parity-check matrix H clearly has rank r, so $C = \mathcal{H}_r(\mathbf{F}_q)$ has dimension $n - r$. To show that C has $d(C) = 3$, notice simply that, as before, no two columns are dependent and that there are three that are dependent if $r \geq 2$: if v and w are two vectors in U such that $\langle v \rangle \neq \langle w \rangle$, then there is a column corresponding to $\langle (v + w) \rangle$ and the three given columns will be linearly dependent.

To show that C is perfect we argue as in Theorem 2.5.1: consider the spheres of radius 1 about the codewords. Since we know that each sphere contains no other codeword, the number of vectors of $\mathbf{F}_q^n$ covered by these spheres is

$$q^{n-r}(1 + (q - 1)n) = q^{n-r}(1 + (q^r - 1)) = q^n,$$

giving all the vectors of $\mathbf{F}_q^n$. $\square$

Remark We have abused notation here and, indeed, above when we defined $\mathcal{H}_r$: the q-ary Hamming code depends on the choices of the representatives of the 1-dimensional subspaces of U, and on their ordering. But, whatever the choices, the codes defined are equivalent and, in the binary case, isomorphic. Obviously, $\mathcal{H}_r = \mathcal{H}_r(\mathbf{F}_2)$.

The orthogonal code has H as its generator matrix and has dimension r. Just as for the binary case, we can again say something about the non-zero vectors of $\mathcal{H}_r(\mathbf{F}_q)^\perp$. Once again the simplest way to see this is to use the coordinate functions of Proposition 2.2.1: since to form the columns we have taken transposes, the columns of H are representatives of the 1-dimensional subspaces of the dual space U^t, and evaluating a linear functional at a vector in U is simply given as a matrix product of a row vector and a column vector — i.e. as an inner product when the proper identification is made. The choice of representatives simply means that the set S of coordinate functionals of Proposition 2.2.1 is a set of $n = (q^r - 1)/(q - 1)$ linear functionals such that if $f \in S$ then $af \notin S$ for any $a \in \mathbf{F}_q, a \neq 1$. Then $u \in U$ is encoded in the usual way:

$$u \mapsto (uf_1, uf_2, \ldots, uf_n).$$

Any entry, uf_i, is 0 if and only if the vector u is in the kernel of f_i. The n kernels are precisely the $(r - 1)$-dimensional subspaces of U. Since any non-zero u is in exactly $(q^{r-1} - 1)/(q - 1)$ kernels, there will be that number of zero entries and $(q^r - 1)/(q - 1) - (q^{r-1} - 1)/(q - 1)$ non-zero entries, i.e. every non-zero vector of $\mathcal{H}_r(\mathbf{F}_q)^\perp$ has weight q^{r-1}. Thus $\mathcal{H}_r(\mathbf{F}_q)^\perp$ is a $[(q^r - 1)/(q - 1), r, q^{r-1}]$ code and any two distinct vectors are at distance q^{r-1} from one another.

Remark The reader familiar with projective geometry will understand that we were really working projectively above without actually saying so. In Chapter 3 we discuss projective geometries *ab initio*. To work projectively in the binary case simply means discarding the zero vector and thus the binary Hamming codes were more easily discovered — the homogeneous coordinates being unique.

Exercise 2.5.3 (1) Prove that the supports of the vectors of weight 3 of $\mathcal{H}_r(\mathbf{F}_q)$ form a design with parameters 2-$((q^r - 1)/(q - 1), 3, q - 1)$.

(2) Prove that an $[n, k, 2e + 1]$ code over $\mathbf{F}_q$ is perfect if and only if the supports of the minimum-weight vectors form a design with parameters $(e + 1)$-$(n, 2e + 1, (q - 1)^e)$.

(3) Prove that the supports of the minimum-weight vectors of the binary Golay code, a perfect $[23, 12, 7]$ code, form a Steiner system with parameters 4-$(23, 7, 1)$.

(4) Prove that the minimum-weight vectors of any linear perfect code generate the code.

Remark There are perfect single-error-correcting codes that are non-linear. The first were found by Vasil'ev [284]. The determination of all the linear perfect codes has a long and distinguished history and, despite the fact that the two Golay codes were discovered very early and that an enormous effort was made to find others or prove that they had all been found, it was not until the 1970s that the story ended when Tietäväinen [275] showed that all were known. The reader should consult van Lint [187] for a complete account.

2.6   Cyclic codes

We now come to a rather large, and very important, class of codes, the *cyclic codes*. They were introduced by Prange (see [241]) very early in the history[6] of this subject and their importance was immediately recognized. Peterson [235] made them the centrepiece of the first textbook on the subject of error-correcting codes. These codes have the property that a certain device, called a *shift register*, can effectively be used for encoding. We will have no need to discuss shift registers here and we refer the reader to the accounts in almost any of the general books on coding theory that we have referenced, in particular MacWilliams and Sloane [203, Chapter 3, p. 89], van Lint [188, Chapter 11], and Pless [238, Chapter 5].

Our definition may appear to be broader than the standard definition, but it will soon be seen to be a natural one and every code that is cyclic in the standard sense will be cyclic in ours.

Definition 2.6.1 *Let C be a linear code of block length n. Then C is* **cyclic** *if* $\mathrm{Aut}(C)$ *contains a cycle of length n.*

It may very well happen that a code is cyclic in more than one way, i.e. $\mathrm{Aut}(C)$ may have many cycles of length n. The *standard* definition for a cyclic code orders the basis for C and requires that a *particular* n-cycle

---

[6]See also Huffman [141]. The book in which this paper appears is interesting historically; many of the important early workers in the field — such as Elias, Slepian and Golay — were present at the Symposium and their comments on the papers read are recorded.

be in Aut(C): take C to be a subspace of F^n and require that any cyclic shift of any codeword is also a codeword. Thus C is cyclic if whenever $(a_0, a_1, \ldots, a_{n-1}) \in C$ then $(a_{n-1}, a_0, \ldots, a_{n-2}) \in C$, where we are using $0, 1, 2, \ldots, n-1$ instead of $1, 2, \ldots, n$ for the coordinate positions, for reasons that will become apparent shortly. This means, of course, that the cycle $\sigma = (0, 1, 2, \ldots, n-1)$, and all its powers, are in Aut(C), since

$$(a_{n-1}, a_0, \ldots, a_{n-2}) = (a_{\sigma^{-1}(0)}, a_{\sigma^{-1}(1)}, \ldots, a_{\sigma^{-1}(n-1)}).$$

Now we make use of the ring $F[X]$ of polynomials over the field F. Let $(a_0, a_1, \ldots, a_{n-1}) \in F^n$ correspond to the polynomial

$$a_0 + a_1 X + \cdots + a_{n-1} X^{n-1} = \sum_{i=0}^{n-1} a_i X^i.$$

This correspondence obviously defines an isomorphism between the vector space F^n and the space of polynomials of degree at most $n-1$. But this latter vector space can be given a ring structure by viewing it as a factor ring of the polynomial ring — using the ideal generated by any polynomial of degree n. The polynomial to use in the current context is $X^n - 1$ and we identify our vector space with the factor ring $F[X]/(X^n - 1)$, where $(X^n - 1)$ denotes the principal ideal of $F[X]$ generated by the polynomial $X^n - 1$. Note that since the polynomial ring is a principal ideal ring, so is $F[X]/(X^n - 1)$. We shall abuse the notation and use X for the image of the polynomial X in $F[X]/(X^n - 1)$. Now the vector $(a_{n-1}, a_0, \ldots, a_{n-2})$ corresponds to

$$X(a_0 + a_1 X + \cdots + a_{n-1} X^{n-1}) = (a_0 + a_1 X + \cdots + a_{n-1} X^{n-1})X$$

and we see that a cyclic code corresponds to an ideal of the factor ring. In fact we have the following:

Proposition 2.6.1 *A subset S of $F[X]/(X^n - 1)$ corresponds to a cyclic code of length n over F (through the correspondence defined above) if and only if S is an ideal.*

Proof: Suppose S corresponds to a cyclic code $C \subseteq F^n$. Since C is linear, S is a subspace of $F[X]/(X^n - 1)$; since S is cyclic, $XS \subseteq S$ as we saw above. Thus $rS \subseteq S$ for any $r \in F[X]/(X^n - 1)$ and hence S is an ideal. This argument is reversible, giving the stated converse. $\square$

Thus we can identify a cyclic code C with an ideal in $F[X]/(X^n - 1)$. Notice that this is precisely the situation we had in Section 2.2 where we spoke of the group algebra, $F[G]$. There G was a regular automorphism group of C. Here we have $G = \langle g \rangle$ where g is the cycle of length n, and the element $a_0 + a_1 g + \ldots + a_{n-1} g^{n-1}$ corresponds precisely to the polynomial in X with the same coefficients. Effectively, $F[G]$ is $F[X]/(X^n - 1)$ and the ideals in $F[G]$ are all cyclic codes. The apparently different approaches are essentially equivalent and we will for the most part use the polynomial approach since the vast machinery developed for polynomial rings can then be exploited. The polynomial ring is not the only approach to cyclic codes; some engineers prefer using the discrete Fourier transform — which we will also employ — and Theorem 2.7.1 below should make it clear why this is so. For this other approach the reader should consult Blahut [41] or Schaub [255]. Blake and Mullin [42, Chapter 4] have given a detailed description of the structure of the group algebra and codes with abelian automorphism groups.

Now we can employ further properties of the ring $F[X]/(X^n - 1)$ to identify the cyclic codes: recall that an ideal I in a commutative ring R is *principal* if it is spanned by a single element a of R, i.e. every element of I can be written in the form ar where $r \in R$. The ring R is a *principal ideal ring* if every ideal in R is principal. Since $F[X]$ is a principal ideal domain – see virtually any book on algebra, for example, MacLane and Birkoff [201] — $F[X]/(X^n - 1)$ is a principal ideal ring and, moreover, the natural map puts the ideals of $F[X]/(X^n - 1)$ into one-to-one correspondence with the ideals of $F[X]$ that contain the ideal $(X^n - 1)$ and hence into one-to-one correspondence with the divisors of $X^n - 1$. If $a(X)$ is a polynomial in $F[X]$ then the ideal of $F[X]/(X^n - 1)$ spanned by the element $a(X) + (X^n - 1)F[X]$ will simply be denoted by $(a(X))$, by abuse of language, and the elements of $(a(X))$ are understood to be reduced modulo the polynomial $X^n - 1$, the point here being that $F[X]/(X^n - 1)$, like the set of integers modulo n, has a standard set of representatives — and we exploit that fact by always expressing an element of $F[X]/(X^n - 1)$ by a polynomial in X of degree less than n. The proof of the following theorem is now obvious.

Theorem 2.6.1 *Let C be a cyclic code in F^n and let I be the corresponding ideal in $F[X]/(X^n - 1)$. Then I contains a unique monic polynomial $g(X)$ of smallest degree, $I = (g(X))$, and $g(X)$ divides $X^n - 1$.*

Definition 2.6.2 *The monic polynomial of smallest degree of a cyclic code I is called the* **generator polynomial** *of I.*

Now the proof of the following corollary should be equally obvious.

Corollary 2.6.1 *If the monic polynomial $g(X)$ divides $X^n - 1$, then $g(X)$ is the generator polynomial for the cyclic code $(g(X))$.*

The requirement that $g(X)$ divides $X^n - 1$ is crucial, of course, and if an arbitrary polynomial is taken as a generator of an ideal, the greatest common divisor with $X^n - 1$ needs to be computed in order to find the generator polynomial of the cyclic code given by the ideal.

Example 2.6.1 Let $F = \mathbf{F}_2$ and $n = 3$. The polynomial $g(X) = 1 + X^2 = (1 + X)^2$ in $F[X]$ when viewed in $F[X]/(X^3 - 1)$ generates the ideal $(1 + X)$ since $\gcd(1 + X^2, X^3 - 1) = 1 + X$. Since the factorization of $X^3 - 1$ in $\mathbf{F}_2[X]$ is $(X + 1)(X^2 + X + 1)$ there are only four binary cyclic codes of length 3: besides $\{0\}$ and $\mathbf{F}_2^3$, we have $\mathbf{F}_{2^j}$ generated by $1 + X + X^2$ and the even-weight subcode generated by $1 + X$.

Of course any ideal, necessarily principal, of $F[X]/(X^n - 1)$ will produce a cyclic code of block length n, but knowledge of the *generator* polynomial gives the dimension of the code, as the next theorem will show. Since $F[X]$ is a principal ideal domain it is a unique factorization domain and the finitely many ideals of $F[X]/(X^n - 1)$ are found by factoring $X^n - 1$ in that domain. There is an algorithm, discovered first by Prange [242] and then later and independently by Berlekamp [34], who improved and popularized it, for factoring polynomials over finite fields; tables of such factorizations exist. We will not need such detailed information here. For a full discussion of factoring polynomials over finite fields see Lidl and Niederreiter [185, Chapter 4].

Theorem 2.6.2 *Suppose* $g(X) = g_0 + g_1 X + \cdots + g_{r-1} X^{r-1} + X^r$ *is a divisor of* $X^n - 1$ *and that* C *is the cyclic code generated by* $g(X)$*. Then* C *has dimension* $n - r$ *and*

$$
G = \begin{pmatrix}
g_0 & g_1 & g_2 & \cdots & g_{r-1} & 1 & 0 & \cdots & 0 \\
0 & g_0 & g_1 & \cdots & g_{r-2} & g_{r-1} & 1 & \cdots & 0 \\
\vdots & \vdots & \ddots & \vdots & \vdots & \vdots & \vdots & \ddots & \vdots \\
0 & \cdots & \cdots & g_0 & \cdots & \cdots & g_{r-2} & g_{r-1} & 1
\end{pmatrix}.
$$

is a generator matrix for C.

Proof: The $k = n - r$ rows of G are obviously linearly independent and correspond to cyclic shifts of $g(X)$, namely $g(X), g(X)X, \ldots, g(X)X^{k-1}$. We must show that this is a spanning set for C. Set $X^n - 1 = g(X)h(X)$. Now $h(X)$ is a monic polynomial of degree k and hence any multiple of $g(X)$ can be written in the form $b(X) = q(X)g(X) + r(X)[X^n - 1]$ with $q(X)$ of degree at most $k-1$. (For example, $X^k g(X) = [X^k - h(X)]g(X) + X^n - 1$.) This shows that the rows of G span C. $\square$

Example 2.6.2 Let $F = \mathbf{F}_2$, $n = 7$. Then

$$X^7 - 1 = (1 + X)(1 + X + X^3)(1 + X^2 + X^3).$$

Setting $g(X) = 1+X+X^3$ and $C = (g(X))$, C is a binary code of dimension 4 with generator matrix

$$\begin{pmatrix} 1 & 1 & 0 & 1 & 0 & 0 & 0 \\ 0 & 1 & 1 & 0 & 1 & 0 & 0 \\ 0 & 0 & 1 & 1 & 0 & 1 & 0 \\ 0 & 0 & 0 & 1 & 1 & 0 & 1 \end{pmatrix}.$$

In fact this is once again the Hamming code $\mathcal{H}_3$ and we will shortly see why.

Now we turn to the orthogonal code $C^\perp$ of a cyclic code $C = (g(X))$. Clearly, $C^\perp$ is also cyclic and we can find a unique monic generator for it that divides $X^n - 1$. Given C as above, write

$$g(X)h(X) = X^n - 1. \tag{2.10}$$

Then $h(X)$, of degree k if $g(X)$ has degree $r = n - k$, is called the *check polynomial* for C — for the following reason:

Lemma 2.6.1 *If* $C = (g(X))$ *and* $h(X) = (X^n - 1)/g(X)$, *then*

$$C = \{c(X)|c(X)h(X) \equiv 0 \pmod{X^n - 1}\}.$$

Proof: Let $c(X) \in C$. Then $c(X) \equiv a(X)g(X) \pmod{X^n - 1}$ so that

$$c(X)h(X) \equiv a(X)g(X)h(X) \equiv 0 \pmod{X^n - 1}.$$

Conversely, if $c(X)h(X) \equiv 0$, then writing $c(X) = q(X)g(X) + r(X)$ in $F[X]$, where $0 \le \deg(r) < \deg(g)$, we have $c(X)h(X) = q(X)g(X)h(X) + r(X)h(X) \equiv 0$ and so $r(X)h(X) \equiv 0$. Now $\deg(h) = n - \deg(g)$, so $\deg(r) + \deg(h) < n$, and hence $r(X) = 0$, and $c(X) \in C$. $\square$

Definition 2.6.3 *If* $a(X) \in F[X]$ *where* $a(X) = a_0 + a_1 X + \cdots + a_r X^r$, *define the* **reciprocal polynomial** $\bar{a}(X)$ *to be* $\bar{a}(X) = X^r a(X^{-1})$, *i.e.*

$$\bar{a}(X) = a_r + a_{r-1} X + \cdots + a_0 X^r.$$

Theorem 2.6.3 *If* $C = (g(X))$ *and* $h(X) = (X^n - 1)/g(X)$, *then* $C^\perp$ *has generator* $\overline{h}(X)$ *and a parity-check matrix for* C *is*

$$
H = \begin{pmatrix}
h_k & h_{k-1} & h_{k-2} & \cdots & h_0 & 0 & \cdots & 0 \\
0 & h_k & h_{k-1} & \cdots & h_1 & h_0 & \cdots & 0 \\
\vdots & \vdots & \ddots & \vdots & \vdots & \vdots & \ddots & \vdots \\
0 & \cdots & & \cdots & h_k & \cdots & \cdots & h_1 & h_0
\end{pmatrix},
$$

where $h(X) = h_0 + h_1 X + \cdots + h_k X^k$ *and* $k = n - r$.

Proof: Assume $n - k > 0$, for otherwise $C^\perp = \{0\}$ and $C = F^n$. Consider $C = \{c(X) | c(X)h(X) \equiv 0 \pmod{X^n - 1}\}$. Write $c(X) = c_0 + c_1 X + \cdots + c_{n-1}X^{n-1}$. Now $c(X)h(X) = q(X)[X^n - 1]$ has degree at most $n - 1 + k$ and hence $q(X)$ has degree at most $k - 1$; it follows that in the product $c(X)h(X)$ the coefficients of $X^k, \ldots, X^{n-1}$ are 0. Thus

$$
\begin{array}{ccccccc}
c_0 h_k & + & c_1 h_{k-1} & + & \cdots & + & c_k h_0 & = & 0 \\
c_1 h_k & + & c_2 h_{k-1} & + & \cdots & + & c_{k+1} h_0 & = & 0 \\
\vdots & & \vdots & & \vdots & & \vdots & & \vdots \\
c_{n-k-1} h_k & + & c_{n-k} h_{k-1} & + & \cdots & + & c_{n-1} h_0 & = & 0.
\end{array}
$$

Thus $c = (c_0, c_1, \ldots, c_{n-1})$ is orthogonal to each of the $n - k$ vectors obtained from the vectors $(h_k, \ldots, h_0, 0, \ldots, 0)$ by cyclic shifts. The matrix H has the correct rank, so it is a parity-check matrix. We need thus only show that $\overline{h}(X) | (X^n - 1)$: $\overline{h}(X) = X^k h(X^{-1})$ and $X^n - 1 = g(X)h(X)$. Thus $X^{-n} - 1 = g(X^{-1})h(X^{-1}) = (1 - X^n)/X^n$, so that

$$
1 - X^n = X^{n-k} g(X^{-1}) X^k h(X^{-1}) = \overline{g}(X)\overline{h}(X).
$$

This completes the proof. $\square$

Example 2.6.3 In Example 2.6.2 where $F = \mathbf{F}_2$, $n = 7$, and $g(X) = 1 + X + X^3$, we have $h(X) = (X^7 - 1)/g(X) = 1 + X + X^2 + X^4$. Thus $\overline{h}(X) = 1 + X^2 + X^3 + X^4$, $k = 4$ and $n - k = r = 3$, and

$$
H = \begin{pmatrix}
1 & 0 & 1 & 1 & 1 & 0 & 0 \\
0 & 1 & 0 & 1 & 1 & 1 & 0 \\
0 & 0 & 1 & 0 & 1 & 1 & 1
\end{pmatrix}.
$$

This is, evidently, a parity-check matrix for the Hamming code, verifying that $C = \mathcal{H}_3$.

Exercise 2.6.1 Prove that the two binary cyclic $[7, 4, 3]$ codes given by the generator polynomials $1 + X + X^3$ and $1 + X^2 + X^3$ are isomorphic. Show that their intersection is the code $\mathbf{F}_{2^j}$ and hence that the two Fano planes obtained from these two codes have no lines in common. Finally prove that if there are three $2\text{-}(7, 3, 1)$ designs on the same point set then there must be a block common to two of them.

Note: We have not discussed here *idempotent generators* for cyclic codes, since we do not really require them (but see Section 7.8). Any of the standard texts referred to can be consulted for this important topic — and also for a discusssion of cyclotomic cosets.

2.7   Transforms

There are many discrete transforms and some are intimately related to coding theory. We begin with one of the simplest, a so-called *Hadamard transform*. Let $V = \mathbf{F}_2^n$ and equip it with the standard inner product, $(x, y) = \sum_1^n x_i y_i$, where $x = (x_1, \ldots, x_n)$ and $y = (y_1, \ldots, y_n)$. We consider the function space R^V where R is any commutative ring.

For $f \in R^V$, the transform $\hat{f} \in R^V$ is defined by

$$\hat{f}(u) = \sum_{v \in V} (-1)^{(u,v)} f(v). \qquad (2.11)$$

The name of the transform comes from the fact that the non-singular $2^n \times 2^n$ matrix that effects the transform, namely the matrix whose rows and columns are indexed by V with the entry in row u and column v being $(-1)^{(u,v)}$, is a Hadamard matrix. We devote Chapter 7 to a coding-theoretical discussion of these matrices; the matrix here is a very special Hadamard matrix, a Sylvester matrix, and it is intimately related to a binary code — in fact a first-order Reed-Muller code (see Chapter 5).

We shall not need the inverse transform — and, indeed, in the generality of our definition it may fail to exist since R is merely a commutative ring — but the following fact will prove useful:

Lemma 2.7.1 *With the notation of the paragraph above and C any binary code,*

$$\sum_{u \in C} \hat{f}(u) = |C| \sum_{v \in C^\perp} f(v).$$

Proof:

$$\sum_{u \in C} \hat{f}(u) \;=\; \sum_{u \in C} \sum_{v \in V} (-1)^{(u,v)} f(v)$$

$$= \sum_{v \in V} f(v) \sum_{u \in C} (-1)^{(u,v)}.$$

If $v \in C^\perp$, $(u,v) = 0$ and the inner sum is $|C|$. If $v \notin C^\perp$, then (u,v) takes the values 0 and 1 equally often as u varies over C (since the set of vectors u of C for which $(u,v) = 0$ is a subgroup of C of index two) and thus the inner sum is 0. This gives the result. $\square$

We need to generalize the above transform from $\mathbf{F}_2$ to an arbitrary finite field, $\mathbf{F}_q$, where $q = p^e$ is a power of the prime p. The -1 in the definition of the transform, which is a primitive square root of 1, must be replaced by a primitive p^{th} root of unity, ω. We assume that the commutative ring R, the target of our functions, contains such an ω — a typical such R being the complex numbers, with $\omega = e^{2\pi i/p}$. In fact, we will henceforth assume that R is a domain; for the later applications R will simply be a polynomial ring in one or more variables over the complex numbers, thus ensuring that we will have a supply of primitive roots of unity. The inner product will take values in $\mathbf{F}_q$, but a standard tactic solves this problem: we replace the inner product (u,v) by its trace, where the trace is the usual one from $K = \mathbf{F}_q$ to $F = \mathbf{F}_p$ and is defined by: $\text{Tr}_{K/F}(a) = a + a^p + \cdots + a^{p^{e-1}}$ when $q = p^e$. The trace is an F-linear transformation from K onto F.

The proof of the following proposition is essentially that of Lemma 2.7.1 above; thus note that for v not in the orthogonal of the q-ary code C in question, (u,v) takes on every value in $\mathbf{F}_q$ equally often as u varies over C, and hence the trace takes on every value between 0 and $p-1$ equally often; put another way, the map $u \mapsto \text{Tr}(u,v)$ is an F-linear functional and takes on every value in F equally often whenever $v \notin C^\perp$. Thus the inner sum is 0, since $1 + \omega + \cdots + \omega^{p-1} = 0$.

Proposition 2.7.1 *Let R be as above, with ω the primitive p^{th} root of unity, and suppose C is any code over $\mathbf{F}_q$ where q is a power of the prime p. Let* Tr *denote the trace from $\mathbf{F}_q$ to $\mathbf{F}_p$. Set $V = \mathbf{F}_q^n$, equipped with the standard inner product. For $f \in R^V$ define $\hat{f}$ by*

$$\hat{f}(u) = \sum_{v \in V} \omega^{\text{Tr}(u,v)} f(v). \tag{2.12}$$

Then

$$\sum_{u \in C} \hat{f}(u) = |C| \sum_{v \in C^\perp} f(v).$$

We turn next to probably the most celebrated of all discrete transforms, the *discrete Fourier transform*. Usually it is defined over the field of complex numbers but all that is required is the presence of a proper root of unity and we proceed in this generality, simply assuming that we have a field R possessing a primitive n^{th} root of unity, ω, thus ensuring that an inverse of the integer n is in R.

Definition 2.7.1 *If $u = (u_0, \ldots, u_{n-1})$ is any vector in R^n then its* **Fourier transform** *is the vector $\hat{u} = (\hat{u}_0, \ldots, \hat{u}_{n-1})$ where*

$$\hat{u}_k = \sum_{i=0}^{n-1} \omega^{ik} u_i$$

for $k = 0, \ldots, n-1$, where ω is a primitive n^{th} root of unity.

Clearly the map $u \mapsto \hat{u}$ is a linear transformation from the free R-module R^n into itself and, in the standard basis, it is effected by a Vandermonde matrix,

$$\begin{pmatrix} 1 & 1 & 1 & \cdots & 1 \\ 1 & \omega & \omega^2 & \cdots & \omega^{(n-1)} \\ 1 & \omega^2 & \omega^4 & \cdots & \omega^{2(n-1)} \\ \vdots & \vdots & \vdots & & \vdots \\ 1 & \omega^{n-1} & \omega^{2(n-1)} & \cdots & \omega^{(n-1)^2} \end{pmatrix}.$$

The inverse is given by

$$u_k = \frac{1}{n} \sum_{i=0}^{n-1} \omega^{-ik} \hat{u}_i,$$

which is why we needed the inverse of n in our ring R. Were it not for this scaling factor, $1/n$, the inverse of the transform would itself be a Fourier transform — using the primitive root ω^{-1} rather than ω and, in fact, had we been willing to use the scaling factor $1/\sqrt{n}$ in the definition this would be precisely the case. Thus any result about the discrete Fourier transform is true, with the proper modification, of its inverse.

Example 2.7.1 If

$$e = (0, 0, \ldots, 1, 0, \ldots, 0),$$
$$\underbrace{\qquad\qquad}_{i+1}$$

then $\hat{e} = (1, \omega^i, \omega^{2i}, \ldots, \omega^{(n-1)i})$.

Notice that if we associate to the vector $(u_0, \ldots, u_{n-1})$ the polynomial

$$u(X) = \sum_{i=0}^{n-1} u_i X^i$$

then the weight of the Fourier transform is $n - \ell$ where ℓ is the number of roots the polynomial has amongst the nth roots of unity. This rather trivial observation has rather interesting consequences.

We collect together some further properties of the discrete Fourier transform in the theorem which follows. As we shall see it is very useful in algebraic coding theory. It was first introduced into this theory by Mattson and Solomon [212] and the usage was systematized by Blahut [40, 41] who was the first to recognize the connection, via the transform, of linear complexity with Hamming weight; in fact Massey [209] has called the last assertion of Theorem 2.7.1 below Blahut's Theorem. Our development follows Schaub [255].

Before we begin we must define the *linear complexity* of a sequence. The concept is of importance for periodic sequences and their generation by a shift register, but we need not go into these matters for our purposes.

Definition 2.7.2 *If $u_0, \ldots, u_m, \ldots$ is an infinite sequence of elements of a ring R, then the* **linear complexity** *of the sequence is the smallest integer L such that there are elements $a_1, \ldots, a_L$ in R with*

$$u_i + \sum_{j=1}^{L} a_j u_{i-j} = 0$$

for $i \geq L$. If no such L exists we say that the linear complexity is infinite.

It should be clear to the reader that given a finite sequence, $u_0, \ldots, u_{n-1}$, and repeating it over and over to form a periodic sequence, gives an infinite sequence of linear complexity at most n. It is in this sense that we will refer to the linear complexity of a finite sequence, and, similarly, to the linear complexity of a vector $(u_0, u_1, \ldots, u_{n-1})$ of R^n. It is also clear that any constant sequence, other than the all-zero sequence, has linear complexity equal to 1; it is customary to assign the all-zero sequence linear complexity 0. Clearly the linear complexity of a sequence is not altered if each term is multiplied by a fixed invertible element of the ring R.

Suppose now that R is a field. For any finite sequence $x_0, \ldots, x_{n-1}$ we form the $n \times n$ **circulant** of the sequence:

$$C(x) = \begin{pmatrix} x_0 & x_1 & x_2 & \ldots & x_{n-1} \\ x_1 & x_2 & x_3 & \ldots & x_0 \\ \vdots & \vdots & \vdots & & \vdots \\ x_{n-1} & x_0 & x_1 & \ldots & x_{n-2} \end{pmatrix}.$$

It is an easy exercise to verify that the linear complexity of the sequence is the rank of the circulant.

Example 2.7.2 The finite sequence $1, a, a^2, \ldots, a^{n-1}$ has linear complexity equal to 1 for any a satisfying $a^n = 1$; in particular (see Example 2.7.1) the linear complexity of the Fourier transform of any of the standard basis elements of R^n is equal to 1. It follows from the fact that linear complexity is unchanged under multiplication by an invertible element that, if R is a field, any vector in R^n has linear complexity equal to 1 provided its Fourier transform has weight 1. Blahut's theorem, Theorem 2.7.1 (6) below, is a generalization of this fact.

Theorem 2.7.1 *(1) If the vector $(\hat{u}_0, \hat{u}_1, \ldots, \hat{u}_{n-1})$ is the Fourier transform of the vector $(u_0, u_1, \ldots, u_{n-1})$, then the Fourier transform of the vector $(u_0, \omega u_1, \ldots, \omega^{n-1} u_{n-1})$ is the cyclic shift, $(\hat{u}_1, \hat{u}_2, \ldots, \hat{u}_0)$.*

(2) The Fourier transform of the cyclic shift $(u_{n-1}, u_0, \ldots, u_{n-2})$ is the vector $(\hat{u}_0, \omega \hat{u}_1, \ldots, \omega^{n-1} \hat{u}_{n-1})$.

(3) The convolution product is mapped to the component-wise product by the Fourier transform: if

$$w_i = \sum_{j=0}^{n-1} u_{i-j} v_j,$$

where the subscripts are taken modulo n, then $\hat{w}_k = \hat{u}_k \hat{v}_k$.

(4) Let r and s be integers with $rs \equiv 1 \pmod{n}$. Then if $(u_0, \ldots, u_{n-1})$ has Fourier transform $(\hat{u}_0, \ldots, \hat{u}_{n-1})$, the transform of $(v_0, \ldots, v_{n-1})$, where $v_i = u_{ri}$, is $(\hat{v}_0, \ldots, \hat{v}_{n-1})$ where $\hat{v}_i = \hat{u}_{si}$ — again taking the subscripts modulo n.

(5) In terms of the polynomial version of codes, ω^k is a root of $u(X)$ if and only if $\hat{u}_k = 0$ and ω^{-k} is a root of $\hat{u}(X)$ if and only if $u_k = 0$.

(6) The weight of a vector is the linear complexity of its Fourier transform and the linear complexity of a vector is the weight of its Fourier transform.

Proof: The proofs of all but the last of these properties are almost immediate from the definition of the Fourier transform. For the last, first note that our general principle concerning the inverse transform allows us to prove only one of the assertions. We proceed to do so.

Let $\hat{u}$ be the Fourier transform of u. Then by (1), the circulant $C(\hat{u})$ is the matrix MDM, where M is the Vandermonde matrix which effects the Fourier transform and D is the diagonal matrix with $u_0, u_1, \ldots, u_{n-1}$ on the diagonal, where, of course, $u = (u_0, u_1, \ldots, u_{n-1})$. Since the linear complexity of $\hat{u}$ is the rank of the circulant which is the rank of MDM we have the assertion because, M being non-singular, the rank of MDM is the rank of D which is, of course, $\mathrm{wt}(u)$. $\square$

2.8 Cyclic codes through roots

Cyclic codes can be described through the roots — in some extension field — of a generator polynomial rather than through the polynomial itself, and this may be a more convenient approach in some situations. Before describing this process, we need to recall a few facts about minimal polynomials of field elements over specific subfields: consult Lidl and Niederreiter [185] for the background.

Set $F = \mathbf{F}_q$, where $q = p^r$, and suppose $a \in F$; the polynomial $m(X) \in \mathbf{F}_p[X]$ is the *minimal polynomial* of a (over $\mathbf{F}_p$) if it is the monic polynomial of smallest degree in $\mathbf{F}_p$ with a as a root. Let us make a convention that the prime field is understood to be the subfield, as this is all we need here. Notice that for any $a \in F$, $a^q = a$, so that a is a root of the polynomial $X^q - X \in \mathbf{F}_p[X]$. The following properties of $m(X)$ can be easily verified:

- $m(X)$ is irreducible;

- if $f(a) = 0$ for $f(X) \in \mathbf{F}_p[X]$, then $m(X)$ divides $f(X)$;

- $m(X)$ divides $X^q - X$ and $\deg(m(X)) \leq r$;

- if $m(X)$ is primitive, i.e. has a root that is a generator of $\mathbf{F}_q^\times$, then $\deg(m(X)) = r$.

If a is a root of a polynomial $f(X) \in \mathbf{F}_p[X]$ then a^{-1} is a root of $\overline{f}(X)$, the reciprocal of $f(X)$; since a generates $\mathbf{F}_q^\times$ if and only if a^{-1} does, $f(X)$ is primitive if and only if $\overline{f}(X)$ is.

Example 2.8.1 Take $F = \mathbf{F}_{16}$ and let $F = \mathbf{F}_2[Y]/(1 + Y^3 + Y^4)$. Then both $f(X) = 1 + X^3 + X^4$ and its reciprocal $\overline{f}(X) = X^4 + X + 1$ are primitive — as can easily be verified by factoring $X^{15} - 1$. If ω is a root of $f(X)$, then $\omega^2, \omega^4, \omega^8$ are the other roots of $f(X)$ and all have $f(X)$ as minimal polynomial. So

$$f(X) = (X - \omega)(X - \omega^2)(X - \omega^4)(X - \omega^8).$$

The roots of the reciprocal are $\omega^7, \omega^{11}, \omega^{13}$ and ω^{14} and these eight elements of $\mathbf{F}_{16}$ are the eight primitive 15th roots of unity. The other irreducible factors of $X^{15} - 1$ are $X - 1, X^2 + X + 1$ and $X^4 + X^3 + X^2 + X + 1$.

The basic outline of the procedure that we will discuss here is as follows: let $a_1, a_2, \ldots, a_r$ be non-zero elements of $\mathbf{F}_q$ (where $q = p^m$), and let $f_1(X), f_2(X), \ldots, f_r(X)$ be, respectively, their minimal polynomials over $\mathbf{F}_p$. Let n be such that $f_i(X)$ divides $X^n - 1$ for each i. Then if $C = \{c(X)|c(X) \in \mathbf{F}_p[X] \text{ and } c(a_i) = 0 \text{ for all } i\}$, then $C = (g(X))$, the cyclic code whose generator polynomial is $g(X)$, the least common multiple of the $f_i(X)$. As an example of the procedure, we can show that the binary Hamming codes are cyclic; in fact, it is not hard to extend this to a proof that $\mathcal{H}_r(\mathbf{F}_q)$ is cyclic as long as n and $q - 1$ are coprime [42, p. 46].

Theorem 2.8.1 *Let $F = \mathbf{F}_{2^m}$, set $n = 2^m - 1$, and let ω be a primitive element for F. If $m(X)$ is the minimal polynomial over $\mathbf{F}_2$ of ω, then $(m(X)) = \mathcal{H}_m$.*

Proof: Since $m(X)$ is the minimal polynomial of ω, the degree is m and $m(X)$ divides $X^n - 1$. Let $C = (m(X))$. Then C is a $[n, n - m]$ binary code and we just need to show that it is $\mathcal{H}_m$. To do this, we look at a parity-check matrix for C and show that it is the same as one for $\mathcal{H}_m$. We know that F is a vector space over $\mathbf{F}_2$ with $\{1, \omega, \omega^2, \ldots, \omega^{m-1}\}$ as a basis. Write $\omega^i = \sum_{j=0}^{m-1} h_{i,j} \omega^j$ for each element $\omega^i \in F^\times$. Construct the matrix H with n columns and m rows with the ith column the coefficients of ω^i, i.e. $(h_{i,0}, h_{i,1}, \ldots, h_{i,m-1})^t$. Then H is a parity-check matrix for $\mathcal{H}_m$ since the columns are distinct.

Now observe that the polynomial $a(X)$ corresponding to the n-tuple $(a_0, a_1, \ldots, a_{n-1})$ satisfies $(a_0, a_1, \ldots, a_{n-1})H^t = 0$ if and only if $a(\omega) = 0$, i.e. if and only if $m(X)$ divides $a(X)$. Thus we do have a parity-check matrix for $C = (m(X))$ as well. $\square$

From this we have also that the even-weight subcode of $\mathcal{H}_m$ has generator matrix $m_1(X) = (1 + X)m(X)$.

A general method for constructing cyclic codes from roots, with a specified minimum distance, was developed by Bose and Ray-Chaudhuri [45] and independently by Hocquenghem [139]: these codes are known as the *BCH codes*, and we give a brief description.

Definition 2.8.1 *Let q be a prime power, let n be a divisor of $q^m - 1$ for some m, and let ω be a primitive n^{th} root of unity in $\mathbf{F}_{q^m}$. Let δ be a positive integer with $\delta < n$. Then a cyclic code of length n over $\mathbf{F}_q$ is a* **BCH code of designed distance δ** *if its generator polynomial $g(X)$ is, for some integer k, the least common multiple of the minimal polynomials of $\{\omega^k, \omega^{k+1}, \ldots, \omega^{k+\delta-2}\}$. In the case where $n = q^m - 1$, the term* **primitive** *BCH code is used; if $k = 1$, the term* **narrow-sense** *is used.*

Theorem 2.8.2 *The minimum distance d of a* BCH *code of designed distance δ, is at least δ.*

Proof: Let c be any non-zero code vector. By construction, the Fourier transform with respect to ω, $\hat{c}$, has $\delta - 1$ consecutive zeros and it therefore has linear complexity at least δ, for otherwise the transform would necessarily be the zero vector, an impossibility since $c \neq 0$. Thus by Theorem 2.7.1 (6) the weight of c is at least δ. $\square$

The dimension of the BCH code is not apparent from the construction and the designed distance, δ, might well be less than the true minimum distance. Conditions under which they are equal are given in Chapter 9 of MacWilliams and Sloane [203].

Remark The BCH bound above has been generalized in several ways by Roos, by Hartmann and Tzeng, and by van Lint and Wilson; in each case it is a question of what certain zero patterns in the Fourier transform imply about the weight of the vector: in the BCH bound that pattern is simply a block of zeros. A unified treatment, using the transform, can be found in Schaub [255].

Example 2.8.2 (1) The binary Hamming codes $\mathcal{H}_m$, for every $m \geq 2$, are BCH codes. Here take $q = 2$, $n = q^m - 1$, and let ω be a primitive element of the field $\mathbf{F}_{2^m}$. Then the (primitive, narrow-sense) BCH code with $\delta = 3$ has generator polynomial $g(X)$ the minimal polynomial of ω (equal to the minimal polynomial of ω^2), i.e.

$$g(X) = (X - \omega)(X - \omega^2)(X - \omega^{2^2})\ldots(X - \omega^{2^{m-1}}).$$

Thus $(g(X)) = \mathcal{H}_m$. Here the true minimum distance is the designed distance. Similarly, binary t-error-correcting BCH codes of length $2^m - 1$, with $t \leq 2^{m-1} - 1$ and dimension $k \leq 2^m - 1 - mt$, can be constructed: see Pless [238, Chapter 7] or MacWilliams and Sloane [203, Chapter 9]. Note that this yields an alternative proof that the binary Hamming codes are cyclic since the sphere-packing bound shows that $(g(X))$ is perfect and hence it must be the Hamming code.

(2) The *Reed-Solomon* codes are (primitive, narrow-sense) BCH codes with $m = 1$, i.e. of length $n = q - 1$ over $\mathbf{F}_q$, where $q \geq 3$. Taking $2 \leq \delta \leq q - 1$ and $k = 1$, since $\omega \in \mathbf{F}_q$, the minimal polynomial of ω^r is simply $X - \omega^r$ for $r = 1, 2, \ldots, \delta - 1$ Thus

$$g(X) = (X - \omega)(X - \omega^2)(X - \omega^3)\ldots(X - \omega^{\delta-1})$$

is the generator polynomial. The matrix

$$\begin{pmatrix} 1 & \omega & \omega^2 & \cdots & \omega^{(q-2)} \\ 1 & \omega^2 & \omega^4 & \cdots & \omega^{2(q-2)} \\ \vdots & \vdots & \vdots & & \vdots \\ 1 & \omega^{\delta-1} & \omega^{2(\delta-1)} & \cdots & \omega^{(q-2)(\delta-1)} \end{pmatrix}$$

has all its entries in $\mathbf{F}_q$ and is clearly a parity-check matrix for the code since it is non-singular — because the Vandermonde matrix is. Thus the dimension of the Reed-Solomon code is $k = n - (\delta - 1)$. The Singleton bound (Theorem 2.1.2) tells us that $d \le n - k + 1 = \delta$ and thus $\delta = d$, the true minimum distance. The orthogonal of a Reed-Solomon code will also be a Reed-Solomon code. These codes have the maximum possible minimum distance for their length and dimension; they are examples of so-called *maximum-distance-separable* codes — or MDS codes. This is a much studied class of codes, with many applications, and we describe them more fully in the next section. They are $[q - 1, k, q - k] = [q - 1, q - d, d]$ q-ary codes.

2.9   MDS codes

The Singleton bound (Theorem 2.1.2) shows that a $[n, k, d]$ q-ary code must satisfy $d \le n - k + 1$. The codes we look at in this section will achieve this bound.

Definition 2.9.1 *An $[n, k, d]$ code over $\mathbf{F}_q$ for which $d = n - k + 1$ is called* **maximum-distance-separable** *and, for brevity, referred to as an* **MDS** *code.*

These codes have also been called *optimal* codes in the literature, but the terminology seems to have hardened and despite reservations we shall retain the usual coding-theoretical terminology.

Example 2.9.1   (1) For any prime power q, if $F = \mathbf{F}_q$, then the codes $F^n, F\mathbf{J}, (F\mathbf{J})^\perp$ are MDS $[n, n, 1], [n, 1, n]$ and $[n, n - 1, 2]$ q-ary codes, respectively. These are the *trivial* MDS codes.

(2) The Reed-Solomon codes of Section 2.8 provide $[q - 1, k, q - k]$ MDS codes over any field $\mathbf{F}_q$, for $q \ge 3$, of length $q - 1$ and minimum weight d satisfying $2 \le d \le q - 1$. It is also quite easily shown that any extended Reed-Solomon code is also MDS, *viz.* a $[q, k, q - k + 1]$ q-ary code: see [203, Chapter 10].

(3) Over $\mathbf{F}_4$, the matrix

$$\begin{pmatrix} 1 & 0 & 1 & 1 \\ 0 & 1 & \omega & \omega^2 \end{pmatrix},$$

where ω is a primitive element for the field, i.e. a primitive cube root of unity, is a generator matrix for a $[4, 2, 3]$ MDS 4-ary code — as is easily verified.

Theorem 2.9.1 *(1) A linear $[n, k, d]$ q-ary code C is MDS if and only if every set of $n - k$ columns of any parity-check matrix for C forms a linearly independent set of vectors of $\mathbf{F}_q^{n-k}$.*

(2) A linear code C is MDS if and only if $C^\perp$ is MDS.

Proof: (1) Let H be a parity-check matrix for C. Then every set of $d - 1$ columns is linearly independent, while there are d dependent columns. If C is MDS then we have $d - 1 = n - k$ and the result. If H has every set of $n - k$ columns linearly independent, then $n - k \leq d - 1$, and the Singleton bound gives the result.

(2) Suppose C is a $[n, k, d]$ MDS code. A parity-check matrix for C is a generator matrix for $C^\perp$, and any non-zero codeword v of $C^\perp$ can be taken to be a row of H. Since every set of $d - 1 = n - k$ columns of H is linearly independent, v cannot have as many as $d - 1$ entries equal to zero. Thus v has weight at least $n - (d - 1) + 1$, i.e. $k + 1$. The Singleton bound again implies that this must be the minimum weight, and thus $C^\perp$ is also MDS. $\square$

Corollary 2.9.1 *A $[n, k, d]$ code C is MDS if and only if every set of k columns of any generator matrix for C is linearly independent.*

This corollary highlights an important geometrical interpretation of MDS codes: the existence of the code is equivalent to the existence of a set of n vectors in a vector space V of dimension k over $\mathbf{F}_q$ with the property that every k-subset is linearly independent and thus forms a basis for V. The interpretation is more striking in projective space:[7]

Definition 2.9.2 *Given integers $n \geq k \geq 3$, an **n-arc** in the projective space $PG_{k-1}(\mathbf{F}_q)$ is a set of n points of the space with the property that no subset of k points are together on a hyperplane of the space.*

---

[7] The reader entirely unfamiliar with projective geometry may want to look at the first few pages of Chapter 3.

Example 2.9.2 For $k = 3$, an n-arc is a set of n points such that no three are together on a line. Thus an n-arc is exactly what we called such a set of points for a 2-design (Definition 1.5.1), regarding $PG_2(\mathbf{F}_q)$ as a projective plane, i.e. a symmetric design.

One of the major problems of algebraic coding theory is, for given redundancy $r = n - k$ and minimum distance d, to find the maximum length n such that there exists a $[n, k, d]$ code over $\mathbf{F}_q$. For MDS codes, where $d = r + 1$, this is now seen to be equivalent to the well-studied geometrical problem of finding arcs of maximal size in projective spaces. Denote by $\max_k(\mathbf{F}_q)$ the maximum value of n for which there exists an n-arc in $PG_{k-1}(\mathbf{F}_q)$; we will show in Chapter 3 that

$$\max_3(\mathbf{F}_q) = \begin{cases} q+1 & \text{for } q \text{ odd} \\ q+2 & \text{for } q \text{ even} \end{cases}.$$

Further results are discussed in MacWilliams and Sloane [203, Chapter 11], and in Hill [134, Chapter 14]. The weight distribution (see Section 2.11) of an MDS code is uniquely determined by its parameters and can be found in the above references.

Exercise 2.9.1   (1) Prove that an $[n, k, d]$ q-ary code is an MDS code if and only if the number of minimum-weight vectors is $(q - 1)\binom{n}{d}$.

(2) Prove that an $[n, k, d]$ q-ary code is an MDS code if and only if the supports of the minimum-weight vectors form a trivial design — i.e. every d-subset is the support of a minimum-weight vector.

2.10   Quadratic-residue codes

We need only consider quadratic-residue codes in the narrowest of senses.[8] They will constitute a class of cyclic codes of prime block length n over a field $\mathbf{F}_q$ where q is a prime power relatively prime[9] to n, an odd prime. We shall assume that q is a non-zero square (or quadratic residue) in $\mathbf{F}_n$, i.e. that there is a solution in $\mathbf{F}_n$ to $X^2 = q$. In practice q is either a prime or a square of a prime and the quadratic-residue codes that have received the most attention have $q = 2, 3$ or 4, the reason being that in these cases the extended codes have systematic gaps in their weight distributions thus making possible the use of Theorem 2.11.2 — see Exercise 2.10.1.

---

[8] For a sophisticated treatment of a vast generalization see Ward [290].

[9] When q is a power of n there are also interesting things to be said; for this case the reader should consult Ward [291].

Since $\mathbf{F}_n^\times$ is cyclic of order $n-1$, the quadratic residues form the subgroup of order $(n-1)/2$ and q is a quadratic residue if and only if $q^{(n-1)/2} \equiv 1 \pmod{n}$. There exists an extension field $\mathbf{F}_{q^k}$ of $\mathbf{F}_q$ that contains an element ω of order precisely n (for example, take $k = (n-1)/2$; set $K = \mathbf{F}_q(\omega)$, the splitting field of $X^n - 1$ over $\mathbf{F}_q$.

Let $\square = \{b^2 | b \in \mathbf{F}_n^\times\}$, i.e. $\square$ consists of the non-zero squares in $\mathbf{F}_n$, and let $\boxslash = \mathbf{F}_n^\times - \square$, i.e. the non-squares in $\mathbf{F}_n$. Thus $|\square| = |\boxslash| = (n-1)/2$. With ω as above, define polynomials in $K[X]$ as follows:

$$g(X) = \prod_{r \in \square} (X - \omega^r), \tag{2.13}$$

$$h(X) = \prod_{s \in \boxslash} (X - \omega^s), \tag{2.14}$$

so that $X^n - 1 = (X - 1)g(X)h(X)$.

In fact, since $q \in \square$, both polynomials are in $\mathbf{F}_q$: for example, if ζ is a root of $g(X)$, then so is ζ^q. Now writing $g(X) = \prod_i (X - \zeta_i) = \sum_m a_m X^m$, the coefficients a_m are symmetric functions in the roots ζ_i, and whenever ζ_i occurs, so does ζ_i^q. It follows that $a_m^q = a_m$ for each m and hence $a_m \in \mathbf{F}_q$. Put another way, both polynomials are invariant under the Galois group of $K/\mathbf{F}_q$ since $rq \in \square$ whenever $r \in \square$ and $sq \in \boxslash$ whenever $s \in \boxslash$.

Definition 2.10.1 *Let n be an odd prime and q a prime power that is relatively prime to n. Assume that q is a quadratic residue modulo n. Then the cyclic codes*

$$\begin{aligned} \mathcal{Q}(n,q) &= (g(X)), \\ \mathcal{Q}_*(n,q) &= ((X-1)g(X)), \\ \mathcal{N}(n,q) &= (h(X)), \\ \mathcal{N}_*(n,q) &= ((X-1)h(X)) \end{aligned}$$

are the **quadratic-residue codes** *of length n over $\mathbf{F}_q$, where $\mathcal{Q}(n,q)$ and $\mathcal{N}(n,q)$ are $[n,(n+1)/2]$ codes, and $\mathcal{Q}_*(n,q)$ and $\mathcal{N}_*(n,q)$ are $[n,(n-1)/2]$ codes.*

Clearly,

$$\mathcal{Q}(n,q) \supset \mathcal{Q}_*(n,q) \text{ and } \mathcal{N}(n,q) \supset \mathcal{N}_*(n,q),$$

the former sometimes called the augmented quadratic-residue codes, and the latter the expurgated quadratic-residue codes. In the binary case the expurgated code is the even-weight subcode of the augmented code and, in the general case, the kernel of the transformation taking $(a_0, \ldots, a_{n-1})$ to $\sum a_i$.

In fact there is no real distinction between $\mathcal{Q}$ and $\mathcal{N}$ since their definition depends on the choice of the primitive root ω, and changing from ω to ω^s, where s is a non-residue, interchanges the two codes. Thus, it is enough to consider only $\mathcal{Q}$ when considering properties of quadratic-residue codes. The distinction is necessary, however, when discussing the orthogonals as the theorem below indicates.

Theorem 2.10.1 Let $\mathcal{Q} = \mathcal{Q}(n, q)$ and $\mathcal{N} = \mathcal{N}(n, q)$. If $n \equiv -1 \pmod 4$, then $\mathcal{Q}^\perp = \mathcal{Q}_*$; if $n \equiv 1 \pmod 4$, then $\mathcal{Q}^\perp = \mathcal{N}_*$. Further, $d(\mathcal{Q}) = d(\mathcal{N}) \geq \sqrt{n}$.

Proof: From Theorem 2.6.3 we know that $\mathcal{Q}^\perp$ is cyclic with generator polynomial the reciprocal polynomial of $(X - 1)h(X)$. It is an easy matter to check that the results stated do hold, with the observation that -1 is a square in $\mathbf{F}_n$ if and only if $n \equiv 1 \pmod 4$: see Section 7.8.

For the second part of the theorem, which is commonly known as the **square-root bound**, we remark again that $\mathcal{Q}$ and $\mathcal{N}$ are actually isomorphic codes, via a permutation of the coordinate positions induced by $x \mapsto x^s$, where s is any non-square in $\mathbf{F}_n$. So suppose $c(X)$ is a minimum-weight codeword of weight d in $\mathcal{Q}$; then if s is a non-square in $\mathbf{F}_n$, $b(X) = c(X^s)$ is a minimum-weight codeword (of weight d) in $\mathcal{N}$. Clearly $c(X)b(X) \in \mathcal{Q} \cap \mathcal{N}$, so it is a multiple of

$$\prod_{r \in \square} (X - a^r) \prod_{s \in \cancel{\square}} (X - a^s) = \prod_{j=1}^{n-1} (X - a^j) = \sum_{j=0}^{n-1} X^j.$$

This latter polynomial has n non-zero coefficients (and is in fact $\jmath$), but the number of non-zero coefficients in $c(X)b(X)$ is at most d^2 and thus we have $n \leq d^2$, unless $c(X)b(X)$ is the zero-vector, in which case $X - 1$ would have to be a factor of $c(X)$ or $b(X)$. This is ruled out by Corollary 2.10.2 below since it implies that the minimum-weight vectors cannot be in the smaller code.[10] $\square$

Observe that $\mathcal{Q} = \mathbf{F}_q \jmath \oplus \mathcal{Q}_*$ and similarly for $\mathcal{N}$. We want to extend the codes to preserve the orthogonality relations of the above theorem. For the codeword $(a_0, a_1, \ldots, a_{n-1})$ of $\mathcal{Q}$, if the new coordinate is $a_\infty = \gamma \sum_{i=0}^{n-1} a_i$, and for the codeword $(b_0, b_1, \ldots, b_{n-1})$ of $\mathcal{N}$, the new coordinate is $b_\infty = -\gamma \sum_{i=0}^{n-1} b_i$, then this is achieved provided that the vector $(1, 1, \ldots, 1, n\gamma)$

---

[10]Mattson [211] pointed out that the proof of the square-root bound in MacWilliams and Sloane [203] needed to be completed by this last observation

is isotropic, i.e. orthogonal to itself, or else orthogonal to the corresponding vector in the other quadratic-residue code. Hence γ is determined up to sign by requiring that $1 \pm n\gamma^2 = 0$ when $n \equiv \mp 1 \pmod 4$. For a complete discussion see Assmus and Mattson [18]. For the binary case it is the usual extended code, and thus we abuse the notation by denoting the extended codes by $\widehat{Q}$ and $\widehat{N}$ in general. Thus we have the following

Corollary 2.10.1 *For $n \equiv -1 \pmod 4$, both $\widehat{Q}$ and $\widehat{N}$ are self-dual $[n + 1, (n+1)/2]$ codes. For $n \equiv 1 \pmod 4$, we have $\widehat{Q}^\perp = \widehat{N}$ and $\widehat{N}^\perp = \widehat{Q}$.*

The minimum-weight vectors of quadratic-residue codes and their use in creating designs have received a great deal of attention; for a rather complete guide to what is known about the minimum weights the reader should consult Newhart [220].

The quadratic-residue codes of greatest interest are those that are binary or ternary, i.e. for which $q = 2$ or 3. The quadratic-reciprocity law tells us, of course, when this is possible. The following lemma and its elementary proof indicate when $q = 2$ is a possibility.

Lemma 2.10.1 *If p is an odd prime, then 2 is a quadratic residue modulo p if and only if $p \equiv \pm 1 \pmod 8$.*

Proof: Working modulo p, let $a = 1 \times 2 \times \ldots \times (p-1)/2$. Multiply each term by 2 to get $b = 2^{(p-1)/2}a = 2 \times 4 \times \ldots \times (p-1)$. Now b has $(p-1)/2$ factors as written, of which $\lfloor (p-1)/4 \rfloor$ occur in a, i.e. are less than or equal to $(p-1)/2$, where a has $\lfloor (p-1)/4 \rfloor$ even factors and $(p-1)/2 - \lfloor (p-1)/4 \rfloor$ odd factors. For the odd factors: if $k \leq (p-1)/2$, then $p-k \geq (p+1)/2$ and is even, i.e. $p-k$ is even and greater than $(p-1)/2$. So $2^{(p-1)/2}a = a(-1)^{(p-1)/2 - \lfloor (p-1)/4 \rfloor}$ so that $2^{(p-1)/2} = (-1)^{(p-1)/2 - \lfloor (p-1)/4 \rfloor}$ in $\mathbf{F}_p$. Now 2 is a square if and only if $2 \in \langle \omega^2 \rangle$ (where ω is a primitive element for $\mathbf{F}_p$), i.e. if and only if $2^{(p-1)/2} = 1$. This is thus equivalent to $(p-1)/2 - \lfloor (p-1)/4 \rfloor$ being even, which reduces to $p \equiv \pm 1 \pmod 8$. $\square$

Example 2.10.1 (1) The Hamming code $\mathcal{H}_3$: take $q = 2$ and $n = 7$. Then $\square = \{1, 2, 4\}$ and $\boxtimes = \{3, 5, 6\}$, so q is a square in $\mathbf{F}_7$. Let a be the image of Y in $\mathbf{F}_8 = \mathbf{F}_2[Y]/(1 + Y + Y^3)$, a primitive seventh root of unity. Then

$$g(X) = (X - a)(X - a^2)(X - a^4) = 1 + X + X^3,$$

which is the minimal polynomial for $a \in \mathbf{F}_8$.

(2) The ternary Golay code, $\mathcal{G}_{11}$: take $q = 3$ and $n = 11$. If $\mathbf{F}_{3^5}^{\times} = \langle \omega \rangle$, then $a = \omega^{22}$ is a primitive 11th root of unity and $K = \mathbf{F}_3(a) = \mathbf{F}_{3^5}$. We have $\square = \{1, 3, 4, 5, 9\}$ and

$$g(X) = (X - a)(X - a^3)(X - a^4)(X - a^5)(X - a^9)$$

is either $2 + X^2 + 2X^3 + X^4 + X^5$ or $2 + 2X + X^2 + 2X^3 + X^5$, depending on the choice of ω. Moreover, $\mathcal{G}_{11}$ is an $[11, 6, d]$ ternary code; in fact, $d = 5$ because of the square-root bound and orthogonality considerations. For an alternative description of this important code that includes a discussion of its automorphism group see Corollary 7.4.3.

(3) Take $q = 4$ and $n = 5$. Here the primitive fifth root of unity is in $\mathbf{F}_{2^4}$. The $[5, 3]$ quadratic-residue code has minimum weight 3 by the square-root bound and hence it is an MDS code.

The three codes in the above example together with the binary Golay code, which is also a quadratic-residue code, are the only quadratic-residue codes whose groups of equivalences are larger than is guaranteed by the Gleason-Prange theorem, to which we now turn.

The *affine group* of the field $\mathbf{F}_n$ is the group of bijections given by $x \mapsto ax + b$ and is denoted, in general, by $AGL_1(\mathbf{F}_n)$: see Section 3.3 for a discussion of the affine groups in general. As a group it is the semi-direct product of the cyclic group of order n, $\mathbf{F}_n^+$, and the cyclic group of order $n - 1$, $\mathbf{F}_n^{\times}$. The subgroup consisting of those transformations for which a is a quadratic residue leaves, as we have remarked, $\mathcal{Q}_*$, $\mathcal{N}_*$, $\mathcal{Q}$ and $\mathcal{N}$ invariant. Thus it is a subgroup of each of the automorphism groups and, in fact, $AGL_1(\mathbf{F}_n)$ acts on the entire apparatus by interchanging the two quadratic-residue codes. Now this latter group extends to the projective line $\mathbf{F}_n \cup \{\infty\}$ as the linear fractional group, $PGL_2(\mathbf{F}_n)$ with the subgroup extending to $PSL_2(\mathbf{F}_n)$. These linear fractional transformations are of the form

$$x \mapsto \frac{ax + b}{cx + d}$$

where $ad - bc = 1$. The content of the Gleason-Prange theorem is that this group has a *projective* representation on both $\widehat{\mathcal{Q}}$ and $\widehat{\mathcal{N}}$. To prove the theorem we need a monomial transformation whose permutation part sends x to $-1/x$ with the usual interpretation of $1/0$ as ∞ and $1/\infty$ as 0. That is not difficult to construct and the reader should consult Assmus and Mattson [18] for the details. In fact, in terms of the discussion following Definition 2.2.3, the group G is merely $\{\pm 1\}$. Here then is the result:

Theorem 2.10.2 (Gleason-Prange) *The group of* $\{\pm 1\}$*-equivalences of the extended quadratic-residue codes contains a group* P *whose permutation part is* $PSL_2(\mathbf{F}_n)$*.*

The theorem has an immediate and interesting consequence:

Corollary 2.10.2 *The minimum weight of* Q *is one less than the minimum weight of* Q_**.*

Proof: $PSL_2(\mathbf{F}_n)$ acts transitively on the projective line and hence a minimum-weight vector of Q_*, viewed as the subcode of $\widehat{Q}$ with $a_\infty = 0$, can be mapped to a code vector of $\widehat{Q}$ with $a_\infty \neq 0$, yielding of code vector of Q of weight 1 less — and conversely. $\square$

Exercise 2.10.1 (1) Let n be a prime with $n \equiv -1 \pmod 4$ and set $d = d(Q)$. Show that the square-root bound, namely that $d^2 \geq n$, can be improved to $d(d - 1) \geq n - 1$. Hint: if $\sum_{i=0}^{n-1} a_i X^i$ is in Q, then $\sum_{i=0}^{n-1} a_i X^{-i}$ is in $\mathcal{N}$.

(2) Let n be a prime with $n \equiv -1 \pmod{12}$. Show that 3 is a quadratic residue modulo n, that $\widehat{Q}(n, 3)$ is self-dual, and that every vector in this extended quadratic-residue code has weight that is a multiple of 3.

(3) Let n be a prime with $n \equiv -1 \pmod 8$. Show that $\widehat{Q}(n, 2)$ is self-dual and doubly-even (i.e. that all vectors have weights that are multiples of 4). Show that the minimum weight of Q is congruent to -1 modulo 4. Use the square-root bound and the sphere-packing bound to show that $\widehat{Q}(23, 2)$ has minimum weight 8.

(4) Use the square-root bound and the above results to show that $Q(11, 3)$ is a perfect $[11, 6, 5]$ code and that $Q(23, 2)$ is a perfect $[23, 12, 7]$ code.

(5) Use the Gleason-Prange theorem to show that every weight class of any extended quadratic-residue code will yield a 2-design, and that if $n \equiv -1 \pmod 4$, every weight class will yield a 3-design.

2.11 Weight enumerators

In this section we describe an invariant of a linear code, *viz.* its weight enumerator, that is more subtle than the parameter set $[n, k, d]$. We also derive the MacWilliams relations and discuss the Assmus-Mattson theorem.

For an arbitrary code (i.e. possibly non-linear) and an arbitrary code vector, we want to know, for each integer i, how many code vectors there are at distance i from the chosen code vector.[11] For linear codes this information is independent of the code vector chosen and hence depends only on how many code vectors there are of each given weight. This information is summarized by the weight enumerator which we now define:

Definition 2.11.1 *Let C be a linear code. Then the* **weight enumerator** *of C is the polynomial*

$$W_C(Z) = \sum_{c \in C} Z^{\mathrm{wt}(c)}.$$

Clearly, the coefficient of Z^i is the number of code vectors of weight i, and if C has block length n and there are A_i code vectors of weight i in C, then

$$W_C(Z) = \sum_{i=0}^{n} A_i Z^i.$$

Frequently the homogeneous form of the weight enumerator is given, i.e.

$$W_C(X, Y) = \sum_{c \in C} X^{\mathrm{wt}(c)} Y^{n-\mathrm{wt}(c)} = \sum_{i=0}^{n} A_i X^i Y^{n-i};$$

both forms will be employed here. A list of the non-zero A_i is usually called the **weight distribution** of the code.

Weight enumerators are not at all easy to determine, but for some classes of codes, for example the Hamming codes and the MDS codes, they are known. For an example of just how sophisticated such determinations become a dedicated reader may wish to consult Schoof and van der Vlugt [257].

The weight enumerators of the codes of the projective planes of orders 2, 3 and 4 are rather easy to determine by hand, but those of the planes of orders 5 and 7 have not yet been determined. Prange [243] found, very early in the history of the subject (and by computer), the weight distribution of the binary code of the plane of order 8 — which we reproduce[12] in Table 2.1 — but order 9, even in the desarguesian case, seems out of range of current computers.

---

[11]In fact, we would like this information not only for code vectors but also for each of the vectors in the ambient space, for this information allows the computation of the probability of the chosen vector being received — given the proper channel assumptions. For linear codes this amounts to determining the weight

$$A_0 \ = \ 1$$
$$A_9 \ = \ 73$$
$$A_{16} \ = \ 2628$$
$$A_{21} \ = \ 56064$$
$$A_{24} \ = \ 784896$$
$$A_{25} \ = \ 1379700$$
$$A_{28} \ = \ 6671616$$
$$A_{29} \ = \ 10596096$$
$$A_{32} \ = \ 29369214$$
$$A_{33} \ = \ 36301440$$
$$A_{36} \ = \ 49056000$$

Table 2.1: The weight distribution for the plane $PG_2(\mathbf{F}_8)$

Surprisingly, however, the weight enumerator of a given code determines the weight enumerator of its orthogonal in a quite simple manner. This is the content of the MacWilliams relations[13] which we now establish:

Theorem 2.11.1 (MacWilliams) *The weight enumerators of a q-ary linear code and its orthogonal are related by the following equation:*

$$W_{C^\perp}(X,Y) = \frac{1}{|C|} W_C(Y - X, Y + (q-1)X).$$

Proof: The proof is an easy consequence of Proposition 2.7.1. Let the ring R of that proposition be the ring of polynomials in X and Y over, say, the complex numbers, although any field with a primitive p^{th} root of unity, ω, would do, where q is a power of p. For any vector v of the ambient space V of C and $C^\perp$, set $f(v) = X^{\text{wt}(v)} Y^{n-\text{wt}(v)}$. We first compute $\hat{f}$. To this

---

enumerators of each of the code's cosets, a daunting task.

[12] This binary code contains the all-one vector and hence $A_{73-i} = A_i$; thus Table 2.1 lists only the non-zero A_i with $i \le 36$. Note that the minimum weight is 9 and that the vectors of weight 9 give the 73 lines of the plane.

[13] The reader might be confused by the plural, seeing only one equality sign in the theorem, but here "relations" refers to the fact that one can find, for each integer j, an expression for the number of vectors of weight j in $C^\perp$ in terms of the A_i — see Exercise 2.11.1

end, define the function w on $F = \mathbf{F}_q$ as follows: $w(0) = 0$, $w(a) = 1$ for $a \neq 0$. Then, for $v = (v_1, \ldots, v_n)$, $\text{wt}(v) = \sum_{i=1}^{n} w(v_i)$. Now

$$\hat{f}(u) = \sum_{v \in V} X^{\text{wt}(v)} Y^{n-\text{wt}(v)} \omega^{\text{Tr}(u,v)}$$

and, in terms of the function w, this is

$$\sum_{v_1 \in F, \ldots, v_n \in F} X^{w(v_1)+\cdots+w(v_n)} Y^{(1-w(v_1))+\cdots+(1-w(v_n))} \omega^{\text{Tr}(u_1 v_1 + \cdots + u_n v_n)}$$

or

$$\prod_{i=1}^{n} \sum_{v_i \in F} X^{w(v_i)} Y^{1-w(v_i)} \omega^{\text{Tr}(u_i v_i)}.$$

When $u_i = 0$ the inner sum is $Y + (q-1)X$ and when $u_i \neq 0$ it is $Y - X + q/p(1 + \omega + \cdots + \omega^{p-1})X = Y - X$. Hence,

$$\hat{f}(u) = (Y - X)^{\text{wt}(u)} (Y + (q-1)X)^{n-\text{wt}(u)}$$

and the proposition yields the result. $\square$

The same technique yields a MacWilliams relation for the *complete weight enumerator*: here f takes values in a ring of polynomials in q variables, one for each element of F, and is defined by $f(v) = \prod_{a \in F} X_a^{e_a(v)}$, where $e_a(v)$ denotes the number of coordinates of the vector v that have the value a. The above proof and its extension to the complete weight enumerator is due to Gleason: see [20]. For a ternary code, for example, taking the variables to be $X = X_0, Y = X_1$ and $Z = X_{-1}$ the relation is

$$W_{C^\perp}(X, Y, Z) = \frac{1}{|C|} W_C(X + Y + Z, X + \omega Y + \omega^2 Z, X + \omega^2 Y + \omega Z).$$

The complete weight distribution of the ternary code of the projective plane of order 3 can be found in Table 2.2. Here $A_{i,j,k}$ is the coefficient of $X^i Y^j Z^k$ and only the non-zero coefficients are listed. Once again, the 26 vectors of weight 4 are simply $\pm v^L$ for the 13 lines L of the plane. More interesting, perhaps, is the fact that the vectors of weight 9 correspond to the 13 affine subplanes and the 234 ovals of the plane (see Assmus and Sachar [22] and Assmus and van Lint [15]).

Exercise 2.11.1 If $W_C(Z) = \sum A_i Z^i$ and $W_{C^\perp}(Z) = \sum B_i Z^i$ are the weight enumerators of C and $C^\perp$ show that

$$\sum_{i=0}^{n-\nu} A_i \binom{n-i}{\nu} = q^{k-\nu} \sum_{i=0}^{\nu} B_i \binom{n-i}{n-\nu}$$

for $\nu = 0, 1, \ldots, n$, where C is an $[n, k]$ q-ary code.

$$A_{13,0,0} = 1$$
$$A_{9,4,0} = A_{9,0,4} = 13$$
$$A_{7,3,3} = 156$$
$$A_{6,1,6} = A_{6,6,1} = 78$$
$$A_{6,3,4} = A_{6,4,3} = 234$$
$$A_{4,9,0} = A_{4,0,9} = 13$$
$$A_{4,6,3} = A_{4,3,6} = 234$$
$$A_{3,7,3} = A_{3,3,7} = 156$$
$$A_{3,4,6} = A_{3,6,4} = 234$$
$$A_{1,6,6} = 78$$
$$A_{0,9,4} = A_{0,4,9} = 13$$
$$A_{0,13,0} = A_{0,0,13} = 1$$

Table 2.2: The complete weight distribution for the plane $PG_2(\mathbf{F}_3)$

Clearly, knowledge of the number of vectors of each weight there are in a code is crucial in determining whether or not the supports of these vectors could form a design. Notice that if S is the support of a vector in a code over $\mathbf{F}_q$ then it is the support of at least $q-1$ such vectors, in fact precisely $q - 1$ such vectors if the minimum weight of the code is $|S|$. For $q = 2$ the supports are in one-to-one correspondence with the code vectors. The Assmus-Mattson theorem which we state below must take account of the possibility that a set might be the support of more than $q-1$ code vectors, and that accounts for its technical statement in the general case. Since its proof in the 1960s, the theorem has become a cornerstone of the body of results, built during the following decades, that relates algebraic coding theory to the theory of designs. The theorem gives conditions on the weight enumerators of a code and its orthogonal that are sufficient to ensure that the supports of the vectors of minimum weight (and other weights also) yield a t-design where t is a positive integer less than the minimum weight.

Theorem 2.11.2 (Assmus-Mattson) *Let C be a q-ary code of block length n with minimum weight d, and let $d^\perp$ denote the minimum weight of $C^\perp$. Let $w = n$ when $q = 2$ and otherwise the largest integer w satisfying*

$$w - \left(\frac{w+q-2}{q-1} \right) < d,$$

defining $w^\perp$ similarly. Suppose there is an integer t with $0 < t < d$ that satisfies the following condition: for $W_C^\perp(Z) = B_i Z^i$ at most $d - t$ of $B_1, B_2, \ldots, B_{n-t}$ are non-zero. Then for each i with $d \le i \le w$ the supports of the vectors of weight i of C, provided there are any, yield a t-design. Similarly, for each j with $d^\perp \le j \le \min\{w^\perp, n - t\}$ the supports of the vectors of weight j in $C^\perp$, provided there are any, form a t-design.

Remark   (1) In order to apply the theorem, $W_C^\perp(Z)$ must have few non-zero coefficients; fortunately that occurs systematically in certain self-dual codes over $\mathbf{F}_2$, $\mathbf{F}_3$ and $\mathbf{F}_4$. It is in these cases that the theorem has been most useful.

(2) If the theorem applies for $q = 2$, all the weight classes in both C and $C^\perp$ yield t-designs. Although for $C^\perp$ this does not follow directly from the theorem, a result of Shaughnessy [261, Theorem 1] (see also MacWilliams and Sloane [203]) shows that, over $\mathbf{F}_2$, whenever all the weight classes of C yield t-designs the same is true for $C^\perp$.

(3) For $C = C^\perp$ and q either 2, 3 — or 4 with an hermitian inner product — there are systematic gaps in the weight distribution and the maximum minimum weight can be computed; self-dual codes that have the maximum minimum weight are said to be "extremal" and they have received a great deal of attention. For example, for block length 24 and $q = 2$ — with all code vectors of weight divisible by 4 — there is only one extremal code, the binary Golay code, whose minimum weight is 8; for block length 24 and $q = 3$ the extremal minimum weight is 9 and there are two such codes; here all the code vectors have weight divisible by 3. For block length 32 and $q = 2$ with all code vectors of weight divisible by 4, the extremal minimum weight is still 8 and there are five such codes. For an application of these coding-theoretical facts to design theory see Section 7.11, where appropriate references are given. There are only finitely many extremal codes but they have not yet been enumerated. The first binary case where existence is currently undecided is that of a $[72, 36, 16]$ binary doubly-even (i. e. all weights divisible by 4) code.

We will not give the proof of the theorem but instead refer the reader to Assmus and Mattson [18].

There have been several "generalizations" of the theorem and elaborations as well. For example, the generalization to spherical designs can be found in Delsarte, Goethals and Seidel [84, Theorem 8.2]. The first elaborations were given by Venkov and, building on Venkov's work, by Koch; for an account of this work, which is concerned with doubly-even extremal codes,

the reader should consult Koch [174]. The most recent account (with a full
set of references and a new proof) can be found in Calderbank, Delsarte and
Sloane [61], but the interested reader may also wish to consult Safavi-Naini
and Blake [252, 253]. One of the curious aspects of this theorem is that
it has produced many non-trivial 5-designs but never a t-design for $t > 5$,
despite the fact that non-trivial designs with $t > 5$ exist for all values of t.

For the exercises below the reader may wish to refer to Exercise 2.10.1.

Exercise 2.11.2   (1) Show that the minimum-weight vectors of $\widehat{\mathcal{Q}}(11, 3)$
yield a 5-design and solve the MacWilliams relations to show that the
parameters are 5-$(12, 6, 1)$.

(2) Use the square-root bound to show that $\widehat{\mathcal{Q}}(23, 3)$ has minimum weight
9 and apply the Assmus-Mattson theorem to show that the minimum-
weight vectors yield a 5-design.

(3) Show that the minimum-weight vectors of $\widehat{\mathcal{Q}}(23, 2)$ yield a 5-design
and solve the MacWilliams relations to show that the parameters are
5-$(24, 8, 1)$.

(4) Use the square-root bound to show that $\widehat{\mathcal{Q}}(47, 2)$ has minimum weight
12 and use the Assmus-Mattson theorem to show that the code vectors
of each weight yield 5-designs.

(5) Suppose $n \equiv 1 \pmod 4$ and $p \in \square$. Show that if

$$(a_0, \dots, a_{n-1}, a_\infty) \in \widehat{\mathcal{Q}}(n, p^2),$$

then

$$(a_0^p, \dots, a_{n-1}^p, a_\infty^p) \in \widehat{\mathcal{N}}(n, p^2)$$

and hence show that $\widehat{\mathcal{Q}}$ and $\widehat{\mathcal{N}}$ are self-dual under the hermitian inner
product: $(x, y) = \sum x_i y_i^p$. In particular show that for $n \equiv 5 \pmod 8$
all vectors in $\widehat{\mathcal{Q}}(n, 2^2)$ have even weight.

Chapter 3

The geometry of vector spaces

3.1 Introduction

The projective and affine geometries of finite-dimensional vector spaces over finite fields provide the deepest source for the theory of designs. Recreational mathematics, the configurational aspects of classical geometry, and the needs of statisticians in the designing of experiments provide other sources, but much of the motivation and direction for the study of designs comes from geometry and it gives us also a wealth of examples that are easily grasped and well understood. This body of examples and theory has grown so extensive that it has earned a place and a name as a mathematical discipline: it is generally referred to as *finite geometry*.

Chapter 5 discusses the codes — and their important properties — of the designs coming from finite geometries. In Chapter 6 we employ these codes to discuss a theory of projective planes. For now we use the finite geometries to construct our first and most important infinite classes of designs. We will obtain designs from both projective and affine geometries and, as might be expected, they will be rather closely related.

Let F be a field and V a finite-dimensional vector space over F. Then $PG(V)$ denotes the **projective geometry** of V. Its elements are the subspaces of V and its structure is given by set-theoretical containment. Similarly, $AG(V)$ denotes the **affine geometry** of V. Its elements are the cosets, $\mathbf{x}+U$, of subspaces U of V, where $\mathbf{x}$ is any vector in V, and again the structure is given by set-theoretical containment. The "geometry" of these structures arises through viewing containment as an incidence relation. We

will not be considering the general *synthetic* definitions of projective and
affine geometry, except in the case of planes (see Chapter 6). An account
of the synthetic development attuned to our purposes here can be found
in Dembowski [85, Section 1.4] but, of course, in many other standard
references also.

We clearly cannot do justice to the vast subject of the geometry of vector
spaces in a single chapter of a monograph. We will give a basic outline that
will be sufficient for our development and refer the reader to the literature
for more complete treatments. For a detailed discussion of the geometry of
vector spaces the reader should consult the books of Artin [5], Baer [26],
Gruenberg and Weir [112], Hahn and O'Meara [114], Hirschfeld [136, 137],
and Kaplansky [152], or standard reference books with more emphasis on
the geometry such as Todd [278] and Veblen and Young [286]. Dembowski
[85, Chapter 1] also discusses most of the results and concepts we need
and gives a full set of references. Besides Artin and Hahn and O'Meara,
Huppert [145] has a treatment of the classical groups involved. The reader
looking for an elementary treatment should consult [112] or [152] and the
reader looking for the most modern and sophisticated treatment [114].

3.2 Projective geometry

If V is a vector space of dimension $n + 1$ over F, then the **projective
dimension** of $PG(V)$ is defined to be n. Since the dimension of a vector
space, together with the field F over which it is defined, determine it up
to isomorphism, V is frequently suppressed in the notation and $PG(V)$ is
denoted by $PG_n(F)$, to reflect its projective dimension (and, of course, the
field F).[1] Similarly, the **projective dimension** of a subspace is defined
to be 1 less than the dimension of the subspace (as a vector space over F).

Thus the **points** of $PG(V)$ are the 1-dimensional subspaces of V, the
lines are the 2-dimensional subspaces of V, the **planes** the 3-dimensional
subspaces of V, and the **hyperplanes** the n-dimensional subspaces of V.
Since the zero subspace is incident with every element of $PG(V)$ and every
element of $PG(V)$ is incident with V, neither $\{0\}$ nor V play a significant
role and they are usually ignored. Frequently, and as long as no confusion
arises, when working with projective geometry the projective dimension is
referred to as the *dimension* and if a distinction with the dimension as
a vector space is required, the term *rank* might be used for the latter.
In either terminology, the dimension formula for subspaces of V holds for

---

[1]To be precise, $PG_n(F)$ is the specific projective space $PG(F^{n+1})$ and each
point of the geometry then has standard "homogeneous" coordinates.

projective dimension as well, i.e. for U and U' subspaces of V,

$$\dim(U) + \dim(U') = \dim(U + U') + \dim(U \cap U'). \qquad (3.1)$$

Exercise 3.2.1 (1) Let P and Q be distinct points of $PG(V)$. Show that there is a unique line in $PG(V)$ that contains both P and Q.

(2) Let $n \geq 2$ and let H be a hyperplane of $PG_n(F)$. If U is a subspace of dimension t, show that $U \cap H$ has dimension t or $t - 1$, the former if and only if U is contained in H.

(3) Let L and M be distinct lines in $PG(V)$. Show that either L and M meet in a unique point, in which case L and M lie together in a unique plane in $PG(V)$, or L and M are disjoint but contained together in a unique subspace of (projective) dimension 3 in $PG(V)$.

(4) Show that the incidence structure consisting of points and lines in $PG_2(F)$ satisfies the conditions:

- any two points are on a unique line;

- any two lines meet in a unique point;

- there are four points, no three of which are collinear.

The three conditions on points and lines deduced in Exercise 3.2.1 (4) constitute the synthetic definition of a *projective plane* — not necessarily finite. Chapter 6 is devoted to finite projective planes; we shall treat them there as symmetric designs with $\lambda = 1$. We denote the unique line through two points P and Q either by PQ or by $P + Q$, depending on the context, and use geometrical terminology in general: thus we speak of triangles, collinearity, concurrency, pencils of lines, and so on.

Since we shall shortly be using subspaces of a fixed dimension as blocks of a design, it is worthwhile to write down the formula for the number of subspaces of a fixed dimension in case F is finite. Thus, if $F = \mathbf{F}_q$ and V is n-dimensional, the number of subspaces of V of dimension k, where $0 \leq k \leq n$ is

$$\frac{(q^n - 1)(q^n - q) \dots (q^n - q^{k-1})}{(q^k - 1)(q^k - q) \dots (q^k - q^{k-1})}.$$

This is easily seen since the numerator counts the number of ways to choose an ordered set of k linearly independent vectors from a vector space of dimension n and the denominator gives the count of the number of bases of a k-dimensional space.

Exercise 3.2.2 If V is of dimension n, U is a subspace of dimension m, and k an integer with $0 \le m < k \le n$, show that the number of subspaces of V of dimension k that contain U is

$$\frac{(q^n - q^m)(q^n - q^{m+1}) \ldots (q^n - q^{k-1})}{(q^k - q^m)(q^k - q^{m+1}) \ldots (q^k - q^{k-1})}.$$

Given two projective geometries, $PG(V)$ and $PG(W)$, we must decide which functions to admit as morphisms. Clearly, a function from the elements of $PG(V)$ to the elements of $PG(W)$ should preserve incidence if it is to be admissible. Given such a function

$$\varphi : PG(V) \to PG(W)$$

preserving incidence merely means that whenever U and U' are subspaces of V with $U \subseteq U'$, then $U\varphi \subseteq U'\varphi$ (where we are writing maps between spaces on the right). We will not need quite this generality; our morphisms will always be isomorphisms and hence we make the following definition.

Definition 3.2.1 *If V and W are finite-dimensional vector spaces, then $PG(V)$ and $PG(W)$ are* **isomorphic** *if there is a bijection*

$$\varphi : PG(V) \to PG(W)$$

such that, for $U, U' \in PG(V)$, $U \subseteq U'$ if and only if $U\varphi \subseteq U'\varphi$. If $W = V$, then such a map φ is called an **automorphism** *or* **collineation** *of $PG(V)$.*

Since the projective dimension of $PG(V)$ is equal to the length of the longest chain, $U_1, U_2, \ldots, U_k$, of elements of $PG(V)$ satisfying $U_1 \subset U_2 \subset \ldots \subset U_k$, it follows that isomorphic geometries have the same projective dimension. That is, V and W must be of the same dimension and, provided they are vector spaces over the same field, they must be isomorphic as vector spaces. Clearly an invertible linear transformation from V to W will induce an isomorphism of the geometries, but something slightly more general will also do this, *viz.* a so-called **semilinear** transformation. We explain the matter and the fundamental theorem of projective geometry in the setting of vector spaces over the *same* field F but alert the reader to the fact that stronger results, giving a field isomorphism, exist. These results are obvious in our case since an isomorphism of geometries over finite fields determines the number of 1-dimensional subspaces and hence the order of the finite field—which determines it up to isomorphism.

Definition 3.2.2 *Let F be a field and let V and W be vector spaces over F. A* **semilinear transformation** *of V into W is given by a map*

$$T : V \to W$$

together with an associated automorphism, $\alpha(T)$, of the field F. The map T is additive, i.e. $(\mathbf{v} + \mathbf{u})T = \mathbf{v}T + \mathbf{u}T$ for all $\mathbf{v}, \mathbf{u} \in V$, and $(a\mathbf{v})T = a^{\alpha(T)}(\mathbf{v}T)$ for all $a \in F$ and $\mathbf{v} \in V$.

A semilinear transformation is thus a linear transformation of V into $^{\alpha}W$, where this latter vector space is simply W with the action of F "twisted" by $\alpha = \alpha(T)$. It is obvious that a semilinear transformation carries subspaces into subspaces, preserving containment, and thus induces an incidence-preserving map on the projective geometries. It is an isomorphism of the projective spaces whenever T is an isomorphism of the additive structures, the inverse being given by T^{-1}, with the associated automorphism of F being $\alpha(T)^{-1}$. Notice that the composition of semilinear transformations is again semilinear and, in fact, $\alpha(ST) = \alpha(S)\alpha(T)$. It follows that when $V = W$ the semilinear isomorphisms form a group and that the map sending T to $\alpha(T)$ defines a homomorphism into the Galois group of F (here the automorphism group of F). The kernel is clearly the group of invertible linear transformations of V; i.e. a semilinear transformation is linear precisely when the associated automorphism is the identity.

In terms of bases, given ordered bases $\mathbf{v}_1, \mathbf{v}_2, \dots, \mathbf{v}_m$ and $\mathbf{w}_1, \mathbf{w}_2, \dots, \mathbf{w}_n$ of V and W, respectively, then if $(\mathbf{v}_i)T = \sum_{j=1}^{n} a_{ij}\mathbf{w}_j$, $A = (a_{ij})$ and $\alpha = \alpha(T)$, then

$$T : (x_1, x_2, \dots, x_m) \mapsto (x_1^{\alpha}, x_2^{\alpha}, \dots, x_m^{\alpha})A,$$

where, as usual, we have used the bases to identify V with F^m and W with F^n. In matrix form, the composition of two semilinear transformations, (α, A) and (β, B), is $(\alpha\beta, A^{\beta}B)$, where A^{β} denotes the matrix (a_{ij}^{β}). Since any matrix A and any automorphism α clearly yield a semilinear transformation, the map sending T to $\alpha(T)$, in the case where $V = W$, is a homomorphism *onto* the Galois group of F.

Thus, for a given vector space V, the group of semilinear isomorphisms of V contains $GL(V)$, the group of invertible linear transformations of V, as a normal subgroup, the quotient being the Galois group of F, in our case a cyclic group. Moreover, choosing a basis of V gives an isomorphism of the group of semilinear isomorphisms of V with the semi-direct product of the Galois group of F and $GL(V)$, a fact which is obvious from the matrix formulation given above. The group of semilinear isomorphisms will be denoted by $\Gamma L(V)$. Clearly every semilinear isomorphism of V induces an isomorphism of $PG(V)$. The scalar transformations (i.e. those that send **v** to a**v** for some fixed $a \in F$) induce the identity isomorphism and it is not too difficult to show that they are the only semilinear isomorphisms that do. The subgroup of scalar transformations is the centre of $GL(V)$ and a normal subroup of $\Gamma L(V)$; the quotient groups are denoted, respectively, by $PGL(V)$ — the **projective general linear group** — and $P\Gamma L(V)$ — the **projective semilinear group**. If V is n-dimensional and a basis has been chosen these groups become matrix groups modulo scalar matrices and are denoted by $PGL_n(F)$ and $P\Gamma L_n(F)$, respectively. Each of these groups acts as a permutation group on the elements of $PG(V)$, the action on the points of $PG(V)$ being doubly-transitive. The reader should be aware that the standard notation has $PGL_n(F)$ acting on $PG_{n-1}(F)$; similarly for the semilinear group.

All the collineations of $PG(V)$ are induced by semilinear transformations; this is the content of the following classical *fundamental theorem of projective geometry*:

Theorem 3.2.1 *Let V be a vector space of dimension at least 3. Then $P\Gamma L(V)$ is the full automorphism group of $PG(V)$.*

There are well-established proofs of this theorem readily available: see Artin [5, Chapter II], for example, or, for a slightly more modern account, Hahn and O'Meara [114, Chapter 3]. Also note that the theorem starts with planes; the projective line consists merely of points and the lack of any incidences allows an arbitrary permutation to be admitted as an automorphism.

Amongst the automorphisms of $PG_n(\mathbf{F}_q)$ there is always an element of order $v = (q^{n+1} - 1)/(q-1)$ that permutes the points of the geometry in a single cycle of this length, called a **Singer cycle** (after Singer [265]). This element is constructed as follows: first recall that the finite field $K = \mathbf{F}_{q^{n+1}}$ can be viewed as a vector space of dimension $n + 1$ over the field $F = \mathbf{F}_q$, with basis $1, \omega, \omega^2, \ldots, \omega^n$, where ω is a primitive element of K. Using the given field structure, it is clear that multiplication by ω induces a linear transformation on the vector space $V = K$. Recalling that the field F has ω^v as primitive element, it is easy to see that this linear transformation induces an automorphism of $PG(V)$ that acts as a cycle of length v on the v points of the geometry. Examples of this construction are given in Chapter 5; the matrix S of (5.2) gives one way to define a Singer cycle acting on the points of the projective space.

If V is a vector space over the field F, then V^t will denote the **dual space**, i.e. the vector space of linear functionals on V — linear transformations from V to the 1-dimensional vector space F. For a subspace U of

V, define
$$U^\perp = \{f | f \in V^t, (\mathbf{u})f = 0 \text{ for all } \mathbf{u} \in U\}. \qquad (3.2)$$

Since V is canonically the dual space of V^t we use the same notation for subspaces of V^t. Since $U \subseteq (U^\perp)^\perp$, observing that the dimension of $U^\perp$ is $\dim(V) - \dim(U)$, we have that $U = U^{\perp\perp}$ whenever U is a subspace of either V or V^t.

Moreover, $U \subseteq U'$ implies that $(U')^\perp \subseteq U^\perp$ and thus $\perp$ gives a function (called the **annihilator map**) from $PG(V)$ to $PG(V^t)$ that reverses incidence. If U has projective dimension k, its image has projective dimension $n - 1 - k$, where n is the projective dimension of $PG(V)$. We make the following definition, in analogy with Definition 1.2.3:

Definition 3.2.3 *If V and W are finite dimensional vector spaces, then $PG(V)$ and $PG(W)$ are* **anti-isomorphic** *if there is a bijection*

$$\varphi : PG(V) \to PG(W)$$

such that, for $U, U' \in PG(V)$, $U \subseteq U'$ if and only if $U'\varphi \subseteq U\varphi$. If $W = V$, then such a map φ is called an **anti-automorphism** *or* **correlation** *of $PG(V)$ and a* **polarity** *if the correlation has order 2.*

If φ is a correlation, a subspace U of V is called **isotropic** if $U \cap U\varphi \neq \{0\}$ and, in analogy with terminology for designs (Definition 1.2.3), isotropic points and hyperplanes will be called **absolute**.

In order to view $\perp$ as a correlation (in fact a polarity), $PG(V)$ and $PG(V^t)$ must be identified. This is done, classically, via a non-singular bilinear or hermitian form. Even in the absence of such a form, we do have that $PG(V)$ is anti-isomorphic to $PG(V^t)$. Note also that the argument that shows that isomorphic geometries come from vector spaces of the same dimension applies, *mutatis mutandis*, to show that if φ is an anti-isomorphism from $PG(V)$ to $PG(W)$, then $\dim(V) = \dim(W)$ and that $\dim(U) = \dim(V) - \dim(U\varphi)$.

In order to compute, a basis must be chosen for V. There is a natural basis — called the **dual basis** — for V^t: let $\mathbf{v}_0, \mathbf{v}_1, \ldots, \mathbf{v}_n$ be a basis for V, where $\dim(V) = n + 1$. The dual basis for V^t is denoted by $\mathbf{v}_0^t, \mathbf{v}_1^t, \ldots, \mathbf{v}_n^t$ and is defined via

$$(\mathbf{v}_i)\mathbf{v}_j^t = \begin{cases} 1 & \text{if } i = j \\ 0 & \text{if } i \neq j \end{cases}.$$

Then, for any $\mathbf{v} \in V$ and $f \in V^t$, if $\mathbf{v} = \sum_i a_i \mathbf{v}_i$, and $f = \sum_i \mathbf{v}_i^t b_i$,[2]

$$(\mathbf{v})f = \sum_i a_i b_i$$

---

[2] Since we are viewing V as a left vector space over F, we view V^t as a right

which is the matrix product of $(a_0, \ldots, a_n)$ and $(b_0, \ldots, b_n)^t$, where here the "exponent" t denotes transpose. The fact that this is also the inner product of the vectors $(a_0, \ldots, a_n)$ and $(b_0, \ldots, b_n)$ simply reflects the fact that *one* way to identify $(F^{n+1})^t$ with F^{n+1} is to use the standard inner product.

The dual basis now allows us to define **homogeneous coordinates** for $PG(V)$, whereby points are denoted by row vectors and hyperplanes by column vectors. This is because points of $PG(V)$ map under $\perp$ to hyperplanes of $PG(V^t)$ — and conversely. Thus if H is a hyperplane of $PG(V)$, then $H^\perp$ is a point of $PG(V^t)$, i.e. a 1-dimensional subspace of V^t spanned by a vector $(u_0, u_1, \ldots, u_n)^t$, a column vector. A point P, where $P = \langle (w_0, w_1, \ldots, w_n) \rangle$, is on H if and only if

$$(w_0, w_1, \ldots, w_n)(u_0, u_1, \ldots, u_n)^t = \sum_{i=0}^{n} w_i u_i = 0,$$

which can be regarded as the equation for the hyperplane. Similarly, if U is a subspace of projective dimension k, then $U^\perp$ has projective dimension $n-1-k$, and U consists of the set of points that satisfy $n-1-k$ independent homogeneous equations.

Example 3.2.1 In $PG_2(F)$ the line containing the points with homogeneous coordinates $(1, 0, 1)$ and $(0, 1, 1)$ has homogeneous coordinates $(1, 1, -1)^t$.

3.3 Affine geometry

Recall that $AG(V)$, where V is a vector space over a field F, consists of all cosets, $\mathbf{x} + U$, of all subspaces U of V with incidence defined through the natural containment relation.[3] In this case the dimension is the same as that of the vector space—for rather obvious geometric reasons. The dimension of a coset is that of the defining subspace U, and if the latter has dimension r, we will also refer to a coset of U as an **r-flat**. Thus the **points** are all the vectors, including $\mathbf{0}$, the **lines** are 1-dimensional cosets, or 1-flats, the **planes** are the 2-dimensional cosets, or 2-flats, and so on, with the **hyperplanes** the cosets of dimension $n - 1$ — where V

---

vector space over F—which is essential when F is a non-commutative ring. In our case Wedderburn's theorem assures us there can never be any difficulty, but we will frame matters in such a way that the reader should have no *notational* difficulties in examining the more general case.

[3] Note that $\mathbf{x} + \{\mathbf{0}\} = \{\mathbf{x}\}$, but we will normally omit the braces.

is of dimension n over F. We also write $AG_n(F)$ for $AG(V)$, in analogy with the projective case, but once again view it concretely as $AG(F^n)$ so that points come equipped with their coordinates. The affine geometry of these cosets is defined by the containment relation which specifies that, if $M = \mathbf{x} + U$ and $N = \mathbf{y} + W$ are cosets in $AG(V)$, then M *contains* N if $M \supseteq N$, from which it follows that W is a subspace of U. The affine geometry $AG(M)$ is, by definition, the set of cosets of V that are contained in M together with the induced incidence relation. Of course, this is quite clear when M is a subspace but if $M = \mathbf{x} + U$ with $\mathbf{x} \notin U$ the reader should have no difficulty seeing that $AG(M)$ is isomorphic to $AG(U)$ since every element of $AG(M)$ can be written in the form $\mathbf{x} + U'$ for some subspace U' of U. As in the projective case we will use standard geometric terminolgy — in particular the notion of **parallelism**:

Definition 3.3.1 *The cosets* $\mathbf{x} + U$ *and* $\mathbf{y} + W$ *in* $AG(V)$ *are* parallel *if* $U \subseteq W$ *or* $W \subseteq U$.

Clearly then cosets of the same subspace are parallel, and cosets of the same dimension are parallel if and only if they are cosets of the same subspace.

Exercise 3.3.1 If M and N are cosets in $AG(V)$, show that

(1) if $M \subseteq N$ then M and N are parallel;

(2) if $M \not\subseteq N$ and $N \not\subseteq M$ and M and N are parallel, then $M \cap N = \emptyset$;

(3) parallelism is an equivalence relation on the set of all cosets of a fixed dimension of V;

(4) hyperplanes $((m-1)$-flats$)$ in $AG_m(F)$ are parallel if and only if they are equal or intersect in the empty set;

(5) the incidence structure consisting of points and lines in $AG_2(F)$ satisfies the following:

 • every two points are on a unique line;
 • if P is a point and L is a line with P not on L, then there is a unique line parallel to L that contains P;
 • there are three non-collinear points.

The conditions on an incidence structure of points and lines given in Exercise 3.3.1 (5), constitute the synthetic definition of an **affine plane** (not necessarily finite): see Chapter 6. Any affine plane is obtained from a projective plane by removing a line and all the points on the line; this is a residual design in the finite case. The embedding theory given below gives the more general result for geometries of any dimension.

As in the projective case, both $GL(V)$ and $\Gamma L(V)$ act on the geometry, but now we also have V itself acting via translation. The underlying action of the **affine general linear group**, $AGL(V)$, and the **affine semilinear group**, $A\Gamma L(V)$, is given as follow: for $T \in \Gamma L(V)$ and $\mathbf{v} \in V$, the map $(T, \mathbf{v})$ is defined by

$$\mathbf{x}(T, \mathbf{v}) = \mathbf{x}T + \mathbf{v}$$

for each $\mathbf{x} \in V$. Such maps preserve cosets and thus act on $AG(V)$. Composition is given by $(S, \mathbf{v})(T, \mathbf{w}) = (ST, \mathbf{v}T + \mathbf{w})$ and it follows that these affine groups are semi-direct products of the linear and semilinear groups (respectively) with the additive group of V, the action of the linear and semilinear groups on V being the natural one. The permutation action on the points of $AG(V)$, i.e. on the vectors in V, is doubly-transitive and, if $F = \mathbf{F}_2$, it is triply-transitive.

Given an ordered basis $\mathbf{v}_1, \mathbf{v}_2, \ldots, \mathbf{v}_n$ for V, if $(T, \mathbf{v})$ is an element of $A\Gamma L(V)$, and $\mathbf{v} = \sum_i b_i \mathbf{v}_i$, define the matrix A via $\mathbf{v}_i T = \sum_j a_{ij} \mathbf{v}_j$, and let α be the field automorphism associated with T. Then

$$(T, \mathbf{v}) : (x_1, \ldots, x_n) \mapsto (x_1^\alpha, \ldots, x_n^\alpha)A + (b_1, \ldots, b_n).$$

Moreover, given any triple $(\alpha, A, (b_1, \ldots, b_n))$ where α is an automorphism of the field F, A is an $n \times n$ matrix with entries from F and $(b_1, \ldots, b_n)) \in F^n$, the formula above defines an element of $A\Gamma L(V)$ and, in fact, with the obvious multiplication of the triples,

$$(\alpha, A, (b_1, \ldots, b_n))(\beta, B, (c_1, \ldots, c_n)) = (\alpha\beta, A^\beta B, (b_1^\beta + c_1, \ldots, b_n^\beta + c_n)),$$

we have an isomorphism of $A\Gamma L(V)$ with this explicitly given group, which we denote by $A\Gamma L_n(F)$. Similarly we write $AGL_n(F)$ for the affine linear group and $ASL_n(F)$ for the affine special linear group — when they are given explicitly.

In analogy with the projective case, there is a **fundamental theorem of affine geometry** which states that for $n \geq 2$, $\mathrm{Aut}(AG_n(F)) = A\Gamma L_n(F)$. This is the same theorem, in effect, as the fundamental theorem for projective geometry, if we consider the way in which affine geometries are embedded in projective geometries.

When studying the following statement of the *fundamental embedding theorem* the reader might well visualize the projection of subspaces of a given vector space onto a hyperplane not containing the origin (rather than the embedding). Here is the result:

Theorem 3.3.1 *Let V be a vector space over F, H a subspace of codimension 1 (i.e. a hyperplane) and $\mathbf{x}$ a vector in V that is not in H. Set $PG(V)^H = \{U | U \in PG(V), U \nsubseteq H\}$. Define a map*

$$\varphi : AG(\mathbf{x} + H) \to PG(V)$$

by $M \mapsto \langle M \rangle$ for any coset $M \in AG(\mathbf{x} + H)$. Then φ is an injection with

$$AG(\mathbf{x} + H)\varphi = PG(V)^H.$$

Further, the inverse map φ^{-1} satisfies

$$U\varphi^{-1} = U \cap (\mathbf{x} + H),$$

for any $U \in PG(V)^H$.

The proof of this result is quite direct from the definitions; it can be found in Gruenberg and Weir [112]. Note that the choice of the hyperplane H and vector $\mathbf{x}$ that produce the embedding is not crucial since for another choice, K and $\mathbf{y}$, it is clear that $AG(\mathbf{x} + H)$ is isomorphic to $AG(\mathbf{y} + K)$ and, moreover, H and K are equivalent under the projective group. It is precisely the failure of such an isomorphism for a projective plane given synthetically that leads to more affine than projective planes in the general case.

The fundamental theorem thus allows us to move from projective geometry to affine geometry — and back again. The embedded affine geometry of dimension n is obtained from a projective geometry of dimension n by removing a hyperplane and all the subspaces contained in it. The points and subspaces remaining form the affine geometry. In an affine geometry of dimension n, once a basis is chosen for the vector space, any r-flat can be given by a set of $n - r$ independent linear equations.

Example 3.3.1 In terms of homogeneous coordinates, let H be the hyperplane $(1, 0, \ldots, 0)^t$ of V. Then the points of $PG(V)^H$ are all those with first coordinate $x_0 \neq 0$ and so, in homogeneous coordinates, we can take $x_0 = 1$. Dropping the coordinate x_0 gives the n coordinates of the vectors (i.e. points) of the n-dimensional affine space.

3.4 Designs from geometries

To define incidence structures from $PG(V)$ and $AG(V)$ we need to choose point sets and block sets; the incidence relation will be that of the geometry, namely containment. In every case the point set of our design will be the set of points of the geometry: for projective spaces the 1-dimensional subspaces of V and for affine spaces the vectors of V. For blocks we will take all the subspaces (or cosets in the affine case) of a fixed dimension. In every case the double-transitivity of the group involved will assure us that we are dealing with a 2-design. Were we to take as points the subspaces of some fixed dimension and as blocks the subspaces of some other (possibly equal) dimension, we would have merely a 1-design in general. We will have no need here for this more general construction.

Thus, for example, we can consider the design of points and lines, the design of points and planes, or the design of points and hyperplanes of a geometry and be assured of a 2-design. The parameters will depend on both the dimension of the geometry and the cardinality of the finite field. By fixing one of these and letting the other vary we obtain numerous infinite families of designs. Each of these designs will have an automorphism group containing $P\Gamma L(V)$ or $A\Gamma L(V)$ in the projective or affine case, respectively. Except for isolated cases the parameters will admit many designs other than these classical designs, and a large amount of effort has gone into classifying the classical designs amongst those with the same parameters.

Perhaps the most interesting case is that of dimension 2. In the projective case, $PG_2(\mathbf{F}_q)$ produces the design of points and lines of a 3-dimensional vector space over a finite field, a classical *projective plane*. It is a symmetric design with parameters $2\text{-}(q^2+q+1, q+1, 1)$. For q prime only one such design is known, but even for $q = p^2$ where p is prime, the number of such designs tends to infinity with p. Moreover, the synthetic definition of a projective plane permits the existence of $2\text{-}(n^2+n+1, n+1, 1)$ designs for order n an arbitrary positive integer. It is still not known whether or not n can be divisible by two distinct primes.

In the affine case, $AG_2(\mathbf{F}_q)$ produces the design of points and lines of a 2-dimensional vector space, i.e. a classical *affine plane*. It is a $2\text{-}(q^2, q, 1)$ design. Once again the synthetic definition permits q to be replaced by an arbitrary positive integer which is the order of the design.

These two cases are not really different since an affine plane of order n given synthetically embeds uniquely in a projective plane of order n, called its *projective completion*, and a projective plane of order n together with a given line produces an affine plane of order n whose projective completion is the original projective plane. It can happen, however, that a given projective plane produces many inequivalent affine planes. These matters will be discussed more thoroughly in Chapter 6.

Projective planes are symmetric designs, i.e. have the same number of points as blocks (see Chapter 4); more generally, the design of points and

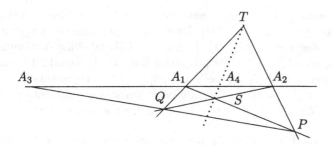

Figure 3.1: Harmonic conjugates

hyperplanes of a projective geometry produces a symmetric[4] design. If the finite field has q elements and the geometry has projective dimension n, then this design of points and hyperplanes is a symmetric design with parameters (see the notation of Section 4.2)

$$\left(\frac{q^{n+1} - 1}{q - 1}, \frac{q^n - 1}{q - 1}, \frac{q^{n-1} - 1}{q - 1} \right)$$

and order q^{n-1}. As we shall see, some of the most interesting questions of design theory involve symmetric designs.

3.5 The cross ratio

Theorem 3.5.1 *Let L be a line in $PG_2(F)$, where F is any field, and let A_1, A_2, A_3 be three distinct points on L. Choose any point P not on L and any point Q on A_3P with $Q \neq P, A_3$. Construct the quadrilateral $PQST$ as follows: $S = A_1P \cap A_2Q$, $T = A_2P \cap A_1Q$, and let ST meet L at A_4. Then A_4 is independent of the choice of P and Q.*

Proof: Refer to Figure 3.1. Choose a basis e_1, e_2, e_3 for the vector space such that $A_1 = \langle e_1 \rangle$, $A_2 = \langle e_2 \rangle$, $P = \langle e_3 \rangle$ and $Q = \langle e_1 + e_2 + e_3 \rangle$. It follows that $A_3 = \langle e_1 + e_2 \rangle$, $S = \langle e_1 + e_3 \rangle$, and $T = \langle e_2 + e_3 \rangle$. Now the point $\langle e_1 - e_2 \rangle$ is on L and on ST and hence must be A_4. Thus A_4 depends only on the choice of A_1, A_2, A_3 on L. $\square$

We introduce now **parametric coordinates** for the projective line $PG(V)$, where V is a 2-dimensional vector space over F: let e_1, e_2 be a basis for V. If $P \in PG(V)$ is a point, then either $P = \langle e_1 \rangle$ or it is of

---

[4]It was for this reason that Dembowski attempted to replace the unfortunate term "symmetric" with "projective".

the form $\langle a\mathbf{e}_1 + \mathbf{e}_2 \rangle$, where $a \in F$, and this accounts for all the points of the line. The element $a \in F$ uniquely determines a point on the projective line; the remaining point, $\langle \mathbf{e}_1 \rangle$, is labelled ∞. This labelling determines the parametric coordinates for the projective line. In the standard homogeneous coordinates of $PG_1(F)$ we have simply picked representatives of the form $(a, 1)$ with $\infty = (1, 0)$, the only point without a representative of the form $(a, 1)$. We are now in a position to define the *cross ratio*:

Definition 3.5.1 *If P_1, P_2, P_3, P_4 are four points on a projective line L with at least the first three points distinct, then the* **cross ratio** *$(P_1, P_2; P_3, P_4)$ is the parametric coordinate of P_4 where the basis for V is chosen so that $P_1 = \langle \mathbf{e}_1 \rangle$, $P_2 = \langle \mathbf{e}_2 \rangle$ and $P_3 = \langle \mathbf{e}_1 + \mathbf{e}_2 \rangle$.*

Further, P_4 is the **harmonic conjugate** *of P_3 with respect to P_1, P_2 if $(P_1, P_2; P_3, P_4) = -1$.*

Example 3.5.1 The point A_4 of Theorem 3.5.1 is the harmonic conjugate of A_3 with respect to A_1, A_2.

Theorem 3.5.2 *Let P_1, P_2, P_3 be distinct points of the projective line $PG(V)$ with parametric coordinates p_1, p_2, p_3 with respect to some fixed basis for the vector space V. If P_4 is a point with parametric coordinate p_4, then*

$$(P_1, P_2; P_3, P_4) = \frac{(p_3 - p_1)(p_4 - p_2)}{(p_4 - p_1)(p_3 - p_2)}.$$

Proof: With respect to the given basis, $P_i = \langle (p_i, 1) \rangle$, for $i = 1, 2, 3, 4$ (unless one of the $p_i = \infty$ for some i, in which case we can still write $\langle (p_i, 1) \rangle = \langle (\infty, 1) \rangle = \langle (1, 0) \rangle$ by treating ∞ in the usual way). Solve for x and y such that

$$x(p_1, 1) + y(p_2, 1) = (p_3, 1),$$

to get $x = (p_3 - p_2)/(p_1 - p_2)$ and $y = (p_3 - p_1)/(p_2 - p_1)$. Choose a new basis $\mathbf{e}_1, \mathbf{e}_2$ for V such that $\mathbf{e}_1 = x(p_1, 1)$, and $\mathbf{e}_2 = y(p_2, 1)$, so that $P_1 = \langle \mathbf{e}_1 \rangle$, $P_2 = \langle \mathbf{e}_2 \rangle$ and $P_3 = \langle \mathbf{e}_1 + \mathbf{e}_2 \rangle$. To find $P_4 = \langle a\mathbf{e}_1 + b\mathbf{e}_2 \rangle$, solve for

$$(p_4, 1) = a\mathbf{e}_1 + b\mathbf{e}_2 = ax(p_1, 1) + by(p_2, 1),$$

which gives

$$\frac{a}{b} = \frac{y(p_4 - p_2)}{x(p_1 - p_4)},$$

and the required result. □

Note: The action of the group $PGL_2(F)$ on $PG_1(F)$ is isomorphic to that of the subgroup of the symmetric group on the set $F \cup \{\infty\}$ that preserves

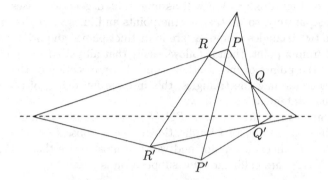

Figure 3.2: Desargues' theorem

the cross ratio. Computation with the matrices is in the usual way: the point with parametric coordinate x goes to the point with parametric coordinate $(ax+b)/(cx+d)$ with, once again, ∞ being treated in the customary manner.

3.6 Desargues' theorem

The geometries we have defined from vector spaces all satisfy the configurational theorem of Desargues, which states that whenever two triangles of the geometry have vertices that are in perspective from a point, then their edges are in perspective along a line. The theorem actually characterizes those projective planes that arise as 2-dimensional projective geometries over division rings; hence the name "non-desarguesian" for those planes that do not. We will give a proof here that $PG(V)$ satisfies Desargues' theorem (i.e. is **desarguesian**) for any vector space, but merely refer the reader to the converse, particularly relevant to us for planes, that any geometry that satisfies Desargues' theorem must arise from a vector space: see Artin [5], Baer [26, Chapter VII], or Lüneburg [195, Chapter 4].

Theorem 3.6.1 *Let Δ and Δ' be two triangles of $PG_n(F)$. Suppose the vertices of Δ are P, Q and R and those of Δ' are P', Q' and R'; let the edges of Δ opposite P, Q and R be, respectively, L, M and N with L', M' and N' defined correspondingly for Δ'. Then the lines $P + P'$, $Q + Q'$ and $R + R'$ meet in a point if and only if the points $L \cap L'$, $M \cap M'$ and $N \cap N'$ lie on a line.*

Proof: Refer to Figure 3.2.[5] We will assume that degenerate cases can be dealt with separately, so we take all the points and lines to be distinct. Further, if the two triangles Δ and Δ' are in distinct planes and if they are in perspective from a point, then it follows easily that all paired sides must meet and that the points of intersection lie on the intersection of the two distinct planes containing the triangles, this intersection being, of course, a line. The dual statement also follows.

Thus we can restrict attention to $PG_2(F)$ and because of duality we need only prove one of the two implications. We suppose that $P + P'$, $Q + Q'$ and $R + R'$ meet in a point and we then must prove that $L \cap L'$, $M \cap M'$ and $N \cap N'$ are collinear. Our supposition is that

$$(P + P') \cap (Q + Q') \subset (R + R')$$

and we must show that

$$(L \cap L') + (M \cap M') \supset (N \cap N').$$

Consider any non-zero vector $\mathbf{v} \in N \cap N' = (P + Q) \cap (P' + Q')$. Then $\mathbf{v} = \mathbf{x} + \mathbf{y} = \mathbf{x}' + \mathbf{y}'$, where $\mathbf{x} \in P$, $\mathbf{x}' \in P'$, $\mathbf{y} \in Q$, $\mathbf{y}' \in Q'$. Thus $\mathbf{x} - \mathbf{x}' = \mathbf{y}' - \mathbf{y} \in (P + P') \cap (Q + Q') \subset (R + R')$, whence $\mathbf{x} - \mathbf{x}' = \mathbf{y}' - \mathbf{y} = \mathbf{z} - \mathbf{z}'$, where $\mathbf{z} \in R$, $\mathbf{z}' \in R'$. So $\mathbf{v} = \mathbf{x} + \mathbf{y} = (\mathbf{x} - \mathbf{z}) + (\mathbf{y} + \mathbf{z}) \in (M \cap M') + (L \cap L')$, since $\mathbf{x} - \mathbf{z} = \mathbf{x}' - \mathbf{z}'$ and $\mathbf{y} + \mathbf{z} = \mathbf{y}' + \mathbf{z}'$. $\square$

3.7   Quadrics and hermitian varieties

In this section we will review briefly the results we will need concerning quadratic and hermitian forms. For a fuller discussion the reader should consult some of the many standard texts dealing with the subject: for example, Artin [5, Chapter III], Dieudonné [90, 91], Hahn and O'Meara [114], Hirschfeld [136, 137], Kaplansky [152], or Weyl [295],

Let V and W be finite-dimensional vector spaces over a field F.[6] We want to relate V and W via a form in such a way that for $V = W$ our vector space will be endowed with further structure in much the same way that the dot product endows real vector spaces with added structure and the unitary (or, what is the same thing, hermitian) dot product endows complex vector spaces with added structure.

---

[5] LaTeXpicture of Desargues' theorem due to P. J. Cameron.

[6] Once again F could be a division ring, in which case V would be a left vector space over F and W would be a right vector space over F. We will couch the definitions in such a way that they will be correct in this more general setting— in fact, even for a non-commutative ring with V and W left and right modules, respectively.

Our principal aim is to define conics and unitals in the classical projective planes. We shall see later that the conics of $PG_2(\mathbf{F}_q)$, where q is odd, have a purely combinatorial characterization due to Segre; the unitals of $PG_2(\mathbf{F}_{q^2})$ do not, but we note here that the coding-theoretical approach does allow one to characterize those unitals defined by hermitian forms for arbitrary q and that, for q even, ovals do appear in the coding theory of the plane.

We will consistently denote the form by (,) to indicate the resemblance to the classical cases and, for $\mathbf{v} \in V$ and $\mathbf{w} \in W$, $(\mathbf{v}, \mathbf{w})$ will be an element of F. Here is the formal definition:

Definition 3.7.1 *A map* (,) $: V \times W \rightarrow F$ *is said to be a* **form on** (V, W) *if it is linear in both variables; i.e. if*

$$(\mathbf{v} + \mathbf{v}', \mathbf{w}) = (\mathbf{v}, \mathbf{w}) + (\mathbf{v}', \mathbf{w}) \text{ and } (a\mathbf{v}, \mathbf{w}) = a(\mathbf{v}, \mathbf{w})$$

for all $\mathbf{v}, \mathbf{v}' \in V$, $\mathbf{w} \in W$ *and* $a \in F$ *and*

$$(\mathbf{v}, \mathbf{w} + \mathbf{w}') = (\mathbf{v}, \mathbf{w}) + (\mathbf{v}, \mathbf{w}') \text{ and } (\mathbf{v}, \mathbf{w}a) = (\mathbf{v}, \mathbf{w})a$$

for all $\mathbf{w}, \mathbf{w}' \in W$, $\mathbf{v} \in V$ *and* $a \in F$. *If* $V = W$ *we shall simply refer to a form on* V.

Remark We have taken the liberty of writing the scalars either on the left or on the right of the vectors, a liberty permitted by commutativity—but see the previous footnote. We shall systematically write linear transformations on the side other than the one the scalars are written on and reverse the side the scalars are on when we take duals. Thus the reader should expect associative laws to occur more often than commutative laws.

Example 3.7.1 For an arbitrary vector space V over F set $W = V^t$ and let $(\mathbf{v}, \mathbf{w})$ denote the value at $\mathbf{v}$ of the functional $\mathbf{w}$. This is the paradigm and the novice reader should bear it in mind whilst reading this section. In fact, *all* examples in the case $W = V^t$ arise by giving a linear transformation L of V into V and setting $(\mathbf{v}, \mathbf{w})$ equal to the value at $\mathbf{v}L$ of the functional $\mathbf{w}$.

Given a form on (V, W), linear transformations

$$L \colon V \rightarrow W^t \text{ and } V^t \leftarrow W \colon R$$

are defined as follows: $(\mathbf{v}L)(\mathbf{w}) = (\mathbf{v}, \mathbf{w}) = (\mathbf{v})(R\mathbf{w})$. That $\mathbf{v}L$ and $R\mathbf{w}$ are functionals and that L and R are linear transformations are, obviously, simple formal consequences of what it means to have a form.

The **left radical** of the form is simply the kernel of the linear transformation L and the **right radical** is the kernel of R. Thus, $\mathbf{v}$ is in the left radical if and only if $(\mathbf{v}, \mathbf{w}) = 0$ for all $\mathbf{w} \in W$ and $\mathbf{w}$ is in the right radical if and only if $(\mathbf{v}, \mathbf{w}) = 0$ for all $\mathbf{v} \in V$.

For a subset S of V denote by $S^\perp$ the subspace of W defined by $S^\perp = \{\mathbf{w} \in W | (\mathbf{s}, \mathbf{w}) = 0 \text{ for all } \mathbf{s} \in S\}$; similarly, for T a subset of W, denote by $T^\perp$ the subspace $\{\mathbf{v} \in V | (\mathbf{v}, \mathbf{t}) = 0 \text{ for all } \mathbf{t} \in T\}$. Clearly $V^\perp$ is the right radical and $W^\perp$ the left. It is easy to see that $S \subseteq (S^\perp)^\perp$ and $T \subseteq (T^\perp)^\perp$. Moreover, $S \subseteq S'$ implies $S'^\perp \subseteq S^\perp$ and $T \subseteq T'$ implies $T'^\perp \subseteq T^\perp$. Now, as a subset of V^t, $R(S^\perp)$ annihilates S and hence, when S is a subspace of V, it can be viewed in the dual of the quotient space V/S, i.e. in $(V/S)^t$. Similarly, $(T^\perp)L$ can be viewed in $(W/T)^t$ when T is a subspace of W. In particular, $\dim R(S^\perp) \leq \dim(V/S)^t = \dim(V/S) = \dim V - \dim S$ when S is a subspace of V and, provided the right radical is zero, we have that

$$\dim S^\perp \leq \dim V - \dim S.$$

On the other hand, if R is a surjection then each $f \in V^t$ is of the form $R\mathbf{w}$ for some $\mathbf{w} \in W$; therefore R maps $S^\perp$ onto $(V/S)^t$ and it follows that

$$\dim R(S^\perp) \geq \dim V - \dim S.$$

Thus when R is an isomorphism we have, for all subspaces S of V,

$$\dim S^\perp = \dim V - \dim S.$$

Similar remarks apply, of course, to L.

When both R and L are isomorphisms we shall call the form **non-singular**. If $V = W$ and the form is **symmetric**, i.e. if $(\mathbf{v}, \mathbf{w}) = (\mathbf{w}, \mathbf{v})$ for all $\mathbf{v}$ and $\mathbf{w}$, then clearly R is an isomorphism if and only if L is and then the (common) radical is $\{0\}$. A symmetric, non-singular form on V will be called an **orthogonal form**. Similarly, the form is **alternating** if $(\mathbf{v}, \mathbf{v}) = 0$ for all $\mathbf{v}$. From this follows immediately that it is also **skew-symmetric**, i.e. $(\mathbf{v}, \mathbf{w}) = -(\mathbf{w}, \mathbf{v})$ for all $\mathbf{v}$ and $\mathbf{w}$. A non-singular alternating form is called a **symplectic form**.

Now suppose that F has an involutory map σ defined on it such that $\sigma : a \mapsto a^\sigma$, where $a^{\sigma^2} = a, (a + b)^\sigma = a^\sigma + b^\sigma$, and $(ab)^\sigma = b^\sigma a^\sigma$ for all $a, b \in F$. (We are here implicitly assuming that $a^\sigma \neq a$ for some $a \in F$.) Let V be a vector space over F with the scalars written on the left and let W be the vector space with scalars written on the right given as $W = V$ additively and $\mathbf{w}a = a^\sigma \mathbf{w}$ for $\mathbf{w} \in V$ and $a \in F$. We shall refer to a form on (V, W) as a form on V and instead of symmetry we now assume that

$$(\mathbf{v}, \mathbf{w}) = (\mathbf{w}, \mathbf{v})^\sigma$$

for all $\mathbf{v}, \mathbf{w} \in V$. Once again, L is an isomorphism if and only if R is; a non-singular form satisfying this condition will be called a **hermitian form**. Besides the additivity in both variables we now have that $(a\mathbf{v}, \mathbf{w}) = a(\mathbf{v}, \mathbf{w})$ and $(\mathbf{v}, a\mathbf{w}) = a^{\sigma}(\mathbf{v}, \mathbf{w})$ for all $\mathbf{v}, \mathbf{w} \in V$ and $a \in F$. Of course, "hermitian" depends on the given involution, but in the case we are interested in, that of a finite field, there will be only one choice, *viz.*

$$a^{\sigma} = a^{q}$$

for $F = \mathbf{F}_{q^2}$. (The Galois group of a finite field, $\mathbf{F}_{p^r}$, is of order r and hence there is an involution only when r is even and then, with $r = 2s$, it must be given by sending a to a^{p^s} since the Galois group is cyclic and generated by the automorphism that sends a to a^{p}.)

Given a non-singular form on (V, W), the map $\perp$ establishes a Galois correspondence between the collection of subspaces of V and the collection of subspaces of W, a fact that follows immediately from the dimensional equalities proved above and the obvious lattice-inverting properties of $\perp$. In fact, $\perp$ is a polarity whenever the form is on V. We will thus speak of **orthogonal**, **symplectic** and **unitary polarities** of $PG(V)$ whenever we have an orthogonal, symplectic or hermitian form, respectively, on V.

Given a non-singular form on V, it may happen that, for a subspace S of V, the form restricted to S is singular. Also, if the form is singular on V it still might be non-singular on S; i.e. it can happen that $S \cap S^{\perp} = \{0\}$. Moreover, if so, then $V = S + S^{\perp}$ and the sum is therefore direct as the following argument shows: for $\mathbf{v} \in V$ restrict $\mathbf{v}L$ to S and, using the non-singularity in S, obtain that $\mathbf{v}L = \mathbf{s}L$ on S for some $\mathbf{s} \in S$, which implies that $(\mathbf{v} - \mathbf{s})L$ is zero on S, i.e. $\mathbf{v} - \mathbf{s} \in S^{\perp}$. Consequently, $\dim S^{\perp} = \dim V - \dim S$ whenever $S \cap S^{\perp} = \{0\}$ even for singular forms on V.

Our principal interest here is in unitary and orthogonal polarities, and it is rather easy to see inductively that any hermitian or orthogonal (when the characteristic of the field is not 2) form on V is "diagonalizable". For this we need to see that, unless the form is identically zero, there is a vector $\mathbf{v}$ with $(\mathbf{v}, \mathbf{v}) \neq 0$. Characteristic 2 gives no trouble in the hermitian case, but for orthogonal forms there is a richer theory in this case. For that we need quadratic forms and quadrics, which we shall introduce presently.

For an arbitrary form on (V, W), let $\mathbf{v}_1, \mathbf{v}_2, \ldots, \mathbf{v}_m$ be a basis of V and $\mathbf{w}_1, \mathbf{w}_2, \ldots, \mathbf{w}_n$ a basis of W. It is enough to know all $(\mathbf{v}_i, \mathbf{w}_j)$ in order to know the form since it is linear in both variables. Moreover, given any $m \times n$ matrix B, a form on (F^m, F^n) is obtained by setting

$$((x_1, x_2, \ldots, x_m), (y_1, y_2, \ldots, y_n)) = (x_1, x_2, \ldots, x_m)B(y_1, y_2, \ldots, y_n)^t.$$

If the form is on V we choose only one basis and the square matrix $B =$

$((\mathbf{v}_i, \mathbf{v}_j))$ determines the form. Now the form is diagonalizable if a basis can be chosen in such a way that the matrix B is diagonal.

When the field is finite, one has even more: in the orthogonal case all but one of the diagonal elements can be chosen to be 1 and this immediately implies that up to a natural equivalence given by isometry there are only two forms in each dimension, the one given by the identity matrix and the one given by the matrix whose diagonal elements are $1, 1, \ldots, 1, a$ where a is some non-square — all this, of course, when the characteristic is not 2.

In the hermitian case the square matrix B must satisfy $B = (B^\sigma)^t$ (it is a hermitian matrix) and, in particular, when the basis is chosen so that B is diagonal, the entries on the diagonal must satisfy the equation $x = x^\sigma$. For finite fields the diagonal elements can be taken to be 1 since the map that sends a to a^{q+1} sends $\mathbf{F}_{q^2}$ onto $\mathbf{F}_q$ and, in fact, $\mathbf{F}_{q^2}^\times$ homomorphically onto $\mathbf{F}_q^\times$; this is an important fact that we use repeatedly below. Therefore there is only one hermitian form in each dimension over the field $\mathbf{F}_{q^2}$ and to compute with such a form we simply identify V with $\mathbf{F}_{q^2}$ using the "good" basis and get

$$((x_1, x_2, \ldots, x_m), (y_1, y_2, \ldots, y_m)) = \sum_{i=1}^m x_i y_i^q.$$

Using such a basis, the absolute points of a unitary polarity, termed a **hermitian variety** in $PG_n(\mathbf{F}_{q^2})$, are given by the equation

$$\sum_{i=0}^n X_i^{q+1} = 0. \tag{3.3}$$

In projective dimension 1 this becomes

$$X^{q+1} + Y^{q+1} = 0$$

which is merely the equation $(X/Y)^{q+1} = -1$ implying, since $-1 \in \mathbf{F}_q$, that there are $q+1$ projective points on the variety. It follows that for projective dimension 2 there are $q+1$ points at infinity and solving $X^{q+1}+Y^{q+1}+1 = 0$ for the affine points gives, for each $Y \in \mathbf{F}_{q^2}$ with $Y^{q+1} \neq -1$, $q+1$ solutions and hence $[q^2 - (q+1)](q+1) + (q+1)$ affine points. Thus the hermitian variety has q^3+1 points in dimension 2. These points are the 1-dimensional subspaces determined by the isotropic vectors of the unitary polarity, i.e. those non-zero vectors $\mathbf{v}$ with $(\mathbf{v}, \mathbf{v}) = 0$ for the hermitian form. If V is the 3-dimensional vector space defining the projective plane $PG_2(\mathbf{F}_{q^2})$ and W is any line of this plane (i.e. $\dim W = 2$), then the form restricted to W is either non-singular, in which case W contains $q + 1$ points of the variety, or else the form restricted to W is singular. In the singular case the form cannot be identically zero, since then it would follow that the form on V would be singular. Thus there is a unique point P of the line with $(\mathbf{v}, \mathbf{w}) = 0$ for $\mathbf{v} \in P$ and $\mathbf{w} \in W$. We have just shown that every line contains either one point of the variety or else contains $q + 1$ points. In terms of the unitary polarity of $PG(V)$, the points of the hermitian variety are the **absolute** points of the polarity (those with $P \subset P^{\perp}$) and the lines meeting the variety only once are the **absolute** lines of the polarity (those with $L^{\perp} \subset L$).

Clearly the points of the variety together with those lines that meet the variety in $q+1$ points, the non-absolute lines, form a design with parameters $2\text{-}(q^3 + 1, q + 1, 1)$. Any design with these parameters is called a **unital**. We have just constructed the classical hermitian unital and for completeness' sake make this formal:

Definition 3.7.2 *A unital is* **hermitian** *if it is isomorphic to a* $2\text{-}(q^3 + 1, q+1, 1)$ *design whose points are the absolute points of a unitary polarity of* $PG_2(\mathbf{F}_{q^2})$ *(for some prime-power q) and whose blocks are the non-absolute lines of the polarity, incidence being that of the projective plane.*

More generally, a unital **embedded** in a projective plane of order q^2 consists of a point set of cardinality $q^3 + 1$ of points of the plane with the property that every line of the plane meets the set in one or $q + 1$ points, and has for blocks those lines that meet this set in $q + 1$ points. It is, in fact, enough to assume that every line of the plane that meets the point set twice meets it $q + 1$ times; it then follows immediately that there is a unique tangent at every point of the embedded unital and that the tangents and block-producing lines account for all the lines of the ambient projective plane. Note here that the projective plane need not be desarguesian. For $q = 2$ the unitals of $PG_2(\mathbf{F}_4)$ are precisely the hermitian unitals (and they are isomorphic to affine planes of order 3) but for $q > 2$ there are always others: see Buekenhout [57] and Metz [215]. Not all unitals can be embedded in projective planes: see Brouwer [50]. In fact, unitals with q not a prime power have been discovered by Mathon [210] and by Bagchi and Bagchi [28]. See Section 8.3 for a fuller discussion of classes of unitals.

Before concentrating on the orthogonal case, which will constitute most of the remainder of this section, notice that for symplectic polarities *all* points of $PG(V)$ are absolute. The following example introduces the *classical* generalized quadrangles (see Example 1.2.3).

Example 3.7.2 The absolute points and absolute lines of a symplectic polarity on $PG_3(\mathbf{F}_q)$ form a generalized quadrangle of type $s = t = q$.

Now we turn to the case of orthogonal forms when $F = \mathbf{F}_q$, q odd. Here the set of absolute points of the orthogonal polarity is called a *quadric* — or *conic* in the case of a projective plane. Since our interest is mainly in conics, we will assume that $\dim V = 3$ but note that what we say about the equivalence of the two varieties in question is true in all even projective dimensions. The equations to be satisfied by the coordinates of those points on the conic are

$$X^2 + Y^2 + Z^2 = 0 \quad \text{or} \quad X^2 + Y^2 + aZ^2 = 0$$

depending on the form, where a is a non-square of $\mathbf{F}_q$. Since, in a finite field, every element is a sum of two squares we can write $a = c^2 + d^2$ and the change of variable, $X := cX + dY$, $Y := dX - cY$, $Z := Z$, reduces the second equation above to a times the first and the varieties defined by the two equations coincide. It is customary and more convenient to again change variables and to use either the equation $X^2 + Y^2 - Z^2 = 0$ or the equation

$$Y^2 = XZ. \tag{3.4}$$

This latter equation clearly has one point at infinity, $x = 1$ and $y = z = 0$, and q affine points. (In fact, this equation could also be used to define a conic in characteristic 2, with "conic" then meaning any projectively equivalent configuration.) Every conic has $q + 1$ points in $PG_2(\mathbf{F}_q)$.

Purely coordinate arguments show that every line of the plane is either disjoint from the conic (an exterior line), meets it in one point (a tangent line) or meets it in two (a secant line). In terms of the form, these lines are the 2-dimensional subspaces that contain no isotropic points, one isotropic point, or two isotropic points, respectively. The tangents correspond, therefore, to those lines L with $L^\perp \subset L$, the absolute lines of the orthogonal polarity.

In particular, a conic in $PG_2(\mathbf{F}_q)$ is a set of $q + 1$ points no three of which are collinear. Purely combinatorial arguments, first enunciated by Qvist [244], show that the largest possible set of points of an arbitrary projective plane of odd order n with no three collinear has $n + 1$ points and that then there is a unique tangent through each point of the set. Since the plane is a 2-$(n^2 + n + 1, n + 1, 1)$ design, this also follows from the general result for arcs in 2-designs that we proved in Section 1.5. It came as quite a surprise when Segre [259] showed that in $PG_2(\mathbf{F}_q)$ with q odd such a set had to be a conic. We shall give a proof of Segre's theorem, but first we need to look at the case of even characteristic.

We have noted that an orthogonal form will not give an oval in the plane from the polarity when the characteristic of the field is 2. For this we need a **quadratic form**, which is defined as follows:

Definition 3.7.3 *A* **quadratic form** *on a vector space V over a field F is a mapping $Q : V \to F$ that satisfies*

$$Q(a\mathbf{v}) = a^2 Q(\mathbf{v}), \quad \textit{for all } a \in F, \mathbf{v} \in V$$

and the condition that the map $f : V \times V \to F$ defined by

$$f(\mathbf{v}, \mathbf{u}) = Q(\mathbf{v} + \mathbf{u}) - Q(\mathbf{v}) - Q(\mathbf{u}), \quad \textit{for all } \mathbf{v}, \mathbf{u} \in V$$

is a form on V.

A **quadric** in $PG(V)$ is then a set of projective points given by vectors $\mathbf{v} \in V$ that satisfy $Q(\mathbf{v}) = 0$. When $PG(V)$ is a projective plane, a quadric is also called a **conic**.

Now note first that in *odd* characteristic a non-degenerate quadric, i.e. one in which the bilinear form is non-degenerate, is precisely the set of absolute points of an orthogonal polarity, i.e. what we have already called a quadric: see any of the texts mentioned, or refer to Dembowski [85, Section 1.4.40]. This is not the case for characteristic 2, since then $f(\mathbf{v}, \mathbf{v}) = 0$ for all $\mathbf{v} \in V$. The quadric is said to be non-degenerate in this case if $Q(\mathbf{v}) \neq 0$ whenever $\mathbf{v} \neq 0$ and $f(\mathbf{v}, \mathbf{x}) = 0$ for all $\mathbf{x} \in V$.

In any characteristic, if a basis $\mathbf{e}_1, \ldots, \mathbf{e}_n$ is chosen for V, then the quadratic form can be represented in the usual way by

$$Q(\mathbf{x}) = \sum_{i=1}^{n} Q(\mathbf{e}_i)x_i^2 + \sum_{1 \le i < j \le n} f(\mathbf{e}_i, \mathbf{e}_j)x_i x_j,$$

where $\mathbf{x} = \sum_i x_i \mathbf{e}_i$. In particular, when $n = 3$ and we have a projective plane, the conic can be represented in the way mentioned above.

Non-degenerate quadrics in dimensions 2 and 3 provide special cases of sets of points in projective space known as **ovoids**:

Definition 3.7.4 *A non-empty set $\mathcal{O}$ of points in $PG(V)$ is an* **ovoid** *if every line meets $\mathcal{O}$ in at most two points and, for any $P \in \mathcal{O}$, the union of all lines that meet $\mathcal{O}$ only at P is a hyperplane of $PG(V)$, called a* **tangent hyperplane.**

If V is $(n+1)$-dimensional over $\mathbf{F}_q$, then an ovoid has $q^{n-1}+1$ points and an easy argument (see Dembowski [85]) shows also that $n \le 3$ if $PG_n(\mathbf{F}_q)$ has an ovoid. When $n = 2$, an ovoid is usually called an **oval**: this is then a set of $q + 1$ points of $PG_2(\mathbf{F}_q)$ such that any line meets the set at most twice, and every point of the oval is on a unique tangent. This is what we have called a $(q + 1)$-arc in Section 1.5, but we have a slight ambiguity here in connection with our terminology for ovals in designs — as defined in Definition 1.5.2 — when considering a finite projective plane as a design. For odd order there is no ambiguity, but for even order an oval as defined for designs must have $q + 2$ points. Such an arc has been called a *hyperoval* in the literature in order to avoid confusion, but we will use the term oval as for designs. In fact, the following lemma, due again to Qvist [244] — but see also Dembowski [85, Section 3.2] or Hirschfeld [136, Chapter 8] — shows that any $(n + 1)$-arc in a projective plane of *even* order n can be completed to an $(n + 2)$-arc.

Lemma 3.7.1 *If* Π *is a projective plane of even order* n *and* $\mathcal{A}$ *is an* $(n + 1)$-*arc, then the tangents to* $\mathcal{A}$ *meet at a point,* K, *called the* **nucleus** *or* **knot** *of* $\mathcal{A}$, *and* $\mathcal{O} = \mathcal{A} \cup \{K\}$ *is an oval.*

Proof: Let $P \notin \mathcal{A}$ be any point of Π and let x_i denote the number of lines through P that meet the arc in i points, where $i = 0, 1, 2$. Then simple counting gives

$$
\begin{aligned}
x_0 + x_1 + x_2 &= n + 1 \\
x_1 + 2x_2 &= n + 1,
\end{aligned}
$$

so that $x_0 = x_2$, and $x_1 = n + 1 - 2x_2$. Since n is even, this shows that $x_1 \geq 1$. In particular, every point on a secant is on at least one tangent. Each point of the arc is on $n + 1$ lines, n of which are secants, so it follows that there are precisely $n + 1$ tangents. Each of the $n - 1$ points on a secant that are not on the arc must be on a tangent, and thus no two tangents can meet on a secant. It follows that the tangents are concurrent. $\square$

Example 3.7.3 In $PG_2(\mathbf{F}_{2^t})$, the equation

$$
Y^2 + XZ = 0
$$

defines a conic, the points of which are $(1, 0, 0)$ and $(y^2, y, 1)$ for all $y \in \mathbf{F}_{2^t}$. Tangents to the conic at $(1, 0, 0)$ and $(0, 0, 1)$ are $(0, 0, 1)^t$ and $(1, 0, 0)^t$, respectively, which meet at $(0, 1, 0)$, which is the nucleus.

In odd characteristic, the ovoids are the non-degenerate quadrics: this is due to Segre for $n = 2$ — we give his proof below — and to Barlotti [30] and Panella [231] for $n = 3$. In even characteristic there are ovoids that are not quadrics both for $n = 2$ and $n = 3$. In neither case are the ovoids classified. For $n = 3$ there are only two classes known, those from quadrics (so-called elliptic quadrics) and the Tits ovoids: see [85, Section 1.4]. For $n = 2$ and $q \leq 16$ the ovals are known: up to projective equivalence there is one for $q = 2, 4$ or 8 and two for $q = 16$ (see Hall [120]). For $q \geq 32$ and $q \neq 64$ there are in general many ovals that are not of the form of a conic together with its nucleus — the so-called *regular ovals*. For $q = 64$, until the recent discoveries of Penttila and Pinneri [234], no "sporadic" oval had been constructed. See Hirschfeld [136], or the survey article by Korchmáros [175], for a general discussion.

Exercise 3.7.1 Let Π be the projective plane $PG(V) = PG_2(\mathbf{F}_q)$ where $q = 2^m$. Let $\mathcal{O}$ be any oval in Π. The structure having for point set all the vectors of V and for block set all the cosets $\mathbf{a} + \langle \mathbf{v} \rangle$, where $\langle \mathbf{v} \rangle \in \mathcal{O}$, forms a generalized quadrangle with $s = q - 1$ and $t = q + 1$.

We will now prove Segre's theorem[7] but we note first some elementary facts discovered by Qvist [244] for odd-order planes. Using the same count as given in Lemma 3.7.1, with n odd, it follows that every point off the oval is on evenly many tangents. Since there are only $n + 1$ tangents, this shows that every point on a tangent (other than the point that is on the oval) must be on exactly one more tangent. In particular this argument implies that the $n + 1$ tangents to an oval in a projective plane of odd order n form an oval in the dual plane.

Theorem 3.7.1 (Segre) *Suppose $\mathcal{C}$ is a set of $q + 1$ points of $PG_2(\mathbf{F}_q)$ with the property that no three lie on a line. Then if q is odd, $\mathcal{C}$ is a conic.*

Proof: Let $\mathcal{C}$ be a set of $q + 1$ points of $PG_2(\mathbf{F}_q)$ with no three collinear and suppose that P, Q, and R are points of $\mathcal{C}$. They form the vertices of a triangle, of course, and their tangents, L', M', and N' with L' the tangent at P, etc., are the edges of another triangle when q is odd by the argument given above.

We first show that these two triangles are in perspective (with the notation as in Desargues' theorem 3.6.1). Since $P + Q + R = V$, the 3-dimensional vector space over $F = \mathbf{F}_q$ that defines $PG_2(\mathbf{F}_q)$, we can choose a basis for V such that the homogeneous coordinates of P, Q, and R are, respectively, $(1, 0, 0)$, $(0, 1, 0)$ and $(0, 0, 1)$, and the equations of the lines $L = Q + R$, $M = P + R$ and $N = P + Q$ are, respectively, $X = 0, Y = 0$ and $Z = 0$. Then the equations of the tangent lines are, respectively, $Y = a'Z$, $Z = b'X$ and $X = c'Y$, where a', b' and c' are necessarily non-zero elements of F.

It follows that if O is any other of the $q - 2$ points of $\mathcal{C}$, its homogeneous coordinates (x, y, z) must all be non-zero. Moreover, the line $O + P$ must

---

[7]The theorem was conjectured by Järnefelt and Kustaanheimo and thought to be implausible by Hall—see *Mathematical Reviews* **14** (1953), page 1008.

have an equation of the form $Y = aZ$ with, clearly, $a \neq 0$ and $a \neq a'$. Similarly, $O + Q$ has an equation of the form $Z = bX$ with $b \neq 0$ and $b \neq b'$ and $O + R$ an equation of the form $X = cY$ with $c \neq 0$ and $c \neq c'$; most importantly, for each such O, $abc = 1$ since $a = yz^{-1}$; $b = zx^{-1}$ and $c = xy^{-1}$. As O varies over those points of C other than P, Q or R, each of a, b and c must vary over, respectively, $F^{\times} - \{a'\}$, $F^{\times} - \{b'\}$ and $F^{\times} - \{c'\}$, since each line of the correct form defines such a point O. Thus, since the product of all elements of $F^{\times}$ is -1, $a'b'c' = (-1)^3 = -1$. By adjusting the basis vectors by scalars, we can now ensure that $a' = b' = c' = -1$ and now the two triangles are in perspective from the point $(1, 1, 1)$ and along the line $X + Y + Z = 0$.

We next want to show that not only $(1, 0, 0), (0, 1, 0)$ and $(0, 0, 1)$ but *all* the points of C satisfy the equation $YZ + ZX + XY = 0$ or, in other words, are the absolute points of the orthogonal polarity given, in matrix form, by the orthogonal form

$$\begin{pmatrix} 0 & 1 & 0 \\ 0 & 0 & 1 \\ 1 & 0 & 0 \end{pmatrix}.$$

If O is any other point of C we know from the above that the homogeneous coordinates (x, y, z) are all non-zero and that, if $aX + bY + cZ = 0$ is the tangent (which we will denote by O') to C at O, then $abc \neq 0$ also. Again by the above, it follows that the triangle with vertices O, Q and R is in perspective with the triangle with edges O', M' and N', and so the points $(Q + R) \cap O'$, $(O + R) \cap M'$ and $(O + Q) \cap N'$ are collinear. The homogeneous coordinates of these points are, respectively, $(0, c, -b)$, $(x, y, -x)$ and $(x, -x, z)$; to be collinear these three vectors must be linearly dependent; i.e. the determinant with these homogeneous coordinates as rows must be zero and hence $b(x+y) = c(z+x)$. Similarly, $c(y+z) = a(x+y)$ and $a(z + x) = b(y + z)$. These three equations imply that

$$\frac{y + z}{a} = \frac{z + x}{b} = \frac{x + y}{c}$$

and we are free, by changing a, b and c by a scalar, to assume all three fractions are equal to 1 since, if they are all 0, $2x = 0$, an impossibility when q is odd. But $ax + by + cz = 0$ and hence $(y + z)x + (z + x)y + (x + y)z = 0$ or $2(yz + zx + xy) = 0$. Using once more the fact that q is odd, we have the required equality. $\square$

Remark We have used the fact that in odd-order planes the tangents to an oval are themselves an oval in the dual plane simply to give the proof some visual clarity. In fact, unlike Segre's original proof, the proof given here (which is modelled directly on Segre's) does not use this fact and only at the very end do we invoke the assumption that q is odd.

Finally we need a few facts about collineations of the geometry that preserve a quadric or hermitian variety, the *isometries* of the geometry. In the case of any form (,), the collineations that commute with the relevant polarity are those that are given by a semilinear transformation T with associated automorphism α that satisfy the property

$$(\mathbf{x}T, \mathbf{y}T) = (\mathbf{x}, \mathbf{y})^{\alpha} a,$$

for all $\mathbf{x}, \mathbf{y} \in V$, where a is a fixed element of F that is given by T. The isometries will then fix the relevant variety, and the group of all isometries will be the *orthogonal*, *symplectic* or *unitary group*, respectively. In the case of a quadratic form, required for quadrics in even characteristic, T is an isometry preserving the quadric if

$$Q(\mathbf{v}T) = Q(\mathbf{v})^{\alpha} a,$$

for all $\mathbf{v} \in V$, where, again, a is fixed and given by T. The group is again an *orthogonal group*, defined as before if the characteristic is not 2.

Example 3.7.4 In Example 3.7.3, parametric coordinates for the conic can be defined as for the projective line, by identifying ∞ with $(1, 0, 0)$ and y with $(y^2, y, 1)$: see Hughes and Piper [143, Chapter 2.7] for a full discussion. It is then not difficult to show (see Theorem 2.37 of [143]) that the group preserving the quadratic (or orthogonal) form, which is also the group that fixes the conic, is the projective semilinear group $P\Gamma L_2(\mathbf{F}_q)$, given more conveniently in the parametric form by the mappings:

$$x \mapsto \frac{ax^{\alpha} + b}{cx^{\alpha} + d}, \text{ where } ad - bc \neq 0$$

for $x \in \mathbf{F}_q \cup \{\infty\}$, $a, b, c, d \in \mathbf{F}_q$ and $\alpha \in \mathrm{Aut}(\mathbf{F}_q)$. This group is then triply-transitive on the points of the conic, whether q is even or odd.

Chapter 4

Symmetric Designs

4.1 Introduction

In this chapter we study those designs that have equally many blocks and points. Historically these designs arose both geometrically (see Chapter 3) and through the needs of statisticians. The original statistical concept is worth recalling here.

Starting with two disjoint sets, $\mathcal{P}$ and $\mathcal{B}$ — called varieties and plots, respectively — each of cardinality v, impose a regular bipartite graph of valency k, with $\mathcal{P}$ and $\mathcal{B}$ as the two parts. The graph should satisfy the further condition that given any two distinct varieties there are precisely λ paths of length 2 from one variety to the other. It follows[1] that the same property holds for any two distinct plots (hence the unfortunate term "symmetric design"). There is thus no reason to call the elements of $\mathcal{P}$ varieties and those of $\mathcal{B}$ plots for it might as well have been the reverse. Moreover, the complementary bipartite graph (eliminate the given edges (P, B) and introduce instead those (Q, C) that were not originally edges) is regular with valency $v - k$ and satisfies the added condition also—with its "λ" equal to $(k - \lambda)(k - \lambda - 1)/\lambda$.[2]

Thus symmetric designs come in pairs, a design and its complement, or in fours, if the dual designs are also taken into account. R. H. Bruck first observed that it is wise when dealing with symmetric designs to consider these four designs together.

Such a statistical design on $(\mathcal{P}, \mathcal{B})$ defines an incidence structure by taking incidence to mean being connected by an edge. The regularity ensures

---

[1] See (3) of Theorem 4.2.1 below.
[2] See (5) of Theorem 4.2.1 below.

that every plot "contains" k varieties and, because of the path condition, any two varieties occur together in λ plots; it is precisely this that makes the design suitable for statistical analysis in the comparison of any two varieties. Dropping the statistical terminology, we will call the elements of $\mathcal{P}$ points and the elements of $\mathcal{B}$ blocks. Hence we have a 2-(v, k, λ) design with equally many points and blocks. We will take this to be the definition of a (v, k, λ) design, where we are assuming that $0 < \lambda < k < v - 1$:

Definition 4.1.1 *A 2-(v, k, λ) design is called a* **symmetric design** *if the number of blocks is the same as the number of points.*

There are some variations of terminology: Hughes and Piper [144] call an incidence structure with $b = v$ a **square** structure, and Dembowski [85] speaks of **projective** rather than symmetric designs. The term "symmetric" has become the customary one, and we reluctantly adopt it.

Example 4.1.1   (1) The examples (1), (2), (3) and (5) of Example 1.2.2 are all symmetric designs.

(2) The designs of points and hyperplanes of a projective geometry as defined in Section 3.4 are symmetric designs with parameters as given in that section.

4.2   The parameters

We first show that the assumption that $b = v$ *implies* that the design could not be a non-trivial t-design for $t > 2$. This is not true, of course, if $k = v-1$ and that is why we have excluded this case in the definition.

Proposition 4.2.1 *If $\mathcal{D}$ is a t-(v, k, λ) design with $t \geq 3$ and with equally many points and blocks, then $\mathcal{D}$ is trivial.*

Proof: $\mathcal{D}$ is a 3-(v, k, λ_3) design, by Theorem 1.2.1. Since $b = v$, we have $k = r$, by Equation (1.2). If P is any point of $\mathcal{D}$, then $\mathcal{D}_P$ is a 2-$(v - 1, k - 1, \lambda_3)$ design with $\lambda_1 = r = k$ blocks. If $\mathcal{D}_P$ is non-trivial, then by Fisher's inequality, Corollary 1.4.1, the number of blocks, *viz.* k, is at least as great as the number of points, *viz.* $v - 1$. So $k \geq v - 1$, and since $k \leq v$, it follows that $\mathcal{D}_P$ is trivial. Equation (1.1) then shows that $\mathcal{D}$ is also trivial. $\square$

Terminology: As is customary we refer to a 2-(v, k, λ) design with $0 < \lambda < k < v - 1$ and $b = v$ simply as a (v, k, λ) design.

Theorem 4.2.1 *Let $\mathcal{D} = (\mathcal{P}, \mathcal{B})$ be a (v, k, λ) design of order n. Then*

(1) $k = r$ *and* $n = k - \lambda$;

(2) $\lambda(v - 1) = k(k - 1)$;

(3) the dual structure, $\mathcal{D}^t$, *is a* (v, k, λ) *design;*

(4) the complementary structure, $\overline{\mathcal{D}}$, *is a* $(v, v - k, v - 2k + \lambda)$ *design;*

(5) λ *divides* $n(n-1)$, *and if* $\lambda\mu = n(n-1)$, *then* $\mathcal{D}$ *is a* $(2n+\lambda+\mu, n+\lambda, \lambda)$ *design,* $\overline{\mathcal{D}}$ *is a* $(2n + \lambda + \mu, n + \mu, \mu)$ *design, and*

$$4n - 1 \le v \le n^2 + n + 1.$$

Proof: (1) and (2) follow from the equations following Theorem 1.2.1, but observe that they are virtually immediate from the original statistical definition.

To prove (3), let A be an incidence matrix for $\mathcal{D}$, so that $B = A^t$ is an incidence matrix for $\mathcal{D}^t$. By Theorem 1.4.1, $A^t A = nI_v + \lambda J_v$, so that $A(A^t A) = nA + \lambda A J_v = nA + \lambda k A = nA + \lambda J_v A = (nI_v + \lambda J_v)A$, and since $\mathcal{D}$ is non-trivial, A is non-singular, by Theorem 1.4.1, and we get $AA^t = nI_v + \lambda J_v = B^t B$. Thus $\mathcal{D}^t$ is a (v, k, λ) design as well.

For (4) first note that $\overline{\mathcal{D}}$ is clearly also symmetric, and the value for its λ can be obtained from Pascal's triangle, Figure 1.3, or easily determined by counting appropriate flags, given that any two blocks meet in λ points.

For (5), since λ divides $k(k - 1)$, it also divides $(k - \lambda)(k - \lambda - 1) = n(n - 1)$. Setting $\lambda\mu = n(n - 1)$, we have that $v = 2n + \lambda + \mu$ and that the parameter sets for $\mathcal{D}$ and $\overline{\mathcal{D}}$ are $(2n+\lambda+\mu, n+\lambda, \lambda)$ and $(2n+\lambda+\mu, n+\mu, \mu)$, respectively.

To obtain the bounds on v, notice that $(\lambda-\mu)^2 = v^2 - 4n(v-1) > 0$, since $\lambda \ne \mu$, $n(n - 1)$ never being a perfect square. Thus $v^2 - 4n(v - 1) - 1 \ge 0$, i.e. $(v - 1)(v - 4n + 1) \ge 0$, giving the lower bound, $4n - 1$, for v. For the upper bound, simply notice that for the positive integers λ and μ, we have $\lambda + \mu \le \lambda\mu + 1$, so that $v = 2n + \lambda + \mu \le n^2 + n + 1$. $\square$

We will see later that both the bounds given in (5) are attained for various infinite sets of the parameter n: those at the lower bound are the *Hadamard* designs, which we will discuss in Chapter 7, and those at the upper bound are the *projective planes*, to be discussed in Chapter 6. Clearly, for a given order n there are only finitely many parameter sets possible and they are easily obtained from (5) by factoring $n(n - 1)$ in all possible ways. For further details concerning this approach see Assmus [7].

Exercise 4.2.1 Show that if a symmetric (v, k, λ) design of order n has $v = 4n - 1$ then it is either a $(4n - 1, 2n - 1, n - 1)$ design or its complement, and if it has $v = n^2 + n + 1$ then it is either a $(n^2 + n + 1, n + 1, 1)$ design or its complement.

Projective planes are those symmetric designs that have $\lambda = 1$, i.e. they are the symmetric designs that are also Steiner systems. Symmetric designs with $\lambda = 2$ are known as **biplanes**; we already have an example of a biplane in Example 1.2.2 (2). Unlike the case of $\lambda = 1$ where an infinite number of designs are known to exist, only a finite number of biplanes have so far been constructed.

For a symmetric design any two blocks intersect in the same number of points and this property characterizes the symmetric designs among the 2-designs. Some attention has been paid to those incidence structures where the cardinalities of the intersections constitute a small set of integers; in particular, the notion of a **quasi-symmetric** design has also been introduced: this is a design in which any two blocks can meet in x or y points, where x and y are distinct, non-negative integers. Generalized quadrangles and Steiner 2-designs are examples, with, in these cases, $\{x, y\} = \{0, 1\}$. For a book-length account of quasi-symmetric 2-designs see Shrikhande and Sane [262].

Exercise 4.2.2 Show that if $\mathcal{D}$ is a 2-$(v, k, 1)$ design that is not symmetric then its **line graph**, defined to be the graph with vertices the block set of $\mathcal{D}$ and two vertices adjacent if – as blocks – they intersect, is a strongly regular graph with parameter set $(b, k(r-1), (r-2) + (k-1)^2, k^2)$.

Point sets of size k that meet every block of a symmetric (v, k, λ) design in at least λ points are immediately characterized by the following result from Lander [181, Lemma 5.2].

Proposition 4.2.2 *If a point set of a (v, k, λ) design has cardinality k and meets every block in at least λ points, then it must be a block.*

Proof: Suppose the point set is S. For each block B of the design set $s_B = |S \cap B| - \lambda$. Now $\sum_{B \in \mathcal{B}} s_B = (\sum_{B \in \mathcal{B}} |S \cap B|) - v\lambda$. Since every point of S is on k blocks, counting pairs (B, x) with x a point of $S \cap B$ yields $\sum_{B \in \mathcal{B}} |S \cap B| = k^2$; hence $\sum_{B \in \mathcal{B}} s_B = k^2 - v\lambda = k - \lambda = n$. Counting next the triples (B, x, y) where x and y are distinct points of $S \cap B$ yields $\sum_{B \in \mathcal{B}} (s_B + \lambda)(s_B + \lambda - 1) = k(k-1)\lambda$. Since $\sum_{B \in \mathcal{B}} s_B = n$, this reduces to $\sum_{B \in \mathcal{B}} s_B^2 = n^2$, and these two equalities, together with the fact that the s_B's are non-negative, forces s_B to be n for one B (and 0 for the other blocks). But then $S = B$. $\square$

Turning next to the residual of a symmetric design, we find that this is always a 2-design. In fact the condition is shown to be necessary and sufficient in Hughes and Piper [144], but we only require the sufficiency.

Theorem 4.2.2 *If $\mathcal{D}$ is a (v, k, λ) design and B is a block of $\mathcal{D}$, then the residual design $\mathcal{D}^B$ is a 2-$(v - k, k - \lambda, \lambda)$ design.*

Proof: The number of points is $v - k$, and any two points of $\mathcal{D}^B$ are still on the same number of blocks as before. The number of points of $\mathcal{D}^B$ incident with a block of $\mathcal{D}^B$ is $k - \lambda$, since we have shown that any two blocks of $\mathcal{D}$, and in particular any block of $\mathcal{D}^B$ and B, meet in λ points. $\square$

Symmetric designs and their residuals play an important role in our analysis using coding theory. Classes of symmetric designs are plentiful, and, as we have indicated, we will later examine more closely those meeting both the upper and lower bounds of Part (5) of Theorem 4.2.1 — along with certain other infinite classes.

We now state the famous theorem of Bruck, Ryser and Chowla, but since we will not actually use it, we will not include a proof; many versions of the proof are readily available: see, for example, Hall [122, p. 133], Hughes and Piper [144, p. 56] or Lander [181].

Theorem 4.2.3 (Bruck-Ryser-Chowla) *If there exists a (v, k, λ) design then*

(1) v is even and n is a square;

(2) v is odd and $z^2 = nx^2 + (-1)^{(v-1)/2} \lambda y^2$ has a non-trivial solution in integers x, y, z.

We remark that the proof of the first part is virtually immediate from Theorem 1.4.1 and this part of the theorem was observed by Schützenberger [258]. The second part uses Hasse-Minkowski results on rational equivalence of quadratic forms. The quoted texts, in particular that of Lander [181], discuss the implications that follow from this theorem for the existence of symmetric designs with particular parameters, in particular the possible orders for finite projective planes.

Exercise 4.2.3 Show that Theorem 4.2.3 implies that there can be no projective plane of order 6.

Finally in this section we state an important theorem of Dembowski and Wagner [86] that characterizes the symmetric designs obtained from points and hyperplanes of finite projective geometries (see Section 3.4) through properties of the *lines* of the design (see Section 1.2 and Exercise 1.2.3 (2)). We only state the theorem, since the proof can be consulted in [86] or, for example, in the texts of Lander [181] or of Hughes and Piper [144]. Furthermore we will not have occasion to make use of the theorem but state it because of its importance in characterizing symmetric designs.

Theorem 4.2.4 (Dembowski-Wagner) *If $\mathcal{D}$ is a non-trivial (v, k, λ) design then the following are equivalent:*

(1) $\mathcal{D}$ is a projective plane or the design of points and hyperplanes of a finite projective geometry;

(2) every line meets every block;

(3) every line contains $(v - \lambda)/(k - \lambda)$ points.

4.3   Automorphisms

There is a proof due to Brauer [48, Lemmas 1, 2 and 3], that dates back to 1941, of Block's theorem [43] and Siemons's [264] generalization of it, but which applies only to incidence structures with the same number of blocks as points[3] (and with incidence matrix of full rank over the rational numbers). In particular, however, it applies to symmetric designs and we include the result and proof here.

Theorem 4.3.1 (Brauer) *Let $\mathcal{S} = (\mathcal{P}, \mathcal{B}, I)$ be an incidence structure with the same number of points as blocks and assume the incidence matrix is non-singular. Then every automorphism of $\mathcal{S}$ fixes as many points as blocks and has the same cycle structure on $\mathcal{P}$ as it does on $\mathcal{B}$. Moreover, if G is any subgroup of $\mathrm{Aut}(\mathcal{S})$, then G has the same number of orbits on points as it has on blocks.*

Proof: Let A be a non-singular incidence matrix of the incidence structure. Each automorphism acts as a permutation on $\mathcal{P}$ and on $\mathcal{B}$. Let C and R, respectively, be permutation matrices representing these actions. Then $AC = RA$ and, as A is non-singular, $A^{-1}RA = C$; thus C and R are similar matrices. Since the number of fixed points of a permutation matrix is its trace and similar matrices have the same trace, the fixed point assertion is immediate.

We next show that the permutations are "similar" (i.e. that they are conjugate in the symmetric group or, in other words, that they have the same cycle structure). Let n_i and m_i denote the number of point and block orbits, respectively, of length i. We have just shown that $n_1 = m_1$. We now apply induction on i: suppose $n_i = m_i$ for $1 \leq i < k$. Then if α is an automorphism, α^k fixes $\sum t n_t$ points and $\sum t m_t$ blocks, the sum being taken over all t dividing k. Since α^k has permutation matrices C^k and R^k

---

[3]In fact, Brauer considered the case of a rectangular matrix with linearly independent columns, precisely the case needed for non-trivial 2-designs in his Lemma 2; his interest was in representation theory, not in design theory, however.

on points and blocks, respectively, they are also similar, and hence have the same trace. Thus $\sum tn_t = \sum tm_t$, so that $kn_k = km_k$ by the induction hypothesis; hence $n_k = m_k$.

Finally, if G is a subgroup of $\mathrm{Aut}(\mathcal{S})$, we need only apply Burnside's lemma (the number of orbits of a permutation group on a finite set is the average number of points fixed by the elements of the group:see Exercise 4.3.1 below) and the first assertion to show that the number of point orbits equals the number of block orbits. $\square$

Exercise 4.3.1 Let G be a permutation group on a finite set Ω. For any $g \in G$ let $\mathrm{Fix}(g)$ denote the set of elements of Ω that are fixed by g. By counting the number of elements in the set

$$\{(\alpha, g) | \alpha \in \Omega, g \in G, \alpha^g = \alpha\}$$

in two ways, show that

$$t|G| = \sum_{g \in G} |\mathrm{Fix}(g)|,$$

where t is the number of orbits of G on Ω.

The proofs of the first two assertions are essentially Brauer's, but he was willing to use the easy representation theory of a cyclic group and hence his proofs are shorter. Even for the last assertion he invoked the representation theory rather than the more elementary lemma of Burnside. Wagner [288] gives another proof of the fixed point assertion (which is the crucial point and implies the others) that works for (v, k, λ) designs. It is due to Baer [25] for the case $\lambda = 1$, but applies for any λ: for any $g \in \mathrm{Aut}(\mathcal{S})$, count in two ways the number of elements in the set $\{(x, B) | x \in \mathcal{P}, B \in \mathcal{B}, x \in B \cap B^g\}$. Choosing points first gives $n_1 k + (v - n_1)\lambda$, and choosing blocks first gives $m_1 k + (v - m_1)\lambda$, and so $n_1 = m_1$. Thus we see that the theorem is highly combinatorial; the generalizations of Block and Siemons (see Theorem 1.6.1) and that of Bridges [49] exploit that fact.

4.4 Difference sets

One way to obtain a symmetric design with a given automorphism is to start with a permutation of the point set and then orbit some particular subset of points under the permutation; the entire orbit of sets is then included as part, or all, of the block set. Of course, the point set must be chosen appropriately to ensure that the design conditions are satisfied. Historically the permutation used simply cycled the points. In that case it is easy to see what the condition on the point set should be and the generalization to arbitrary groups follows rather naturally:

Definition 4.4.1 *Let G be a group of order v and let k and λ be positive integers with $1 < k < v - 1$. A (v, k, λ) **difference set** for G is a set D of k group elements with the property that the set of $k(k - 1)$ group elements*

$$\{gh^{-1}|g, h \in D \text{ with } g \neq h\}$$

contain every non-identity element of G exactly λ times.

The terminology comes from the original concept, where G is a finite cyclic group written additively, for then the products gh^{-1} are *differences*, $g - h$. The term "perfect difference set" was also used: see Ryser [250, Chapter 9] for an early bibliography on difference sets. A difference set is called **abelian** or **cyclic** according to whether G is abelian or cyclic. Note also that in some literature the term "quotient set" has been used for emphasis when the group is not necessarily abelian.

Example 4.4.1   (1) If $G = (\mathbf{Z}/7\mathbf{Z})^+$, the additive group of residues modulo 7, then $D = \{1, 2, 4\}$ is a (7,3,1) difference set.

(2) If $G = (\mathbf{Z}/11\mathbf{Z})^+$, $D = \{1, 3, 4, 5, 9\}$ is a (11,5,2) difference set.

(3) If $G = (\mathbf{Z}/13\mathbf{Z})^+$, $D = \{1, 5, 6, 8\}$ is a (13,4,1) difference set.

(4) If $G = (\mathbf{Z}/2\mathbf{Z})^+ \times (\mathbf{Z}/8\mathbf{Z})^+$, then both

$$D = \{(0, 0), (0, 1), (0, 2), (0, 5), (1, 0), (1, 6)\}$$

and

$$E = \{(0, 0), (0, 1), (1, 0), (1, 2), (1, 5), (1, 6)\}$$

are $(16, 6, 2)$ difference sets.

(5) If $G = \mathcal{Q} \times (\mathbf{Z}/2\mathbf{Z})^+$ where $\mathcal{Q}$ is the quaternion group, then

$$D = \{(1, 0), (i, 0), (j, 0), (k, 0), (1, 1), (-1, 1)\}$$

is a $(16, 6, 2)$ difference set, where we use the customary notation for the quaternion group by viewing it as the group of units of the integral quaternions: $\mathcal{Q} = \{\pm 1, \pm i, \pm j, \pm k\}$.

(6) Let G be the elementary abelian group of order 64 taken as the additive group of the vector space $\mathbf{F}_2^6$. Let D be the vectors of weights 1, 2, 5 and 6. Thus, $|D| = \binom{6}{1} + \binom{6}{2} + \binom{6}{5} + \binom{6}{6} = 28$. It is very easy to

see that every non-zero vector in G can be written in 12 distinct ways as a difference (in this case sum) of these 28 vectors; for example J, the vector 111111, can be written six ways as a weight-1 vector plus a weight-5 vector and six ways as a weight-5 vector plus a weight-1 vector. Thus we have a $(64, 28, 12)$ difference set. (We are indebted to Camion for pointing out this example to us. It is, in fact, a variant of an infinite class of such examples introduced in 1960 by Menon [214, Theorem, p. 373].)

For a census of the difference sets with $v = 16$ see Kibler [164]. For a general treatment of difference sets in elementary abelian groups see Camion [65], where the discrete Fourier transform is employed and the connections with coding theory emphasized.

Exercise 4.4.1 Verify that all the quoted sets in Example 4.4.1 are difference sets.

It is obvious from the definition that, for difference sets, $k(k - 1) = \lambda(v - 1)$, which is precisely the congruence condition for symmetric designs. This equation also implies, because of our assumption on k, that $\lambda < k$. In fact, a symmetric design is present; we next show how to get an incidence structure from a difference set and prove that it is in fact a symmetric design with the group G acting regularly—as a group of automorphisms—on points (and therefore on blocks by Brauer's theorem).

Definition 4.4.2 *Let D be a (v, k, λ) difference set in a group G. Define an incidence structure $\mathcal{D}$, called the **development** of D, by $\mathcal{D} = (\mathcal{P}, \mathcal{B})$, where $\mathcal{P}$ is the set of group elements, and*

$$\mathcal{B} = \{gD | g \in G\},$$

*i.e. $\mathcal{B}$ is the set of left **translates** of D, where $gD = \{gd | d \in D\}$. Incidence is membership.*

Theorem 4.4.1 *Let $\mathcal{D}$ be the development of a (v, k, λ) difference set for a group G. Then $\mathcal{D}$ is a (v, k, λ) design, every block is a (v, k, λ) difference set for G, and $G \subseteq \text{Aut}(\mathcal{D})$ with G acting regularly on points and blocks of $\mathcal{D}$.*

Proof: That every block is a (v, k, λ) difference set follows immediately from the definition of the development: the same is in fact true for the right translates. Furthermore, the v blocks are all distinct because $\lambda < k$. The group G is clearly an automorphism group of $\mathcal{D}$, transitive on points and blocks. Thus $\mathcal{D}$ is certainly a 1-design with equally many points and blocks. To show that it is a 2-design, it is simplest to prove that $\mathcal{D}^t$, the dual structure, is a 2-design, from which it will follow that $\mathcal{D}$ is as well. Thus all we must show is that $gD \cap hD$ has cardinality λ whenever $g \neq h$. But the number of points common to gD and hD is the number of solutions (d, e) in D to $gd = he$, i.e. to $g^{-1}h = de^{-1}$, and this is λ by the definition, for $g \neq h$.

Clearly multiplication on the left by elements of G is regular on points and blocks. $\square$

The development of a difference set D is also frequently called the **translate design** of D.

Example 4.4.2   (1) The development of the difference set of Example 4.4.1(1) above is the projective plane of order 2 and the complementary design is the biplane of order 2, i.e. a $(7, 4, 2)$ symmetric design.

(2) The development of the difference set of Example 4.4.1(2) above is an $(11, 5, 2)$ symmetric design, i.e. a biplane of order 3.

(3) The developments of the three difference sets of Example 4.4.1(4) and (5) are all $(16, 6, 2)$ symmetric designs i.e biplanes of order 4; they are distinct and, in fact, there are, up to isomorphism, exactly three $(16, 6, 2)$ designs: see Burau [60] and Hussain [146].

(4) For $q \equiv 3 \pmod 4$ and $G = \mathbf{F}_q^+$, the set D of non-zero squares of $\mathbf{F}_q$ forms a $(q, \frac{1}{2}(q-1), \frac{1}{4}(q-3))$ difference set whose development is a Hadamard design: see Section 7.8 on the Paley construction of Hadamard matrices for a full discussion of these designs.

Remark The biplane of order 2 is easily seen to be unique, and similarly for the biplane of order 3, as shown in Cameron [62]. For a discussion of the latter biplane and the three other biplanes of the examples above, consult Hussain [146]. For a coding-theoretical discussion of the three $(16, 6, 2)$ designs, see Assmus and Salwach [23]; we have taken the two difference sets of Example 4.4.1(4) directly from Turyn's fundamental paper [283].

The design defined above as the **development** of D, is also referred to as a **difference set design**. In case $\lambda = 1$, the difference set is called a **planar** difference set, since then the development is a projective plane.

The converse of the theorem also holds:

Theorem 4.4.2 *If $\mathcal{D}$ is a (v, k, λ) design with a group of automorphisms acting regularly on points, then $\mathcal{D}$ is the development of a difference set for that group.*

Proof: Let G be the subgroup of $\mathrm{Aut}(\mathcal{D})$ that is acting regularly on points. Clearly $|G| = v$. We take the action of G on $\mathcal{P}$ to be on the right so that xg is the point to which the point x is sent by the group element g. Fix a point $x \in \mathcal{P}$ and let B be any block. Set

$$D = \{d | d \in G, xd \in B\}.$$

We need to show that D is a (v, k, λ) difference set for G. Clearly $|D| = k$. For $g \in G$, $g \neq 1$, we need to show that there are λ pairs $d, e \in D$ for which $g = de^{-1}$. Since $g \neq 1$, $x \neq xg$, and so there are λ blocks $B_1, \ldots, B_\lambda$ containing both the points x and xg. Since G is regular on blocks, for each i there is a unique $h_i \in G$ such that $B_i = Bh_i$, for $i = 1, \ldots, \lambda$. Thus xh_i^{-1} and xgh_i^{-1} are in B for each i, so that gh_i^{-1} and h_i^{-1} are in D with $g = (gh_i^{-1})(h_i^{-1})^{-1}$. There are, therefore, at least λ pairs (d, e) with $d \neq e$ and $d, e \in D$ satisfying $g = de^{-1}$. But there are $v - 1$ choices for g and $k(k-1)$ such pairs. Since $\lambda(v-1) = k(k-1)$ there must be precisely λ such pairs for each $g \neq 1$. $\square$

Corollary 4.4.1 *If D is a difference set in a group G then the complement of D is also a difference set in G.*

Proof: This could be proved directly, but it is immediate from Theorem 4.4.2 since G acts regularly on the complementary design also. $\square$

Suppose we are given a difference set D for a group G. If α is an automorphism of the group, then $D\alpha = \{d\alpha | d \in D\}$ is clearly also a difference set in G.[4] Moreover, since $(gD)\alpha = (g\alpha)(D\alpha)$ the automorphism α will induce an automorphism of the development of D precisely when $D\alpha = gD$ for some element $g \in G$. Here is the appropriate definition:

Definition 4.4.3 *If D is a (v, k, λ) difference set for G and $\mu \in \mathrm{Aut}(G)$, then μ is a **multiplier** of D if $D\mu = gD$ for some $g \in G$.*

The reason for the curious term "multiplier" is historical: if D is a difference set in the cyclic group $G = (\mathbf{Z}/v\mathbf{Z})^+$ written additively, then the automorphism group of G is $(\mathbf{Z}/v\mathbf{Z})^\times$, the group of units of the ring

---

[4]Here we write automorphisms on the right so that $g\alpha$ is the image under α of g.

of integers modulo v,[5] with the action given by multiplication. In this classical case, $m \in (\mathbf{Z}/v\mathbf{Z})^\times$ is a multiplier if $Dm = mD = g + D$ for some $g \in (\mathbf{Z}/v\mathbf{Z})^+$, where addition and multiplication are all taking place in the ring $\mathbf{Z}/v\mathbf{Z}$ and $mD = \{md | d \in D\}$.

Example 4.4.3 (1) If $G = (\mathbf{Z}/7\mathbf{Z})^+$ and $D = \{1, 2, 4\}$, then $2D = \{2, 4, 8 \equiv 1\}$. So the map $a \mapsto 2a$ for $a \in G$ is an automorphism of the group and of the design, and acts as a multiplier. Here the multiplier happens to fix the difference set, but had we taken the difference set to be $D = \{0, 1, 3\}$, then $2D = \{0, 2, 6\} = 6 + D \neq D$.

(2) Every element of $\mathrm{Sym}(6)$, acting naturally on $\mathbf{F}_2^6$, is a multiplier (which fixes the difference set) of Example 4.4.1 (6).

It is obvious that the set of multipliers of a difference set D for a group G is a subgroup of $\mathrm{Aut}(G)$. Since every automorphism of a group fixes the identity element of the group, Brauer's theorem immediately implies the following:

Proposition 4.4.1 *Let D be a difference set for a group G and let μ be a multiplier. Then there is at least one difference set in the development of D that is fixed by μ, i.e. there is a $g \in G$ such that $(gD)\mu = gD$.*

When the group G is abelian we have an even stronger result:

Proposition 4.4.2 *If D is a (v, k, λ) difference set for an abelian group G and k is relatively prime to v, then there is a difference set in the development of D that is fixed by every multiplier. Further, the blocks in the development of D that are fixed by every multiplier are in one-to-one correspondence with the set of fixed points in G of the group of multipliers.*

Proof: Let h be the product of all the elements of D. Then the product of all the elements of gD is $g^k h$ since G is abelian. Since k is relatively prime to v the map $x \mapsto x^k$ is an automorphism of G and hence there is a unique $g \in G$ such that the product of the elements of gD is the identity element of G. Thus there is a unique difference set in the development of D, the product of whose elements is the identity element of G and clearly this difference set is fixed by every multiplier. Change notation and let D be this unique element of the development. Then $(gD)\mu = (g\mu)D$ and the last assertion is clear. $\square$

We shall next prove Hall's multiplier theorem, but to do that efficiently it is best to examine difference sets in the group algebra context. We therefore pause to develop the necessary notation and results.

---

[5]Of course, $(\mathbf{Z}/v\mathbf{Z})^\times$ consists precisely of those residues modulo v that are relatively prime to v.

4.5 Group algebras

In this section we will develop the machinery necessary to build more hospitable living quarters for difference sets and the standard geometric codes that we will treat in Chapter 5. Although this is standard algebra, there are certain aspects that do not seem to be treated deeply enough in the standard texts, so we review the material here. In principle, only the integral group algebra need be defined, obtaining the others via the tensor product, but we take the slightly less sophisticated approach and define the group algebra over an arbitrary commutative ring. Here that ring will in fact be either $\mathbf{Z}$ or $\mathbf{F}_q$ for some prime power q.

Thus let R be a commutative ring and let G be a finite group. We use the multiplication in G to impose an algebra structure on the space of all functions from G to R. We adopt for the present the usual notation for these functions: $\sum_{g \in G} a_g g$ is that function that sends the group element g to the ring element a_g. The group algebra itself we denote by $R[G]$. It is a free R-module of rank $|G|$ and the addition in the algebra and the scalar multiplication is that of the module structure (here a left module). The multiplication is an extension of the multiplication in G, i.e.

$$\left(\sum_{g \in G} a_g g \right) \left(\sum_{h \in G} b_h h \right) = \sum_{g \in G, h \in G} (a_g b_h)(gh) = \sum_{k \in G} \left(\sum_{g \in G} a_g b_{g^{-1} k} \right) k.$$

The unit element of the algebra is the function taking the value 1 at the neutral element of the group and the value 0 elsewhere. We will abuse the notation and denote this unit element by 1, as is customary. The group algebra has a canonical involution given by extending the function $g \mapsto g^{-1}$ of G onto G to a function from $R[G]$ to $R[G]$ via

$$\sum_{g \in G} a_g g \mapsto \overline{\sum_{g \in G} a_g g}$$

where

$$\overline{\sum_{g \in G} a_g g} = \sum_{g \in G} a_g g^{-1} = \sum_{g \in G} a_{g^{-1}} g.$$

(If G is an elementary abelian 2-group this is just the identity map.) The map has the usual properties: for any x and y in $R[G]$

$$\overline{\overline{x}} = x, \ \overline{x + y} = \overline{x} + \overline{y}, \text{ and } \overline{xy} = \overline{y}\,\overline{x}.$$

What about those functions that are characteristic functions of subsets of the group G? Once again we will abuse the notation and if $X \subseteq G$,

$X \in R[G]$ will be denoted by $\sum_{g \in X} g$. Thus the element G when viewed in $R[G]$ is simply $\sum_{g \in G} g$. The reader should think of a subset, when viewed in the group algebra, as a name for its characteristic function; that is precisely what g is when viewed in the group algebra as the formal sum with $a_g = 1$ and $a_h = 0$ for $h \neq g$.

Example 4.5.1 A set D being a (v, k, λ) difference set for the group G is expressed by the following equation in the group algebra $\mathbf{Z}[G]$:

$$D\overline{D} = (k - \lambda)1 + \lambda G = n1 + \lambda G$$

where $n = k - \lambda$ is the order of the development of D.

The passage from the finite group to the group algebra is functorial; in particular, if $\alpha : G \to H$ is a homomorphism of groups, it induces a homomorphism of the group algebras $R[G] \to R[H]$ given by $(\sum_{g \in G} a_g g)\alpha = \sum_{g \in G} a_g(g\alpha)$, where we are now using α for the group algebra homomorphism also. Moreover, what is more important for us here, if A is any algebra over the commutative ring R, then any group homomorphism, $\alpha : G \to A^{\times}$, where $A^{\times}$ is the group of units of the algebra A, extends to an algebra homomorphism, $R[G] \to A$, using precisely the same formula.

Suppose next that χ is a character[6] of G, i.e. a homomorphism of G into $F^{\times}$, the multiplicative group of some field F. Now F is canonically a $\mathbf{Z}$-algebra or, what is the same thing, every integer is interpretable uniquely in F; thus χ extends to an algebra homomorphism of $\mathbf{Z}[G]$ into F with, as above, $(\sum_{g \in G} a_g g)\chi = \sum_{g \in G} a_g(g\chi)$. More generally, if F is an R-algebra (for example, if R is a subring of F) the same is true for $R[G] \to F$. If χ is the principal character (i.e. if $g\chi = 1$ for all $g \in G$), then on the group algebra the map is simply $\sum_{g \in G} a_g g \mapsto (\sum_{g \in G} a_g)1$. For any character χ we define $\overline{\chi}$ by $g\overline{\chi} = (g^{-1})\chi$ or, equivalently (on the group algebra), $x\overline{\chi} = \overline{x}\chi$. This is a direct generalization of using conjugation in the classical case, where F is the field of complex numbers, since then $\overline{g\chi}$, the complex conjugate of $g\chi$, is the inverse of $g\chi$, i.e. $(g^{-1})\chi$. Since for any character χ,

$$(gh)\chi = (g\chi)(h\chi) = (h\chi)(g\chi) = (hg)\chi,$$

$\overline{\chi}$ is also a character of G. Moreover, for a fixed field F, the set of all characters into F forms a group whose multiplication is simply given by $g(\chi\eta) = (g\chi)(g\eta)$; the neutral element of this group is the principal character and the inverse of χ is $\overline{\chi}$.

---

[6]In terms of representation theory, we deal only with 1-dimensional characters here.

Of course, if F is not large enough, there may be few characters and even when F is large there may only be the principal character (when G is its own commutator subgroup). But when G is abelian with exponent e and F contains a primitive e^{th} root of unity, the group of characters of G into F is (non-canonically) isomorphic to G. This is a rather easy consequence of the fundamental theorem for finitely generated abelian groups and the obvious fact that for a cyclic group of order m the characters are obtained by sending a generator of the group to any m^{th} root of unity. Defining the inner product of χ and η by

$$(\chi, \eta) = \frac{1}{|G|} \sum_{g \in G} (g\chi)(g\overline{\eta})$$

(where our restrictions on the field imply that $|G| \neq 0$ in F) in this case the orthogonality relations follow easily:

$$(\chi, \eta) = \frac{1}{|G|} \sum_{g \in G} (g\chi)(g\overline{\eta}) = \delta_{\chi, \eta} \tag{4.1}$$

where δ is the Kronecker δ. The proof of this can be reduced to the statement for a cyclic group G of order m, say, with generator a: for if w is an element of order m in F, then the characters of G are the maps χ_i defined by

$$\chi_i : a \mapsto w^i$$

for $i = 0, \dots, m - 1$. Then

$$
\begin{aligned}
(\chi, \eta) = (\chi_i, \chi_j) &= \frac{1}{|G|} \sum_{g \in G} (g\chi)(g\overline{\eta}) \\
&= \frac{1}{m} \sum_{k=0}^{m-1} w^{k(i-j)} = \delta_{i,j}
\end{aligned}
$$

for the usual reason, i.e. the factorization of $X^m - 1$. Once again this is a direct generalization of the classical case.

Example 4.5.2 If D is a (v, k, λ) difference set in a finite abelian group G and χ is any non-principal character, applying χ to the equation expressing the fact that D is a difference set (see Example 4.5.1) gives

$$(D\chi)(D\overline{\chi}) = n,$$

since

$$G\chi = \sum_{g \in G} g\chi = (\chi, \chi_0) = 0$$

where χ_0 is the principal character. (Turyn [283] exploited this fact in the classical case to give his non-existence theorems.)

Note: The orthogonality relations (4.1) hold for the characters of any finite group whose order is not divisible by the characteristic of the field F. Thus also in the example above the group need not be abelian, under this condition.

Assuming that G is abelian and that F is sufficiently large,[7] denote the character group of G by G^*. Let X be the matrix whose rows are indexed by G and columns by G^* and whose entry at position (g, χ) is $g\chi$; similarly, define $\overline{X}$ to have entry $g\overline{\chi}$ in the $(g, \chi)^{\text{th}}$ position. Then the orthogonality relations are expressed by

$$X \overline{X}^t = |G|I$$

where I is the $|G| \times |G|$ identity matrix. Because of our assumption on F, X and $\overline{X}$ are virtually inverses of one another. They serve to define the "discrete Fourier transform" and its inverse: see Definition 2.7.1.

We will not need the classical case and here we take for F the field $\mathbf{F}_{p^m}$, where p has order m modulo e and p is a prime not dividing e, the exponent of G. Each character of G maps $F[G]$ onto F. Moreover, since p does not divide e and hence does not divide $|G|$, X is non-singular and we obtain, by the categorical definition of the direct product, an isomorphism of $F[G]$ onto $F \times F \times \cdots \times F$, the product taken $|G|$ times.

We have described much more than we will at present need concerning group algebras. For now we will simply use the group algebra as a home for the difference sets to be discussed in the next section.

4.6   The codes and the multiplier theorem

As the following results will make clear, the codes over fields of characteristic p dividing the order of a symmetric design will always have some role to play. In Chapter 6 we will give a detailed discussion of the case of projective planes, but here we will discuss the general case and show that, roughly speaking, self-orthogonal codes are obtained. We begin with a simple but useful lemma that facilitates the computation of inner products of code vectors. For any $v \in F^{\mathcal{P}}$ here we will denote the value of v at the point x by v_x.

---

[7] The reader can take F to be $\mathbf{Q}(\zeta)$, where ζ is a primitive e^{th} root of unity, for example $\exp(\frac{2\pi i}{e})$, or $\mathbf{F}_{p^m}$, where p has order m modulo e and p is a prime not dividing e, the exponent of G.

Lemma 4.6.1 *Let $\mathcal{D}$ be a (v, k, λ) design of order n and let p be any prime dividing n. If B and B' are any two blocks of $\mathcal{D}$, then $(v^B, v^{B'}) \equiv k \pmod{p}$ and if $v = \sum_{B \in \mathcal{B}} a_B v^B$ and $w = \sum_{B' \in \mathcal{B}} b_{B'} v^{B'}$ are any two vectors of $C_p(\mathcal{D})$, then*

$$(v, w) = k \left(\sum_{B \in \mathcal{B}} a_B \right) \left(\sum_{B \in \mathcal{B}} b_B \right).$$

Moreover,

$$k(v, w) = \left(\sum_{x \in \mathcal{P}} v_x \right) \left(\sum_{x \in \mathcal{P}} w_x \right).$$

Proof: Since $B \cap B'$ has cardinality k or λ and $\lambda \equiv k \pmod{p}$, the first assertion is obvious. The second follows immediately by linearity. In fact, since $(v, w) = k \sum_{B \in \mathcal{B}, B' \in \mathcal{B}} a_B b_{B'}$ and $\sum_{x \in \mathcal{P}} v_x = \sum_{B \in \mathcal{B}} k a_B$ with a similar assertion for w we have that $k(v, w) = \sum_{B \in \mathcal{B}, B' \in \mathcal{B}} (k a_B)(k b_{B'}) = (\sum_{x \in \mathcal{P}} v_x)(\sum_{x \in \mathcal{P}} w_x)$. $\square$

Theorem 4.6.1 *Let $\mathcal{D}$ be a (v, k, λ) design of order n and let p be any prime dividing n. Then $\dim C_p(\mathcal{D}) \leq (v + 1)/2$. If, further, p divides k, then $C_p(\mathcal{D}) \subseteq C_p(\mathcal{D})^\perp$ and $\dim C_p(\mathcal{D}) \leq v/2$.*

Proof: The first part follows from Theorem 2.4.2, but we can also look at a direct proof in this case. Let B, B' and D be any three blocks of $\mathcal{D}$ and set $C = C_p(\mathcal{D})$. Then $(v^B - v^{B'}, v^D) = 0$ by Lemma 4.6.1. Hence $E_p(\mathcal{D}) \subseteq \text{Hull}_p(\mathcal{D})$. Since

$$\dim(C^\perp) \geq \dim(E_p(\mathcal{D})) \geq \dim(C) - 1,$$

and $\dim(C^\perp) + \dim(C) = v$, we have $\dim(C) \leq (v + 1)/2$.

If p divides k also, the lemma immediately implies that $C \subseteq C^\perp$. In that case, then, $\dim(C) \leq v/2$. $\square$

When p sharply divides n, i.e. when p divides n but p^2 does not, we obtain, with the help of Klemm's theorem (Theorem 2.4.2), the stronger result:

Theorem 4.6.2 *Let $\mathcal{D}$ be a (v, k, λ) design of order n and let p be any prime that sharply divides n. Then*

(1) if p does not divide k, $C_p(\mathcal{D})^\perp \subseteq C_p(\mathcal{D})$ and $\dim C_p(\mathcal{D}) = (v + 1)/2$;

(2) if p divides k, $C_p(\mathcal{D}) \subseteq C_p(\mathcal{D})^\perp$ and $\dim C_p(\mathcal{D}) = (v - 1)/2$.

In both cases the smaller code has codimension 1 in the larger and

$$C_p(\mathcal{D})^\perp = C_p(\overline{\mathcal{D}}).$$

Proof: Set $C = C_p(\mathcal{D})$. If p does not divide k, then by Theorem 2.4.2, $C^\perp \subseteq C$ and $\dim C \geq v/2$. By Theorem 4.6.1, $\dim C \leq (v+1)/2$. Now since p sharply divides n, we know that n is not a square, so by Theorem 1.4.1 (or by Theorem 4.2.3), v is odd, and so $\dim C = (v+1)/2$.

If p divides k, then $C \subseteq C^\perp$ and $\dim C \leq v/2$, implying that $\dim C \leq (v-1)/2$, since v is odd. Using the notation of Theorem 4.2.1, the complementary design, $\overline{\mathcal{D}}$, has order n, and parameters $(v, n + \mu, \mu)$, where $\lambda \mu = n(n-1)$. Since p divides n, p does not divide $(n-1)$ and hence, because p sharply divides n, p does not divide μ. So $\dim(\overline{\mathcal{D}}) = (v+1)/2$ by the above. Now clearly $\jmath \in C_p(\overline{\mathcal{D}})$, and so $C \subseteq C_p(\overline{\mathcal{D}})$, of codimension at most 1; in fact, of codimension exactly 1, by the above argument, and so $\dim C = (v-1)/2$. The last assertion should be clear. $\square$

Remark: The usual proof of Theorem 4.6.2 involves the use of the Smith normal form and invariant factors: Lander [181] has a treatment of this method.

For symmetric designs $\text{Hull}_p(\mathcal{D})$ and $E_p(\mathcal{D})$ are closely related:

Theorem 4.6.3 *Let $\mathcal{D}$ be a (v, k, λ) design and let p be a prime dividing the order n of $\mathcal{D}$. Then $E_p(\mathcal{D}) \subseteq \text{Hull}_p(\mathcal{D})$ and $\text{Hull}_p(\mathcal{D})$ has codimension at most 1 in $C_p(\mathcal{D})$. Further, $C_p(\mathcal{D}) = \text{Hull}_p(\mathcal{D})$ if and only if p divides k but, whenever p does not divide k, $E_p(\mathcal{D}) = \text{Hull}_p(\mathcal{D})$ and this code is of codimension 1 in $C_p(\mathcal{D})$.*

Proof: Set $C = C_p(\mathcal{D})$, $H = \text{Hull}_p(\mathcal{D})$, and $E = E_p(\mathcal{D})$. As above we have $E \subseteq H$, and since E has codimension at most 1 in C, the same is true for H.

Now if p divides k, then $C \subseteq C^\perp$, i.e. $H = C$, and conversely. If p does not divide k, then clearly $H \neq C$, and since E and H both have codimension 1 in C, and $E \subseteq H$, we have $E = H$. $\square$

As far as we know, nothing further can be said about the dimensions of the codes of symmetric designs in the general case. Hamada in [125] made a conjecture that the p-rank of any design with the same parameters as a design obtained from a projective geometry over a field of characteristic p (see Section 3.4) is greater than that of the projective-geometry design unless it is isomorphic to it. Tonchev [280] found counter-examples to this conjecture in finding non-isomorphic designs with the same p-rank as the projective-geometry design of points and planes in $PG_4(\mathbf{F}_2)$. However, the conjecture has neither been proved nor disproved if restricted to *symmetric* designs: see Mackenzie [200] for some discussion of this. We will have something further to say about this conjecture in Section 6.9 in the case of projective planes.

The minimum weight of the code of a (v, k, λ) design is clearly at most k since the blocks themselves (in any design) yield vectors of weight k. But for $\lambda > 1$ it is frequently, but not always, smaller. Moreover, even if the minimum weight is k there may be other vectors of weight k besides those given by the blocks. There is, however, a general result which we now record.

Proposition 4.6.1 *Let $\mathcal{D}$ be a (v, k, λ) design, p a prime dividing $n = k - \lambda$ with $p > \lambda$, and S a set of points of cardinality k with $v^S \in C_p(\mathcal{D})$. Then S is a block.*

Proof: Since $p > \lambda$ we have that p does not divide k. Lemma 4.6.1 now ensures that, for any block B, $k(v^S, v^B) \equiv k^2 \pmod{p}$, and hence that $(v^S, v^B) \equiv k \equiv \lambda \pmod{p}$. Using once again that $p > \lambda$, we have that $|S \cap B| \geq \lambda$, since $(v^S, v^B) \equiv |S \cap B|$, and hence that S is a block by Proposition 4.2.2. $\square$

Remark   (1) The assumption that $p > \lambda$ is necessary here: the two $(16, 6, 2)$ designs whose binary codes are of dimension 7 and 8, respectively, have vectors of weight 6 that are not blocks: see Assmus and Salwach [23].

(2) The assumption that the vector of weight k is a characteristic function (or the scalar multiple of such a function) is also necessary: the $(11, 5, 2)$ design yields a code over $\mathbf{F}_3$ that has weight-5 vectors that are not scalar multiples of characteristic functions of blocks—and in this case 5 is the minimum weight.

We turn now to symmetric designs given by difference sets. As an example of the use of coding-theoretical techniques we give a proof of Hall's multiplier theorem. For a fuller discussion of these matters the reader may wish to consult Lander [181], but see also Pott [240] and the references given there. Hall's original version of the theorem was for cyclic difference sets in planes [116]; it was later generalized in Hall and Ryser [123] to cyclic difference sets in general, and then further generalized in Hall [117]. The theorem given below is a further generalization to abelian difference sets and is a particular case of a version due to Mann [207, Theorem 7.3]. Ott [227] proved a much broader version that is not restricted to the abelian case.

If G is an abelian group of order v with a difference set D and m is a positive integer relatively prime to v then the map given by $g \mapsto g^m$ is an automorphism of G. In many cases this map, which we will denote by α_m, is not only an automorphism of G but also a multiplier for D and then we say that "m is a multiplier".

Theorem 4.6.4 (The multiplier theorem) *Let D be a (v, k, λ) difference set in an abelian group G. Suppose p is a prime dividing $n = k - \lambda$
and assume that $p > \lambda$. Then p is a multiplier.*

Proof: The fact that p divides n and $p > \lambda$ implies, from the parameters, that p is coprime to v. We work in the group algebra $\mathbf{F}_p[G]$. Viewing D in this algebra we have that $D\alpha_p = D^p$ since $x \mapsto x^p$ is an automorphism of the algebra that fixes the coefficients (i.e. $\mathbf{F}_p$). We want to show, therefore, that $D^p = gD$ for some $g \in G$. Viewing the code $C_p(\mathcal{D})$, where $\mathcal{D}$ is the development of D, in the group algebra we have, since the code is generated by all gD, $g \in G$, that it is an ideal of that algebra. In particular, then, $D^p \in C_p(\mathcal{D})$. By Proposition 4.6.1 above, D^p is a block, i.e. $D^p = gD$ for some $g \in G$. $\square$

Remark It is not known whether or not the assumption that $p > \lambda$ is necessary. Certainly it is necessary in the proof we have given, which is, essentially, that of Pott [240]. The theorem we give here is a special case of the broader theorem due to Ott concerning multipliers in groups that are not necessarily abelian: see [227, Theorem 1.1, p. 70].

We conclude this chapter by proving a result of Bagchi and Sastry that yields information about difference sets in elementary abelian 2-groups. But first we make the following observation: let D be a (v, k, λ) difference set in an abelian group G. Then $\varphi : g \mapsto Dg^{-1}$ and $Dh \mapsto h^{-1}$ defines a polarity of the development of D. Here g is an absolute point of φ if $g \in Dg^{-1}$, i.e. if $g^2 \in D$. Thus if v is odd there are precisely k absolute points and if G is an elementary abelian 2-group every point is absolute since we can take D to contain the neutral element of G.

Bagchi and Sastry [27] have shown that in an arbitrary incidence system with a polarity, the characteristic function of the set of absolute points is in the binary code of the system. Dillon [92] has given a simple proof of this result which we state and prove in its matrix form. Note that given a polarity of an incidence system the incidence matrix can be chosen in such a way that the absolute points are distinguished by the fact that there is a 1 on the diagonal of the row corresponding to the point.

Proposition 4.6.2 *Let $A = (a_{ij})$ be an $m \times m$ symmetric matrix with entries from $\mathbf{F}_2$. Then the vector $(a_{11}, a_{22}, \ldots, a_{mm})$ is in the row space of A.*

Proof: Let $\mathbf{x} = (x_1, x_2, \ldots, x_m)$ be any vector in $\mathbf{F}_2^m$. The fact that we are over a field of characteristic 2 yields

$$\mathbf{x} A \mathbf{x}^t = \sum_{i=1}^{m} x_i a_{ii}.$$

Thus $\mathbf{x} A \mathbf{x}^t = 0$ if and only if $\mathbf{x} \in \langle \mathbf{a} \rangle^{\perp}$, where we have set

$$(a_{11}, a_{22}, \ldots, a_{mm}) = \mathbf{a}.$$

But, if C is the row space of A, then $\mathbf{x} \in C^{\perp}$ if and only if $A\mathbf{x}^t = 0$. Thus $C^{\perp} \subseteq \langle \mathbf{a} \rangle^{\perp}$ so that $\langle \mathbf{a} \rangle \subseteq C$ and the proposition follows. $\square$

Corollary 4.6.1 (Bagchi and Sastry) *The incidence vector of the set of absolute points of a polarity of a symmetric incidence structure (and hence of a (v, k, λ) design) is contained in the binary code of the design.*

Now an incidence structure with a polarity is necessarily self-dual and a design given by a difference set in an elementary abelian 2-group possesses a polarity, the one discussed above, all of whose points are absolute. From what we have already proved, therefore, we have the following:

Corollary 4.6.2 *If $\mathcal{D}$ is the design given by a difference set in an elementary abelian 2-group, then the all-one vector is in $E_2(\mathcal{D})$, which is of codimension 1 in $C_2(\mathcal{D}) = C_2(\overline{\mathcal{D}})$. Moreover, $C_2(\mathcal{D}^t) = C_2(\overline{\mathcal{D}}^t)$ is code-isomorphic to $C_2(\mathcal{D}) = C_2(\overline{\mathcal{D}})$.*

Chapter 5

The standard geometric codes

5.1 Introduction

In this chapter we present much of the classic material on the codes of those designs that arise from affine and projective geometries. When the geometry is over the field of two elements, these codes are the Reed-Muller and punctured Reed-Muller codes; otherwise they are various cases of the class of generalized Reed-Muller codes. The description of these codes as precisely the codes of the geometries — with the minimum-weight vectors as the incidence vectors of the blocks of the geometric design in question — is the work of Delsarte [80, 81], Delsarte, Goethals and MacWilliams [83] and Goethals and Delsarte [105]. We have reworked this material and simplified the development. Since several of the proofs are rather technical, we will not give them here; we will, however, indicate exactly how the code of the geometry can be obtained as a generalized Reed-Muller code (primitive or non-primitive subfield code) and illustrate the theory with examples.

There is quite an extensive literature on the generalized Reed-Muller codes: an overall historical description, together with the main references, is given in [83], which refers to the pioneering work of Kasami, Lin and Peterson [153, 154, 155], Rudolph [249] and Weldon [292] — amongst others. The book by Blake and Mullin [42] has a detailed description of the codes based mostly on the quoted works of Delsarte *et al.*; these authors also give a general description of the polynomial codes of Kasami *et al.* [155]. Our motivation here comes from design theory, where the geometries are the starting point and the codes are those that arise as codes of designs, as we have defined them. This is a different starting point from those mentioned above, where the codes and their various decoding algorithms were the salient features, and it is this geometric point of view which has led to the simplifications that have been achieved. An alternative approach to the one we take, an approach which may in time achieve a notational and conceptual clarity, has recently been developed. This approach to generalized Reed-Muller codes involves seeing the codes as ideals in a group algebra and it began with the work of Berman [36]. The work of Berger and Charpin [33], Charpin [67, 68, 69, 70], and Landrock and Manz [182] deals specifically with this method. Some of the important properties of the codes that are of particular relevance to our interest in the designs are still in the process of being explored. Thus we have not indicated how this method may be developed and we refer the reader to the above papers.

The first case, involving the Reed-Muller codes where the field is $\mathbf{F}_2$, is somewhat different and conceptually much simpler than the case for general q. Since the results in the binary case form a self-contained whole, we will deal with this case separately, even though this will lead to a certain amount of repetition, since Reed-Muller codes are, as the terminology implies, special cases of generalized Reed-Muller codes. The generalization is, however, by no means a trivial one — especially when q is not prime — and, happily, for the simpler Reed-Muller and punctured Reed-Muller codes it is extremely easy to establish that they are precisely the codes of the geometric designs we wish to study.

5.2   The Reed-Muller codes

We give in this section a leisurely[1] and elementary account of the Reed-Muller codes that is, in several respects, new: the expert will notice that we do not invoke Lucas's theorem — nor the cyclicity[2] of the punctured Reed-Muller codes — but still prove that they are projective-geometry codes and establish all the essential facts. The new ideas and some further consequences are explored in Rose [247]. For the traditional — and more complex — treatment see van Lint [186] or MacWilliams and Sloane [203, Chapter 13].

Throughout this section F will denote the field $\mathbf{F}_2$. Let V be a vector space of dimension m over F. As usual, we let F^V denote the vector space over F of all functions from V to F; we write $\mathbf{v}$ for a typical vector in

---

[1]For a brief, but still complete, account see [8].

[2]In fact, the cyclicity is an immediate consequence of our new proof that the punctured Reed-Muller codes are the codes of projective designs.

V. The functions in this case may conveniently be thought of as *Boolean functions* and specified by *truth tables* — giving the value of a function at all of its 2^m arguments. As such, the addition in F^V and the component-wise multiplication — defined by

$$(fg)(\mathbf{v}) = f(\mathbf{v})g(\mathbf{v})$$

for $f, g \in F^V$ and $\mathbf{v} \in V$ — have natural logical interpretations:

$$
\begin{aligned}
f + g &\equiv f \text{ or } g \text{ but not both,} \\
fg &\equiv f \text{ and } g, \\
f + g + fg &\equiv f \text{ or } g, \\
1 + f &\equiv \text{ not } f,
\end{aligned}
$$

where 1 is the function that is identically 1. Notice that $f^2 = f = 1f$ for any function f.

Now F^V has dimension 2^m with our chosen basis the characteristic functions of the members of V. Viewing V as F^m it too has a standard basis $\mathbf{e}_1, \mathbf{e}_2, \ldots, \mathbf{e}_m$, where

$$\mathbf{e}_i = \underbrace{(0, 0, \ldots, 1, 0, \ldots, 0)}_{i}.$$

In fact, we were implicitly using this basis above when we mentioned truth tables and gave the logical interpretations of addition and multiplication. Then any $f \in F^V$ can be given as a function of m Boolean variables corresponding to the m coordinate positions: writing the vector $\mathbf{x} \in V$ as

$$\mathbf{x} = (x_1, x_2, \ldots, x_m) = \sum_{i=1}^{m} x_i \mathbf{e}_i,$$

then $f = f(x_1, x_2, \ldots, x_m)$. The "polynomial" x_i is, for example, the linear functional $\mathbf{e}_i^t$, since its value at $(\sum_{j=1}^{m} x_j \mathbf{e}_j)$ is x_i, and the sum and product of these polynomial functions of m variables correspond to the logical interpretation of the Boolean functions as given above. Observe that we are deliberately confusing the polynomial form of the function with its name and this means that the same function will have many names. Thus $x_i^2 = x_i$ for each i, and we obtain *all* the monomial functions via the 2^m monomial functions:

$$\mathcal{M} = \{x_1^{i_1} x_2^{i_2} \ldots x_m^{i_m} \mid i_k = 0 \text{ or } 1; k = 1, 2, \ldots, m\},$$

where we write, as above, 1 for the constant function $x_1^0 x_2^0 \ldots x_m^0$ with value 1 at all points of V; as a code vector it is the all-one vector $\jmath$. The linear combinations over F of these 2^m monomials give *all* the polynomial functions, since we can reduce any polynomial in the x_i modulo $x_i^2 - x_i$, for $i = 1, 2, \ldots, m$. The set $\mathcal{M}$ of 2^m monomials is another basis for the vector space F^V; the following lemma indicates how each of our given basis elements of characteristic functions of the vectors in V is given as a polynomial, i.e. as a sum of elements of $\mathcal{M}$. Notice that the notation in the lemma is that introduced in Definition 1.2.5 for the characteristic function of a subset of V; thus $v^{\mathbf{w}}$ is that function on V that takes the value 1 at $\mathbf{w}$ and the value zero elsewhere.

Lemma 5.2.1 *Set $K = \{1, 2, \ldots, m\}$ and, for $\mathbf{w} = (w_1, w_2, \ldots, w_m) \in V$, let*

$$I_{\mathbf{w}} = \{i \in K | w_i = 1\}.$$

Then

$$v^{\mathbf{w}} = \prod_{k=1}^{m} (x_k + 1 + w_k) = \sum_{K \supseteq J \supseteq I_{\mathbf{w}}} \prod_{j \in J} x_j.$$

Proof: The proof is simple: the first polynomial is easily seen to define the characteristic function of the vector $\mathbf{w}$; and the expansion of this product is clearly the sum on the right. $\square$

Exercise 5.2.1 With the notation of Lemma 5.2.1 show that

$$f = \sum_{\mathbf{w} \in V} g(\mathbf{w}) x_1^{w_1} x_2^{w_2} \ldots x_m^{w_m}$$

for any $f \in F^V$, where g is defined by

$$g(\mathbf{w}) = \sum_{I_{\mathbf{u}} \subseteq I_{\mathbf{w}}} f(\mathbf{u}).$$

Hint: Write $f = \sum_{\mathbf{u}} f(\mathbf{u}) v^{\mathbf{u}}$ and use the lemma.

Example 5.2.1   (1) $v^{(1,1,\ldots,1)} = x_1 x_2 \ldots x_m.$

(2) $v^{(0,0,\ldots,0)} = \sum_{J \subseteq K} \prod_{j \in J} x_j = \sum_{f \in \mathcal{M}} f.$

(3) For $m = 3$, $v^{(1,0,0)} = x_1 + x_1 x_2 + x_1 x_3 + x_1 x_2 x_3.$

(4) The incidence vector of the hyperplane, $X_1 + X_3 = 0$ of $V = F^m$ is the polynomial $1 + x_1 + x_3$.

Here is the definition of the Reed-Muller codes:

Definition 5.2.1 *Let V denote the vector space of dimension m over $F = \mathbf{F}_2$ and let r satisfy $0 \le r \le m$. The **Reed-Muller code of order r**, denoted by $\mathcal{R}(r, m)$, is the subspace of F^V (with basis the characteristic functions on the vectors of V) that consists of all polynomial functions in the x_i of degree at most r, i.e.*

$$\mathcal{R}(r,m) = \left\langle \prod_{i \in I} x_i \,\middle|\, I \subseteq \{1, 2, \ldots, m\}, 0 \le |I| \le r \right\rangle.$$

Example 5.2.2 (1) For $m = 4$, $r = 2$, $\mathcal{R}(2,4)$ has basis

$$1, x_1, x_2, x_1x_2, x_3, x_1x_3, x_2x_3, x_4, x_1x_4, x_2x_4, x_3x_4$$

and is, thus, a $[16, 11]$ binary code. As we shall prove below (see Theorem 5.2.1 and Example 5.2.3) this is the $[16, 11]$ extended Hamming code $\widehat{\mathcal{H}_4}$.

(2) The first-order Reed-Muller code $\mathcal{R}(1, m)$ consists of all linear combinations of the monomials x_i and 1 and hence each codeword, apart from 0 and $\jmath$, is given either by a non-zero linear functional on V or by 1 plus such a functional. Since any non-zero linear functional has 2^{m-1} zeros, every vector of $\mathcal{R}(1, m)$, apart from 0 and $\jmath$, has weight 2^{m-1}. A generator matrix for $\mathcal{R}(1, m)$ using the basis $x_1, x_2, \ldots, x_m, 1$ can be written so that the first $2^m - 1$ columns and m rows are the binary representations of the numbers between 1 and $2^m - 1$, whereas the last column is all 0, apart from a final row where all entries are equal to 1. This is clearly a generator matrix for the orthogonal of the extended Hamming code, i.e.

$$\mathcal{R}(1, m) = (\widehat{\mathcal{H}_m})^{\perp}. \tag{5.1}$$

(See Definitions 2.5.1 and 2.3.3, and the subsequent discussion.) One usually takes a generator matrix of $\mathcal{R}(r, m)$ to be slightly different, however, by putting the all-one vector first and ordering the remaining monomials inductively, as we have done just above for $\mathcal{R}(2, 4)$, the columns being ordered inductively also.

As an immediate consequence of the definition and the linear independence of the functions in $\mathcal{M}$, we have

$$\dim(\mathcal{R}(r, m)) = \binom{m}{0} + \binom{m}{1} + \binom{m}{2} + \cdots + \binom{m}{r}.$$

In particular,

$$\dim(\mathcal{R}(1, m)) = 1 + m.$$

The trivial cases include $\mathcal{R}(0,m) = F\jmath$, $\mathcal{R}(m,m) = F^V$ and the code $\mathcal{R}(m-1,m)$, which is of codimension 1 in F^V and equal to $(F\jmath)^\perp$ — as we will verify below. For the fixed basis of V we have the embeddings:

$$\mathcal{R}(r,m) \subseteq \mathcal{R}(s,m)$$

whenever $0 \leq r \leq s \leq m$.

We have already mentioned that $\mathcal{R}(0,m) = F\jmath$ has as its orthogonal $\mathcal{R}(m-1,m)$. This is a special case of the following result:

Theorem 5.2.1 *For any $m \geq 1$ and any r such that $0 \leq r < m$,*

$$\mathcal{R}(r,m)^\perp = \mathcal{R}(m-r-1,m).$$

Proof: Let $f \in \mathcal{R}(m-r-1,m)$ and $g \in \mathcal{R}(r,m)$. Then f is a polynomial in the x_i's of degree at most $m-r-1$ and g is a polynomial of degree at most r. Thus the product fg has degree at most $m-1$ and hence $fg \in \mathcal{R}(m-1,m)$. Since all vectors of weight 1 are given by polynomials of degree m (by Lemma 5.2.1), all vectors of odd weight are also, there being only one monomial of degree m. Hence fg has even weight and the inner product (f,g) is zero. Thus

$$\mathcal{R}(r,m)^\perp \supseteq \mathcal{R}(m-r-1,m)$$

and since

$$
\begin{aligned}
\dim(\mathcal{R}(m-r-1,m)) &= \tbinom{m}{0} + \tbinom{m}{1} + \cdots + \tbinom{m}{m-r-1} \\
&= 2^m - (\tbinom{m}{m-r} + \cdots + \tbinom{m}{m-1} + \tbinom{m}{m}) \\
&= 2^m - (\tbinom{m}{r} + \cdots + \tbinom{m}{1} + \tbinom{m}{0}) \\
&= 2^m - \dim(\mathcal{R}(r,m)) \\
&= \dim(\mathcal{R}(r,m)^\perp)
\end{aligned}
$$

we have the stated result. $\square$

Example 5.2.3 From (5.1) we get immediately that

$$\mathcal{R}(1,m)^\perp = \widehat{\mathcal{H}_m} = \mathcal{R}(m-2,m).$$

In the next section we will see the connection between the Reed-Muller codes and the codes of the designs of points and flats in affine space over $\mathbf{F}_2$. The codes of the analogous designs from projective spaces over $\mathbf{F}_2$ arise as *punctured* Reed-Muller codes:

Definition 5.2.2 *For* $0 \leq r < m$ *the* **punctured Reed-Muller code of order** r, $\mathcal{R}(r, m)^*$, *is the code obtained from* $\mathcal{R}(r, m)$ *by puncturing at the vector* $\mathbf{0} \in V$.

One could puncture at *any* vector of V and get an isomorphic code since the set of polynomial functions is invariant under translation in V; i.e. if f is a polynomial in the x_i's then so is g where $g = f(x_1 + a_1, \ldots, x_m + a_m)$ for any vector $\mathbf{a} = (a_1, \ldots, a_m) \in V$.

Example 5.2.4 If $m = 3$, $r = 1$, $\mathcal{R}(1, 3)$ is a self-dual $[8, 4, 4]$ binary code, and $\mathcal{R}(1, 3)^*$ is a $[7, 4, 3]$ code, *viz.* the Hamming code $\mathcal{H}_3$. Example 5.2.3 gives the reason for this.

Proposition 5.2.1 *For* $r < m$ *the punctured Reed-Muller code* $\mathcal{R}(r, m)^*$ *is a*

$$[2^m - 1, \binom{m}{0} + \binom{m}{1} + \cdots + \binom{m}{r}]$$

binary code.

Proof: This is almost trivial: the dimension must be that of $\mathcal{R}(r, m)$ since all the vectors in this code are of even weight and the projection cannot, therefore, have a kernel. $\square$

Finally, note that it follows from Theorem 5.2.1 that

$$(\mathcal{R}(r, m)^*)^\perp = \mathcal{R}(m - r - 1, m)^* \cap (\mathbf{F}_2 \jmath)^\perp$$

provided $r < m$. That is, $(\mathcal{R}(r, m)^*)^\perp$ consists of the vectors of $(\mathcal{R}(m - r - 1, m))$ with a zero at $\mathbf{0}$, that coordinate being discarded.

5.3 Geometries and Reed-Muller codes

The set of vectors V is the point set for any design defined from an affine geometry $AG_m(\mathbf{F}_2)$; moreover the binary codes of all the associated designs of points and flats are subspaces of F^V. Similarly, the designs from the projective geometry $PG_{m-1}(\mathbf{F}_2)$ all have point set $V^* = V - \{\mathbf{0}\}$ and F^{V^*} is the ambient space of their binary codes. In this section we indicate how to associate these design codes with the Reed-Muller and punctured Reed-Muller codes of the last section.

Consider the generating elements of $\mathcal{R}(r, m)$: the polynomial x_i as a codeword has value 1 at a point $\mathbf{x}$ in V if the vector $\mathbf{x}$ has a 1 in the coordinate position i and value 0 otherwise. Thus $1 + x_i = v^H$, where H is the hyperplane with the equation $X_i = 0$. Also, x_i is the characteristic function of the complement of this hyperplane, i.e. the $(m - 1)$-flat with equation $X_i = 1$. Similarly, $(1 + x_i)(1 + x_j)$, for $i \neq j$, is the characteristic function of the intersection of two hyperplanes, a subspace of dimension $m - 2$. In general, all the elements of $\mathcal{M}$ are the incidence vectors of flats in the affine geometry and $\mathcal{R}(r, m)$ is spanned by the incidence vectors of these $(m - s)$-flats, for $0 \leq s \leq r$. In order to show that $\mathcal{R}(r, m)$ is the binary code of the design of points and $(m - r)$-flats of $AG_m(\mathbf{F}_2)$, which is our aim, we need to show that the vectors given by the $(m - r)$-flats span $\mathcal{R}(r, m)$. Notice that we already have this result for the first-order Reed-Muller codes, since the linear equations certainly define $(m - 1)$-flats and, furthermore, $\mathcal{R}(1, m)$ has precisely $2(2^m - 1)$ such vectors, the number of $(m - 1)$-flats in $AG_m(\mathbf{F}_2)$. Thus, if $\mathcal{A}$ is the affine design of points and $(m - 1)$-flats, we have that

$$\mathcal{R}(1, m) = C_2(\mathcal{A}).$$

The general case is almost as easy. First of all we have that the flats are in the Reed-Muller code:

Proposition 5.3.1 *The incidence vectors of the $(m - r)$-flats of $AG_m(\mathbf{F}_2)$ are all in $\mathcal{R}(r, m)$.*

Proof: Any $(m - r)$-flat T in $AG_m(\mathbf{F}_2)$ consists of all the vectors (points of the affine space) $\mathbf{x} = (x_1, x_2, \ldots, x_m)$ that satisfy r linear equations,

$$\sum_{j=1}^{m} a_{ij} X_j = b_i, \text{ for } i = 1, 2, \ldots, r \ ,$$

where all the a_{ij} and b_j are in $\mathbf{F}_2$. The polynomial,

$$\prod_{i=1}^{r} \left(b_i + 1 + \sum_{j=1}^{m} a_{ij} x_j \right),$$

has degree at most r and thus is in $\mathcal{R}(r, m)$. Moreover it is clearly the characteristic function v^T of T. $\square$

In fact the degree of the polynomial is exactly r when the equations are independent and the proof actually shows that all the $(m - s)$-flats are in $\mathcal{R}(r, m)$ provided $s \leq r$.

Now observe that the characteristic function of any $(t + 1)$-flat is the sum of the characteristic functions of two t-flats contained in it and thus the binary code of the design of points and $(m - r)$-flats contains, by a trivial induction, the characteristic function of every $(m - s)$-flat for $0 \leq s \leq r$ and hence the code of this design *is* $\mathcal{R}(r, m)$. We reverse the roles of r and $m - r$ and state this formally as

Theorem 5.3.1 *Let $\mathcal{A}$ be the design of points and r-flats of the affine geometry $AG_m(\mathbf{F}_2)$, where $0 \leq r \leq m$. Then the binary code $C_2(\mathcal{A})$ is the Reed-Muller code $\mathcal{R}(m - r, m)$.*

Corollary 5.3.1 *The dimension of the binary code of the design of points and r-flats of $AG_m(\mathbf{F}_2)$ is*

$$\binom{m}{0} + \binom{m}{1} + \cdots + \binom{m}{m-r}.$$

Of course, the characteristic functions of the r-flats are vectors of weight 2^r of $\mathcal{R}(m - r, m)$, but are there others? The answer is "no", a fact that we shall soon prove. Before doing so, however, we introduce a new element into the discussion and depart from the traditional treatment; as we shall presently see, we will simultaneously have a proof of the fact that the minimum weight of $\mathcal{R}(m - r, m)$ is 2^r.

We first show that the binary code of the design of points and r-dimensional subspaces of $PG_{m-1}(F)$ is $\mathcal{R}(m - r - 1, m)^*$. The proof below is remarkably simple, but seems to have been missed.

It is clear, of course, that the code of the design is contained in the punctured Reed-Muller code. Extend the code of the design by an overall parity check and note that this extended code is a subcode of $\mathcal{R}(m-r-1, m)$ and that incidence vectors of the $(r+1)$-dimensional *subspaces* of V generate this extended code. Now every $(r + 2)$-dimensional subspace of V has an incidence vector that is the sum of all the incidence vectors of the $(r + 1)$-dimensional subspaces it contains (this is simply the fact that $\jmath$ is in the code of the design of points and hyperplanes of any projective space). But, over $\mathbf{F}_2$, an $(r + 1)$-flat consists of an $(r + 2)$-dimensional subspace from which the points of an $(r + 1)$-dimensional subspace, the subspace of which it is a coset, have been removed. In other words, in the code of the design it is the sum of the incidence vectors of an $(r + 2)$-dimensional subspace and an $(r + 1)$-dimensional subspace. Thus all the $(r + 1)$-flats of V are in the extended code of the design and it is, therefore, $\mathcal{R}(m - r - 1, m)$. We have proved the following:

Proposition 5.3.2 *The binary code of the design of points and r-dimensional subspaces of $PG_{m-1}(\mathbf{F}_2)$ is the punctured Reed-Muller code $\mathcal{R}(m - r - 1, m)^*$.*

Corollary 5.3.2 *The dimension of the binary code of the design of points and r-dimensional subspaces of $PG_{m-1}(\mathbf{F}_2)$ is*

$$\binom{m}{0} + \binom{m}{1} + \cdots + \binom{m}{m-r-1}.$$

We will now give an *alternative* proof of the above corollary in order to provide insight into Theorem 5.3.2 to follow and, more importantly, into the generalization we discuss in the next section. Although Theorem 3.3.1 is not strictly necessary for what follows, the novice may wish to review it and the material surrounding it before proceeding since the main idea is geometric, involving the passage from the projective to the affine.

Here is the **alternative proof**:

We fix r and prove the result by induction on m. The induction starts at $m = r + 1$ where the result is trivial, the dimension being 1. So suppose the result is true for the design of points and r-dimensional subspaces of $PG_{m-1}(F)$ and let $\mathcal{D}$ be the design of points and r-dimensional subspaces of $PG_m(F)$. Set $C = C_2(\mathcal{D})$. Let W be the underlying $(m+1)$-dimensional vector space defining the m-dimensional projective space and let H be a hyperplane of W. Now $\overline{H} = W - H$ is an m-flat of W giving rise to an affine geometry isomorphic to $AG_m(F)$. Further, $W - \{0\} = (H - \{0\}) \cup \overline{H}$. Every r-dimensional subspace of $PG_m(F)$ either lies in $H - \{0\}$ or else meets $\overline{H}$ in an r-flat of this affine space; moreover every r-flat of this affine space is so obtained. Thus the projection of the code C onto the coordinate places corresponding to $\overline{H}$ is isomorphic to $\mathcal{R}(m - r, m)$ and the kernel of this projection contains $C_2(\mathcal{K})$ where $\mathcal{K}$ is the design of points and r-dimensional subspaces of the $(m - 1)$-dimensional projective space defined by H. By Corollary 5.3.1 and the induction assumption we have that

$$\dim(C) \;\geq\; \binom{m}{0} + \binom{m}{1} + \cdots + \binom{m}{m-r}$$
$$+ \binom{m}{0} + \cdots + \binom{m}{m-r-1}.$$

Using the fact that $\binom{m+1}{k} = \binom{m}{k} + \binom{m}{k-1}$ repeatedly, the right-hand side of the above inequality is

$$\binom{m+1}{0} + \binom{m+1}{1} + \cdots + \binom{m+1}{m+1-r-1}$$

which gives the result since the containment of C in the punctured Reed-Muller code, $\mathcal{R}(m - r, m + 1)^*$, implies that

$$\dim(C) \leq \binom{m+1}{0} + \binom{m+1}{1} + \cdots + \binom{m+1}{m+1-r-1}.$$

This completes the alternative proof. $\square$

The above proof of the corollary is very natural and, indeed, yields more as the following theorem indicates.

Theorem 5.3.2 *Any embedding of $PG_{m-1}(\mathbf{F}_2)$ in $PG_m(\mathbf{F}_2)$ gives rise to the following two short exact sequences:*

$$0 \to \mathcal{R}(m - r - 1, m)^* \to \mathcal{R}(m - r, m + 1)^* \to \mathcal{R}(m - r, m) \to 0$$

and

$$0 \to \mathcal{R}(m-r-1,m) \to \mathcal{R}(m-r,m+1)^* \to \mathcal{R}(m-r,m)^* \to 0.$$

Proof: The first result is implicit in the alternative proof, but note that we have an explicit inductive representation of the geometric codes in question. The second short exact sequence comes from the projection of the code of the design $\mathcal{D}$ onto $H-\{0\}$ — where we are using the notation of the alternative proof. First observe that the intersection of an r-dimensional subspace of $PG_m(\mathbf{F}_2)$ not contained in $H - \{0\}$ is an $(r-1)$-dimensional subspace of that embedded projective space and that every $(r-1)$-dimensional subspace of the embedded space arises in this way. Clearly, then, $C_2(\mathcal{D}) = \mathcal{R}(m-r,m+1)^*$ maps onto $C_2(\mathcal{L}) = \mathcal{R}(m-r,m)^*$, where $\mathcal{L}$ is the design of points and $(r-1)$-dimensional subspaces of $PG_{m-1}(\mathbf{F}_2)$. Observe next that the difference (in this case the sum) of two incidence vectors of parallel r-flats of $AG_m(\mathbf{F}_2)$ coming from intersections of r-dimensional subspaces of $PG_m(\mathbf{F}_2)$ that meet the embedded projective space in the same $(r-1)$-dimensional subspace is in the kernel. Such a sum is an incidence vector of an $(r+1)$-dimensional flat of the affine space and hence $\mathcal{R}(m-r-1,m)$ is in the kernel. A dimensional argument completes the proof, just as it did in the alternative proof. $\square$

As we shall see in the next section, there are analogous short exact sequences for arbitrary finite fields and they will play an equally clarifying role for arbitrary geometric codes, i.e. those arising — in the traditional treatments — from generalized Reed-Muller codes.

We next draw out the consequences of Theorem 5.3.2 and in so doing prove that the minimum weights of the Reed-Muller codes are as we have indicated and determine the nature of the minimum-weight vectors.

Corollary 5.3.3 *The minimum weight of $\mathcal{R}(m-r,m)$ is 2^r and the vectors of minimum weight are the incidence vectors of the r-flats of $AG_m(\mathbf{F}_2)$. The minimum weight of $\mathcal{R}(m-r,m)^*$ is 2^r-1 and the vectors of minimum weight are the incidence vectors of the $(r-1)$-dimensional subspaces of $PG_{m-1}(\mathbf{F}_2)$.*

Proof: Since we know that the incidence vectors of the r-flats are in $\mathcal{R}(m-r,m)$ we clearly have that its minimum weight is at most 2^r and, if we have equality, then the minimum weight of $\mathcal{R}(m-r,m)^*$ is clearly 2^r-1. Conversely, if the minimum weight of $\mathcal{R}(m-r,m)^*$ is 2^r-1 then that of $\mathcal{R}(m-r,m)$ is 2^r since it is obtained from $\mathcal{R}(m-r,m)^*$ by an overall parity check.

We employ the short exact sequences and induction on m and we use the notation of the alternative proof of Corollary 5.3.2. The induction assumption is that the corollary is true for m *and all* $r < m$. Thus we are assuming that $\mathcal{R}(m - s, m)$ has minimum weight 2^s whenever $m - s \leq m$ and that $\mathcal{R}(m - s, m)^*$ has minimum weight $2^s - 1$ whenever $m - s < m$ and that the minimum-weight vectors are as announced.

For $r = 0$ the affine result is trivial and this case does not occur for the punctured codes. So assume $r > 0$ and consider dimension $m + 1$. If $r = m$ the results are trivial from the nature of the first-order Reed-Muller codes and hence we may assume that $0 < r < m$. Now fix an embedding of $PG_{m-1}(\mathbf{F}_2)$ in $PG_m(\mathbf{F}_2)$ as in the alternative proof and suppose v is a minimum-weight vector of the code $C_2(\mathcal{D})$, i.e. of $\mathcal{R}(m - r, m + 1)^*$. If v is zero at the coordinates corresponding to $\overline{H}$, then v can be viewed in $C_2(\mathcal{K})$ and hence, by the induction assumption, has weight $2^{r+1} - 1$ and is the incidence vector of an r-dimensional subspace of $PG_{m-1}(\mathbf{F}_2)$ and hence of $PG_m(\mathbf{F}_2)$. If v is zero at the coordinates corresponding to $H - \{\mathbf{0}\}$ then v can be viewed in $\mathcal{R}(m - r - 1, m)$ and, by the induction assumption, its weight would be at least 2^{r+1}, an impossibility since the minimum weight is bounded above by $2^{r+1} - 1$. We can therefore restrict ourselves to those minimum-weight vectors, neither of whose projections are the zero vector. Thus the weight of v, by induction, is at least $2^r - 1 + 2^r = 2^{r+1} - 1$ and, moreover, such a vector — of weight $2^{r+1} - 1$ — when restricted to $H - \{\mathbf{0}\}$ is the incidence vector of an $(r - 1)$-dimensional subspace of the embedded projective space. To show that v is the incidence vector of an r-dimensional subspace of $PG_m(\mathbf{F}_2)$ we construct an r-dimensional subspace of $PG_m(\mathbf{F}_2)$ whose incidence vector w agrees with v on $H - \{\mathbf{0}\}$ and has at least one 1 in common with v on $\overline{H}$. Then the weight of $v - w$ is easily seen to be less than $2^{r+1} - 1$ and hence $v = w$. This gives the projective result for projective dimension m from which the affine result for dimension $m + 1$ follows since the Reed-Muller codes are invariant under translation in V — as we remarked in the last section — which means it is sufficient to consider only those minimum-weight vectors of the Reed-Muller code with a 1 at $\mathbf{0}$. $\square$

We shall see that the proof, which relies on the exact sequences, generalizes to arbitrary finite fields and considerably simplifies the discussion of the relationship of generalized Reed-Muller codes to the codes of the corresponding projective designs. The reader interested in that case should be certain that the above proof has been mastered.

It should be noted that the code of the projective-geometry design is cyclic and hence so are the punctured Reed-Muller codes. This is, of course, true for any projective-geometry design over any field, due to the existence of Singer cycles, as already mentioned in Chapters 3 and 4. See Exercise 5.3.1 for the determination of the full automorphism groups of the

Reed-Muller and punctured Reed-Muller codes.

In conclusion, and for future reference, we summarize the results obtained on the properties of the Reed-Muller codes and finite geometries over the field of two elements:

Theorem 5.3.3 *Let m be any positive integer.*

(1) If $\mathcal{A}$ is the design of points and r-flats of the affine geometry $AG_m(\mathbf{F}_2)$, where $0 \leq r \leq m$, then the binary code $C = C_2(\mathcal{A})$ is the Reed-Muller code $\mathcal{R}(m-r, m)$. It is a $[2^m, \binom{m}{0} + \binom{m}{1} + \cdots + \binom{m}{m-r}, 2^r]$ binary code and the minimum-weight vectors are the incidence vectors of the r-flats. Further, C contains the incidence vectors of all t-flats for $r \leq t \leq m$.

For $r > 0$ the orthogonal, $C^\perp$, is the Reed-Muller code $\mathcal{R}(r-1, m)$, which is the binary code of the design of points and $(m-r+1)$-flats of the affine geometry $AG_m(\mathbf{F}_2)$.

(2) If $\mathcal{D}$ is the design of points and r-dimensional subspaces of the projective geometry $PG_m(\mathbf{F}_2)$, where $0 \leq r \leq m$, then the binary code $C = C_2(\mathcal{D})$ is the punctured Reed-Muller code $\mathcal{R}(m-r, m+1)^*$. It is a $[2^{m+1} - 1, \binom{m}{0} + \binom{m+1}{1} + \cdots + \binom{m+1}{m-r}, 2^{r+1} - 1]$ binary cyclic code and the minimum-weight vectors are the incidence vectors of the r-dimensional subspaces. Further, C contains the incidence vectors of all t-dimensional subspaces for $r \leq t \leq m$.

The orthogonal code to $\mathcal{R}(m-r, m+1)^*$ is $\mathcal{R}(r, m+1)^* \cap (\mathbf{F}_2\jmath)^\perp$, which is the even-weight subcode of the binary code of the design of points and $(m-r)$-dimensional subspaces of $PG_m(\mathbf{F}_2)$.

Example 5.3.1 (1) In $AG_4(\mathbf{F}_2)$, the design of points and lines has code $\mathcal{R}(3, 4)$, which is the even-weight subcode of F^V, of dimension 15. Its orthogonal is $F\jmath = \mathcal{R}(0, 4)$. The design of points and planes has code $\mathcal{R}(2, 4)$, of dimension 11, with orthogonal the code from the design of points and hyperplanes, of dimension 5, i.e. $\mathcal{R}(1, 4)$.

(2) In $PG_3(\mathbf{F}_2)$, the design of points and lines has code $\mathcal{R}(2, 4)^*$, of dimension 11 and minimum weight 3; it is, of course, a binary Hamming code. This is generally true, and the design given by the minimum-weight vectors of a binary Hamming code is this classical Steiner triple system of points and lines of a projective space over $\mathbf{F}_2$.

Exercise 5.3.1 (1) Show that $AGL_m(\mathbf{F}_2)$ in its natural action on $V = \mathbf{F}_2^m$ yields a group of automorphisms of every Reed-Muller code $\mathcal{R}(r, m)$. Use the triple transitivity of this group to give another proof that the minimum-weight vectors of $\widehat{\mathcal{H}_m}$ form a Steiner quadruple system and, in fact, the design of points and planes of the corresponding affine geometry.

(2) Show that $PGL_m(\mathbf{F}_2)$ in its natural action on $V^* = V - \{\mathbf{0}\}$ yields a group of automorphisms of every punctured Reed-Muller code $\mathcal{R}(r,m)^*$. Use the double transitivity of this group to give another proof that the minimum-weight vectors of $\mathcal{H}_m$ form a Steiner triple system and, in fact, the design of points and lines of the corresponding projective geometry.

That $PGL_m(\mathbf{F}_2)$ is the *full* group of automorphisms of $\mathcal{R}(r,m)^*$ whenever $0 < r < m - 1$ follows from Corollary 5.3.3 and the fundamental theorem of projective geometry, Theorem 3.2.1, and from this it follows that $AGL_m(\mathbf{F}_2)$ is the full group of automorphisms of $\mathcal{R}(r,m)$ whenever $0 < r < m - 1$. One must be careful here and note that it is not the entire projective space that must be preserved, but only part of it, to ensure that the automorphism comes from the general linear group; the interested reader should consult a proof of the fundamental theorem, for example: Hahn and O'Meara [114, Theorem 3.1.12].

5.4   Generalized Reed-Muller codes

Our description of the generalized Reed-Muller codes is based, primarily, on the paper by Delsarte *et al.* [83]. We have, however, reworked the material in several important respects and have introduced a different notation; a dictionary of the correspondence between our notation and theirs is at the end of this chapter (see page 197), to assist those readers who wish to read this fundamental paper. The definitions are based on the polynomial codes introduced by Kasami *et al.* [153, 155]; these authors introduced the primitive generalized Reed-Muller codes in [154] and Weldon [292] introduced the non-primitive generalized Reed-Muller codes and the single-variable treatment of Mattson and Solomon [212]. Our treatment of the Mattson-Solomon polynomial appears to be new in that we view it in a quotient ring that is slightly different from the one traditionally used; this allows us to considerably simplify some of the proofs and, we hope, it also makes for greater clarity.

We start with the so-called *m-variable* approach. This is entirely analogous to our previous approach to the Reed-Muller codes, which are, simply, the generalized Reed-Muller codes for $q = 2$.

Let $q = p^t$, where p is a prime. Set $E = \mathbf{F}_q$ and let V be a vector space of dimension m over E. Again we will denote a general vector in V by $\mathbf{v}$, and we will take V to be the space E^m of m-tuples, with standard basis $\mathbf{e}_1, \ldots, \mathbf{e}_m$, where

$$\mathbf{e}_i = (\underbrace{0, 0, \ldots, 1, 0, \ldots, 0}_{i}).$$

Our codes will be q-ary codes, i.e. codes over E, and the ambient space will be the function space E^V, with the usual basis of characteristic functions of the vectors of V. As in Section 5.2, we can denote the members $f \in E^V$ by functions of the m-variables denoting the coordinate positions in V, i.e. if

$$\mathbf{x} = (x_1, x_2, \ldots, x_m),$$

then $f \in E^V$ is given by

$$f = f(x_1, x_2, \ldots, x_m)$$

and the x_i take values in E. Since every element in E satisfies $a^q = a$, the polynomial functions in the m variables can be reduced modulo $x_i^q - x_i$ (as was done in Section 5.2 for $q = 2$) and we can again form the set $\mathcal{M}$ of q^m monomial functions

$$\mathcal{M} = \{x_1^{i_1} x_2^{i_2} \ldots x_m^{i_m} \,|\, 0 \le i_k \le q - 1, k = 1, 2, \ldots, m\}.$$

For a monomial in $\mathcal{M}$ the degree is the total degree, i.e. $\nu = \sum_{k=1}^m i_k$ and we have that $0 \le \nu \le m(q-1)$. We will shortly show that $\mathcal{M}$ forms another basis (that we will not use for the codes) of E^V — as was done for $q = 2$; we do this in an entirely analogous way by expressing each characteristic function of a vector as a polynomial in members of $\mathcal{M}$:

Lemma 5.4.1 *For* $\mathbf{w} = (w_1, w_2, \ldots, w_m) \in V$,

$$v^{\mathbf{w}} = \prod_{i=1}^m \left(1 - (x_i - w_i)^{q-1} \right).$$

Proof: Since $a^{q-1} = 1$ for any non-zero $a \in E$, $1 - (x_i - w_i)^{q-1} = 0$ whenever $x_i \ne w_i$; thus the polynomial function on the right is clearly the same as the characteristic function on the left. $\square$

Example 5.4.1 The polynomial $1 - (x_i - a)^{q-1}$ is the characteristic function of the $(m-1)$-flat in E^m given by the equation $X_i = a$. Observe that this polynomial is *not* linear unless $q = 2$.

Since E^V has dimension q^m, $\mathcal{M}$ is another basis for E^V — by the above lemma. The space E^V can then be viewed as the space of all polynomials (reduced modulo $x_i^q - x_i$) in the m variables; i.e. all linear combinations with coefficients in E of the monomials in $\mathcal{M}$. We write $\mathcal{N}_E(\nu, m)$, or simply $\mathcal{N}(\nu, m)$ if q is understood from the context, for the subspace — of the space of these polynomials — consisting of those of degree at most ν, where $0 \le \nu \le m(q-1)$. We use this interpretation to define the generalized Reed-Muller codes:

Definition 5.4.1 *Let $E = \mathbf{F}_q$, where $q = p^t$ and p is a prime, and set $V = E^m$. Then for any ν such that $0 \le \nu \le m(q-1)$, the νth **order generalized Reed-Muller code** $\mathcal{R}_E(\nu, m)$ over E is the subspace of E^V (with basis of characteristic functions on the vectors in V) of all reduced m-variable polynomial functions of degree at most ν. Thus*

$$\mathcal{R}_E(\nu, m) = \mathcal{R}_{F_q}(\nu, m) = \left\langle x_1^{i_1} x_2^{i_2} \ldots x_m^{i_m} \mid \sum_{k=1}^{m} i_k \le \nu \right\rangle.$$

Example 5.4.2 (1) For $q = 2$ and $0 \le \nu \le m$, $\mathcal{R}_{F_2}(\nu, m) = \mathcal{R}(\nu, m)$.

(2) For any q, $\mathcal{R}_{F_q}(0, m) = \mathbf{F}_q \jmath = \langle \jmath \rangle$.

The dimension of the generalized Reed-Muller codes can be obtained by counting the number of elements in the basis of monomials:

Theorem 5.4.1 *For any ν such that $0 \le \nu \le m(q-1)$,*

$$\dim(\mathcal{R}_{F_q}(\nu, m)) = \sum_{i=0}^{\nu} \sum_{k=0}^{m} (-1)^k \binom{m}{k} \binom{i - kq + m - 1}{i - kq}.$$

Proof: Use the fact that the number of ways of picking j objects from a set of m objects — with repetitions allowed — is $\binom{j+m-1}{j}$ and then an inclusion-exclusion argument to show that the inner sum is the number of ways of picking i objects from a set of m objects, when no object can be chosen more than $q - 1$ times. Summing on i yields the result. $\square$

Example 5.4.3 (1) By applying the formula or by a direct count we have that

$$\dim(\mathcal{R}_{F_q}(1, m)) = 1 + m$$

just as in the binary case — but note that the function x_i is zero on the hyperplane $X_i = 0$ but takes on many different values off the hyperplane unless $q = 2$. Since every non-constant linear polynomial in the m variables has q^{m-1} zeros, these codes have minimum weight $q^m - q^{m-1} = q^{m-1}(q - 1)$ and all code vectors except the zero vector and non-zero multiples of $\jmath$ have this weight.

(2) By applying the formula or by a direct count we have that, for $0 \le \nu \le q - 1$,

$$\dim(\mathcal{R}_{F_q}(\nu, 1)) = 1 + \nu$$

and, since a polynomial in one variable of degree at most ν can have at most ν distinct roots, the minimum weight in this code is $q - \nu$, there being a polynomial of degree ν with exactly ν distinct roots since $\nu < q$. These codes are MDS codes.

The orthogonal codes are also generalized Reed-Muller codes, as we already saw in the binary case. Here again the situation is entirely analogous to the binary case and we need a lemma which generalizes the result that $\mathcal{R}(m - 1, m)$ consists of even-weight vectors. The result is an easy consequence of the orthogonality relations (4.1); thus the proof depends simply on the fact that $\sum_{i=0}^{n-1} \omega^i = 0$ whenever ω is an nth root of unity different from 1.

Lemma 5.4.2 *If $f \in E^V$ has degree $\nu < m(q - 1)$ as a polynomial in $x_1, x_2, \ldots, x_m$ then*

$$\sum_{\mathbf{w} \in V} f(\mathbf{w}) = 0.$$

In fact, if f is any linear combination of the elements of $\mathcal{M}$ with coefficients in any overfield of E, the same result holds.

Proof: The result is clearly true for any constant function (since the block length of the code is a multiple of, in fact a power of, p, the characteristic of $\mathbf{F}_q$) so we need to prove the assertion only for monomial functions, i.e. elements of $\mathcal{M}$, of positive degree less than $m(q - 1)$. Moreover, if in such a monomial any $i_k = 0$, the sum is again a multiple of q and hence 0. We thus restrict ourselves to those monomials in which every x_i appears. The orthogonality relations (4.1) for the group $E^\times \times \ldots \times E^\times$, using E itself as the field where the characters take their values, yields immediately, taking η as the principal character and χ the character sending $\mathbf{a} = (a_1, \ldots, a_m)$ to $a_1^{i_1} \ldots a_m^{i_m}$, that

$$\sum_{\mathbf{a}} a_1^{i_1} \ldots a_m^{i_m} = 0$$

since there is some k for which $i_k < q - 1$, and since the sum only need be taken over those vectors all of whose entries are non-zero. $\square$

Theorem 5.4.2 *For $\nu < m(q-1)$ and $\mu = m(q-1) - 1 - \nu$,*

$$\mathcal{R}_{F_q}(\nu, m)^{\perp} = \mathcal{R}_{F_q}(\mu, m).$$

Proof: If f has degree ν and g has degree μ, then the product fg has degree less than $m(q-1)$. Thus Lemma 5.4.2 implies that

$$\sum_{\mathbf{w} \in V} f(\mathbf{w})g(\mathbf{w}) = 0$$

and the corresponding codewords are orthogonal. Hence

$$\mathcal{R}_{F_q}(\nu, m)^{\perp} \supseteq \mathcal{R}_{F_q}(\mu, m)$$

and now we need only check the dimensions: the involution of $\mathcal{M}$ that sends $x_1^{i_1} \ldots x_m^{i_m}$ to $x_1^{q-1-i_1} \ldots x_m^{q-1-i_m}$ yields the fact that the number of monomials of degree greater than μ is equal to the number of degree less than or equal to ν and hence the dimension of $\mathcal{R}_{F_q}(\mu, m)$ is $q^m - \dim(\mathcal{R}_{F_q}(\nu, m))$, as required. $\square$

Example 5.4.4 For $q = 2$,

$$\mathcal{R}_{F_2}(\nu, m)^{\perp} = \mathcal{R}(\nu, m)^{\perp} = \mathcal{R}(m - 1 - \nu, m) = \mathcal{R}_{F_2}(m - 1 - \nu, m).$$

The generalized Reed-Muller codes are, in general, codes over non-prime fields and thus could not be the codes of designs coming from affine geometries — unless q happens to be a prime. They do contain the incidence vectors of flats in the geometry, as we will shortly see. In order to demonstrate this, it is convenient, notationally, first to make an observation, important in itself, about the automorphism group of $\mathcal{R}_{F_q}(\nu, m)$:

Theorem 5.4.3 *For $0 \le \nu \le m(q-1)$, the automorphism group of $\mathcal{R}_{F_q}(\nu, m)$ contains the affine group $AGL_m(\mathbf{F}_q)$.*

Proof: Recall that for any code $C \subseteq F^{\mathcal{P}}$, an automorphism of C is a permutation $\sigma \in \text{Sym}(\mathcal{P})$ that preserves C, i.e. for which, if $c \in C$, $c^{\sigma} \in C$, where c^{σ} is defined by $c^{\sigma}(P) = c(\sigma(P))$ for any $P \in \mathcal{P}$.

Now $\gamma \in AGL_m(\mathbf{F}_q)$ is given by

$$\gamma : \mathbf{v} \mapsto A\mathbf{v} + \mathbf{a},$$

where $\mathbf{v}, \mathbf{a} \in V = E^m$ — viewed as column vectors — and A is a non-singular $m \times m$ matrix over E. Thus, for $f \in \mathcal{R}_{F_q}(\nu, m)$, f^{γ} is defined by

$$f^{\gamma}(\mathbf{x}) = f(A\mathbf{x} + \mathbf{a}),$$

and so, clearly, $f^\gamma \in \mathcal{R}_{F_q}(\nu, m)$. $\square$

Note: Recently Berger and Charpin [33] have shown that $AGL_m(\mathbf{F}_q)$ is the *full* automorphism group of these generalized Reed-Muller codes. As we remarked at the end of Section 5.3, for $q = 2$ this result is an easy consequence of the geometric nature of the codes and the fundamental theorem of projective geometry.

Theorem 5.4.4 *For $0 \le r \le m$ and $\nu \ge r(q - 1)$, the generalized Reed-Muller code $\mathcal{R}_{F_q}(\nu, m)$ contains the incidence vector of any $(m - r)$-flat of $AG_m(\mathbf{F}_q)$.*

Proof: Any $(m - r)$-flat in $AG_m(\mathbf{F}_q)$ consists of all the points $\mathbf{x}$ of V satisfying r independent equations

$$\sum_{j=1}^{m} a_{ij} X_j = w_i, \text{ for } i = 1, 2, \ldots, r$$

where all a_{ij} and w_i are in $\mathbf{F}_q$. If the code $\mathcal{R}_{F_q}(\nu, m)$ contains the incidence vector of some t-flat, then it will contain the incidence vector of every t-flat, since the affine group $AGL_m(\mathbf{F}_q)$ acts transitively on t-flats and, as we have just seen, preserves the code. So we need only construct one $(m - r)$-flat that is in $\mathcal{R}_{F_q}(\nu, m)$.

Consider the polynomial

$$p(x_1, \ldots, x_m) = \prod_{i=1}^{r} \left(1 - x_i^{q-1}\right),$$

of degree $r(q - 1)$. Then $p(x_1, \ldots, x_m) = 0$ on V unless $x_i = 0$ for $i = 1, 2, \ldots, r$. Thus the codeword corresponding to $p(\mathbf{x})$ has an entry 1 at points in the $(m - r)$-flat defined by the r equations

$$X_1 = 0, X_2 = 0, \ldots, X_r = 0$$

and an entry 0 at points off the flat. Hence it is the incidence vector of this $(m - r)$-flat. Since $p(x_1, \ldots, x_m) \in \mathcal{R}_{F_q}(\nu, m)$ for $\nu \ge r(q - 1)$, we have the result. $\square$

Just as in the binary case the proof shows more, namely that the generalized Reed-Muller code contains all $(m - s)$-flats for $0 \le s \le r$. Moreover, the subcode of the generalized Reed-Muller code that is generated by the $(m - r)$-flats contains, by the same trivial induction argument used in the binary case, all $(m - s)$-flats for $0 \le s \le r$. Note, however, that when using characteristic functions of t-flats to obtain characteristic functions of

$(t + 1)$-flats one could use coefficients other than 1 provided $q > 2$ and hence obtain vectors that are supported on the $(t + 1)$-flat but are not characteristic functions.

Example 5.4.5 Take $q = 3$ and $m = 2$. The geometry is then $AG_2(\mathbf{F}_3)$, the affine plane of order 3. Let $C = C_3(AG_2(\mathbf{F}_3))$ be the code over $\mathbf{F}_3$ associated with this plane, i.e. the code generated by the incidence matrix of the plane. The incidence vectors of the lines (1-flats) will be in $\mathcal{R}_{F_3}(\nu, 2)$ for $\nu = 2, 3$ and 4. In fact $C = \mathcal{R}_{F_3}(2, 2)$, while $\mathcal{R}_{F_3}(3, 2) = (\mathbf{F}_3 \jmath)^\perp$, as we know, and $\mathcal{R}_{F_3}(4, 2) = \mathbf{F}_3^9$, the entire ambient space.

The geometric designs we are interested in will have codes that are so-called subfield subcodes of appropriate generalized Reed-Muller codes. We have already seen that the design of points and r-flats of $AG_m(\mathbf{F}_q)$, $\mathcal{A}$ say, has its incidence vectors in $\mathcal{R}_{F_q}((m - r)(q - 1), m)$. It follows that if q is a power of the prime p, then $C_p(\mathcal{A})$ is a subcode of the subset of those vectors in $\mathcal{R}_{F_q}((m-r)(q-1), m)$ all of whose entries are in $\mathbf{F}_p$, that subset being a vector space over $\mathbf{F}_p$, i.e. a subfield subcode of the generalized Reed-Muller code.

The work of Delsarte and others established the full connection between the codes of designs coming from geometries and the generalized Reed-Muller codes, i.e. those we have already defined and also the punctured, non-primitive and subfield codes that we will define shortly. The main results are not at all easy to establish in general, but there are some particular cases where less involved results from the combinatorial side help to establish this connection. The above example is one of these cases. More generally, we have the full story for desarguesian affine planes of prime order.

Proposition 5.4.1 *Let p be a prime and let C be the code over $\mathbf{F}_p$ of the desarguesian affine plane $AG_2(\mathbf{F}_p)$. Then $C = \mathcal{R}_{F_p}(p - 1, 2)$. Moreover,*

$$C^\perp = \mathcal{R}_{F_p}(p - 2, 2) \subset C.$$

The code C is a $[p^2, p(p + 1)/2, p]$ code and the minimum-weight vectors are multiples of the incidence vectors of the lines.

Proof: As in the above example, with $r = 1$, the generalized Reed-Muller code $\mathcal{R}_{F_p}(p-1, 2)$ contains the code of the plane. From Theorem 5.4.1 — or by a very simple count — and Theorem 4.6.2, together with the discussion in Section 6.3, it can be seen that the dimension of C is equal to that of $\mathcal{R}_{F_p}(p - 1, 2)$, and thus these codes are equal. The orthogonal is the code $\mathcal{R}_{F_p}(\mu, 2)$ where $\mu = 2(p - 1) - (p - 1) - 1 = p - 2$, which is a subcode of $\mathcal{R}_{F_p}(p - 1, 2)$ and of dimension $p^2 - p(p + 1)/2 = p(p - 1)/2$.

That the minimum weight is p and that the minimum-weight vectors are multiples of the incidence vectors of the lines is a consequence of a general result for *arbitrary* planes of prime order: see Corollary 6.4.1 and the subsequent discussion. $\square$

Just as for the Reed-Muller codes, we can remove a coordinate position to obtain a code of length $q^m - 1$, which will turn out to be cyclic:

Definition 5.4.2 *The νth order punctured generalized Reed-Muller code, where $0 \le \nu < m(q-1)$, denoted by $\mathcal{R}_{F_q}(\nu, m)^*$, is the code of length $q^m - 1$ obtained by deleting the coordinate position $\mathbf{0}$ from $\mathcal{R}_{F_q}(\nu, m)$.*

For $q = 2$, $\mathcal{R}_{F_2}(\nu, m)^* = \mathcal{R}(\nu, m)^*$, the punctured Reed-Muller code. These are also called *shortened* generalized Reed-Muller codes (see van Lint [188]) or *cyclic* generalized Reed-Muller codes (see Blake and Mullin [42]) since our next result will show they are cyclic (which we already know for $q = 2$). Observe also that any coordinate position can be deleted in place of $\mathbf{0}$, since $AGL_m(\mathbf{F}_q)$ acts transitively on the vectors of $V = E^m$.

Theorem 5.4.5 *The automorphism group of $\mathcal{R}_{F_q}(\nu, m)^*$ contains, for any ν, the general linear group $GL_m(\mathbf{F}_q)$. In particular, $\mathcal{R}_{F_q}(\nu, m)^*$ is a cyclic code.*

Proof: The group $GL_m(\mathbf{F}_q)$ is the stabilizer of $\mathbf{0}$ in $AGL_m(\mathbf{F}_q)$, so it obviously acts on $\mathcal{R}_{F_q}(\nu, m)^*$. We can obtain a cyclic group of order $q^m - 1$ acting on it in the following way: consider the field $K = \mathbf{F}_{q^m}$ and let ω be a primitive root in K satisfying an irreducible equation of degree m over E, *viz.* $p(\omega) = 0$, where

$$p(X) = \prod_{j=0}^{m-1} (X - \omega^{q^j}) = \sum_{i=0}^{m} w_i X^i,$$

where $w_i \in E$ for $i = 1, 2, \ldots, m$ with $w_m = 1$. Of course, K as a vector space over $\mathbf{F}_q = E$ is isomorphic to V and multiplication by ω simply cycles the elements of $K^\times = V - \{\mathbf{0}\}$. Using the basis $1, \omega, \ldots, \omega^{m-1}$ multiplication by ω is given by the following $m \times m$ matrix over E, the companion matrix of $p(X)$, i.e.

$$S = \begin{pmatrix} 0 & 0 & \cdots & 0 & -w_0 \\ 1 & 0 & \cdots & 0 & -w_1 \\ 0 & 1 & \cdots & 0 & -w_2 \\ \vdots & \vdots & & \vdots & \vdots \\ 0 & 0 & \cdots & 1 & -w_{m-1} \end{pmatrix}. \tag{5.2}$$

Then S generates a subgroup of order $q^m - 1$ of $GL_m(\mathbf{F}_q)$, operating on vectors of V by

$$S : \mathbf{x} \mapsto S\mathbf{x},$$

where we are viewing the vectors as column vectors. (Actually this map yields a Singer cycle on the projective points — as discussed earlier in Chapter 3.) Thus starting with $\mathbf{e} = \mathbf{e}_1^t = (1, 0, 0, \ldots, 0)^t$, say, we order the coordinate places so that the vector corresponding to an $f \in E^V$ is the $(q^m - 1)$-tuple

$$(f(\mathbf{e}^t), f((S\mathbf{e})^t), f((S^2\mathbf{e})^t), \ldots, f((S^{(q^m-2)}\mathbf{e})^t)).$$

Then the function f^S is in $\mathcal{R}_{F_q}(\nu, m)^*$ provided f is; the code, therefore, is cyclic. $\square$

Corollary 5.4.1 *Provided that $\nu < m(q-1)$, the generalized Reed-Muller codes $\mathcal{R}_{F_q}(\nu, m)$ are extended cyclic codes and of the same dimension as the corresponding cyclic codes $\mathcal{R}_{F_q}(\nu, m)^*$.*

Proof: By Lemma 5.4.2 $f(\mathbf{0}) = -\sum_{\mathbf{w} \neq \mathbf{0}} f(\mathbf{w})$ provided that the degree of f is less than $m(q-1)$. $\square$

5.5   Dimensions and minimum weights

In this section we will derive the dimensions and minimum weights of some of the generalized Reed-Muller codes in which we are interested. We will use the cyclicity of the punctured codes in order to do so, for we are, in fact, interested in certain non-primitive subfield subcodes and we see no other way to establish their dimensions. Since $\mathcal{R}_{F_q}(\nu, m)^*$ is a cyclic code, we can specify it by its generator polynomial or, alternatively, by the roots of its generator polynomial. The reader should understand that when we say $\mathcal{R}_{F_q}(\nu, m)^*$ is a cyclic code we are assuming that ω has been chosen and it is cyclic as in the representation given in the proof of Theorem 5.4.5. We need a lemma and some notation:

Definition 5.5.1 *For q a prime power and u a positive integer with q-ary representation*

$$u = \sum_{i=0}^{\infty} u_i q^i, \text{ where } 0 \leq u_i \leq q - 1,$$

the **q-weight** *of u is $\mathrm{w}_q(u)$ given by*

$$\mathrm{w}_q(u) = \sum_{i=0}^{\infty} u_i.$$

Lemma 5.5.1 *Given $\nu < m(q-1)$, set $\mu = m(q-1) - 1 - \nu$. Suppose $f \in \mathcal{R}_{F_q}(\nu, m)$ and suppose u is a positive integer with $\mathrm{w}_q(u) \leq \mu$. Let L be any extension field of E and let $\mathbf{a} \in L^m$. Using the standard inner product in L^m,*

$$\sum_{\mathbf{w} \in E^m} (\mathbf{w}, \mathbf{a})^u f(\mathbf{w}) = 0.$$

Proof: If $\mathbf{w} = (w_1, w_2, \ldots, w_m)$ with $w_i \in E$ and $\mathbf{a} = (a_1, a_2, \ldots, a_m)$ with $a_i \in L$, then

$$(\mathbf{w}, \mathbf{a}) = \sum_{i=1}^{m} w_i a_i.$$

Let $u = \sum_{k=0}^{s} u_k q^k$. Then

$$\sum_{\mathbf{w} \in E^m} (\mathbf{w}, \mathbf{a})^u f(\mathbf{w}) = \sum_{\mathbf{w} \in E^m} \left(\sum_{i=1}^{m} w_i a_i \right)^{\sum_k u_k q^k} f(\mathbf{w})$$

$$= \sum_{\mathbf{w} \in E^m} \left[\prod_k \left(\sum_i w_i a_i^{q^k} \right)^{u_k} \right] f(\mathbf{w}).$$

All we need to do is to show that the degree, in terms of the variables w_i, of the expression within the square brackets is at most μ, for then this polynomial multiplied by $f(\mathbf{w})$ will have degree less than $m(q-1)$ and Lemma 5.4.2 will give the result.

In the inner summation raised to the power u_k for fixed k, each term has the form $c(\mathbf{a}) w_1^{i_1} \ldots w_m^{i_m}$ where $\sum_j i_j = u_k$. Taking the product of such terms over k will give a sum of terms of the form $c'(\mathbf{a}) w_1^{j_1} \ldots w_m^{j_m}$ where $\sum_l j_l \leq \sum_k u_k = \mathrm{w}_q(u) \leq \mu$. This gives the required result. $\square$

Theorem 5.5.1 *If $K = \mathbf{F}_{q^m}$, $E = \mathbf{F}_q$, and ω is a primitive root of K, then, for $0 \leq u \leq q^m - 2$, ω^u is a root of the generator polynomial of the code $\mathcal{R}_{F_q}(\nu, m)^*$ if and only if $0 < \mathrm{w}_q(u) \leq m(q-1) - 1 - \nu$.*

Proof: Using the notation of Theorem 5.4.5, consider K as a vector space over E. Let $\mathbf{a} = (1, \omega, \ldots, \omega^{m-1})^t \in K^m$, and define the map T:

$$T : \begin{cases} E^m & \to & K \\ \mathbf{e}_i^t & \mapsto & (\mathbf{e}_i^t, \mathbf{a}) \end{cases}$$

for $i = 1, \ldots, m$. Then T extends to a non-singular linear transformation over E and, moreover,

$$T(S^i \mathbf{e}) = (S^i \mathbf{e}, \mathbf{a}) = \omega^i, \tag{5.3}$$

for $i = 0, 1, \ldots, q^m - 2$. Using the customary correspondence between a cyclic code of length n and an ideal in the polynomial ring, we have, for $f \in \mathcal{R}_{F_q}(\nu, m)$, the codeword

$$(f(\mathbf{e}^t), f((S\mathbf{e})^t), \ldots, f((S^{(q^m-2)}\mathbf{e})^t))$$

in $\mathcal{R}_{F_q}(\nu, m)^*$ corresponding to the polynomial

$$f(X) = \sum_{i=0}^{q^m-2} f((S^i\mathbf{e})^t)X^i.$$

Now ω^u is a root of the generator polynomial for $\mathcal{R}_{F_q}(\nu, m)^*$ if and only if $f(\omega^u) = 0$ for all $f \in \mathcal{R}_{F_q}(\nu, m)$. But

$$
\begin{aligned}
f(\omega^u) &= \sum_{i=0}^{q^m-2} f((S^i\mathbf{e})^t)(\omega^u)^i \\
&= \sum_{i=0}^{q^m-2} f((S^i\mathbf{e})^t)(S^i\mathbf{e}, \mathbf{a})^u \\
&= \sum_{\substack{\mathbf{w}\in V \\ \mathbf{w}\neq\mathbf{0}}} f(\mathbf{w})(\mathbf{w}, \mathbf{a})^u \\
&= 0
\end{aligned}
$$

for $0 < w_q(u) \le m(q-1) - 1 - \nu = \mu$, by Lemma 5.5.1. On the other hand the number of integers u with $u = 0$ or $w_q(u) > \mu$ is, since $w_q(u) = m(q-1) - w_q(q^m - 1 - u) > \mu$ if and only if $w_q(q^m - 1 - u) \le \nu$, clearly the dimension of $\mathcal{R}_{F_q}(\nu, m)$ and this shows that we have exactly the roots of the generator polynomial. Thus we have the necessary and sufficient condition on ω^u. $\square$

Corollary 5.5.1 *For $0 \le \nu < m(q-1)$ the code $\mathcal{R}_{F_q}(\nu, m)^*$ is the cyclic code with generator polynomial*

$$g(X) = \prod_{\substack{0 < u < q^m-1 \\ w_q(u) \le \mu}} (X - \omega^u),$$

where ω is a primitive element of $\mathbf{F}_{q^m}$ and $\mu + \nu + 1 = m(q-1)$.

Corollary 5.5.2 *For* $0 \leq \nu < m(q-1)$ *the code* $(\mathcal{R}_{F_q}(\nu, m)^*)^{\perp}$ *is the cyclic code with generator polynomial*

$$g(X) = \prod_{\substack{0 \leq u < q^m - 1 \\ w_q(u) \leq \nu}} (X - \omega^u),$$

where ω *is a primitive element of* $\mathbf{F}_{q^m}$. *Moreover, if* $\nu + \mu + 1 = m(q-1)$, *then*

$$(\mathcal{R}_{F_q}(\nu, m)^*)^{\perp} = (\mathbf{F}_q \jmath)^{\perp} \cap \mathcal{R}_{F_q}(\mu, m)^*.$$

Proof: That the generating polynomial is as asserted follows from an application of Theorem 2.6.3. The second statement then follows from Corollary 5.5.1, with the extra factor $(X - 1)$ placing the code inside $(\mathbf{F}_q \jmath)^{\perp}$. $\square$

Corollary 5.5.3 *For* $\nu < m(q-1)$, *the dimensions of both* $\mathcal{R}_{F_q}(\nu, m)^*$ *and* $\mathcal{R}_{F_q}(\nu, m)$ *are given by*

$$|\{u | 0 \leq u \leq q^m - 1 \text{ and } w_q(u) \leq \nu\}|.$$

Proof: This is simply a restatement of the value of the dimension in terms of the q-weight. $\square$

Example 5.5.1 For $m = 2$ and $q = 3$, $E = \mathbf{F}_3$ and $K = \mathbf{F}_9$. The quadratic $f(Z) = Z^2 + Z - 1$ is a primitive polynomial for K, with primitive root ω. Since $m(q-1) = 4$, $\mathcal{R}_E(1, 2)^*$ and $\mathcal{R}_E(2, 2)^*$ are the only interesting cases. The generator polynomial for $\mathcal{R}_E(2, 2)^*$ is $g(X) = (X - \omega)(X - \omega^3) = X^2 + X - 1$ and the code has dimension 6. A generator matrix is

$$G = \begin{pmatrix} -1 & 1 & 1 & 0 & 0 & 0 & 0 & 0 \\ 0 & -1 & 1 & 1 & 0 & 0 & 0 & 0 \\ 0 & 0 & -1 & 1 & 1 & 0 & 0 & 0 \\ 0 & 0 & 0 & -1 & 1 & 1 & 0 & 0 \\ 0 & 0 & 0 & 0 & -1 & 1 & 1 & 0 \\ 0 & 0 & 0 & 0 & 0 & -1 & 1 & 1 \end{pmatrix}.$$

The extended code, $\mathcal{R}_E(2, 2)$, has a generator matrix that is G augmented by an extra column whose entries are -1's: this is a generator matrix for the code of the affine plane $AG_2(\mathbf{F}_3)$. If the extra column, corresponding to $\mathbf{0}$, is labelled 0, and added as the first column, and the columns of G then labelled 1 to 8, then the plane can be pictured as in Figure 5.1, with incidence matrix as given in Figure 5.2, where the rows are arranged in parallel classes. The columns then correspond to the points

$$(0,0),(1,0),(0,1),(1,-1),(-1,-1),(-1,0),(0,-1),(-1,1),(1,1)$$

which are $\mathbf{0}$ and $S^i e^t$ for $i = 0, 1, \ldots, 7$, with

$$S = \begin{pmatrix} 0 & 1 \\ 1 & -1 \end{pmatrix},$$

the matrix of (5.2). Equivalently, they correspond to the elements of $\mathbf{F}_9$ in the order $0, 1, \omega, \omega^2, \omega^3, \omega^4, \omega^5, \omega^6, \omega^7$. The line $\{0,1,5\}$, for example, has the equation $X_2 = 0$ and the line $\{6, 5, 8\}$ has the equation $X_1 + X_2 + 1 = 0$.
The generator polynomial for $\mathcal{R}_E(1,2)^*$ is

$$f(X) = (X - \omega)(X - \omega^2)(X - \omega^3)(X - \omega^4)(X - \omega^6)$$

and that for $(\mathcal{R}_E(2,2)^*)^\perp$ is

$$(X - 1)f(X) = X^6 + X^5 - X^4 - X^2 - X + 1,$$

so a parity-check matrix for $\mathcal{R}_E(2,2)^*$ is

$$H = \begin{pmatrix} 1 & -1 & -1 & 0 & -1 & 1 & 1 & 0 \\ 0 & 1 & -1 & -1 & 0 & -1 & 1 & 1 \end{pmatrix}.$$

Theorem 5.5.2 *If $\nu = r(q - 1) + s$, where $0 \leq s < q - 1$, then the code $\mathcal{R}_{F_q}(\nu, m)^*$ is a subcode of a BCH code of length $q^m - 1$ over $\mathbf{F}_q$ with designed distance*

$$(q - s)q^{m-r-1} - 1.$$

Proof: From Theorem 5.5.1, for a primitive element ω of $\mathbf{F}_{q^m}$, ω^u is a root of the generator polynomial of $\mathcal{R}_{F_q}(\nu, m)^*$ if and only if $0 < w_q(u) < \mu + 1$. Now

$$\mu + 1 = m(q - 1) - \nu = (m - r)(q - 1) - s = (m - r - 1)(q - 1) + (q - 1 - s),$$

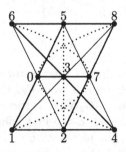

Figure 5.1: The affine plane $AG_2(\mathbf{F}_3)$

$$
\left(
\begin{array}{ccccccccc}
0 & 1 & 2 & 3 & 4 & 5 & 6 & 7 & 8 \\
\hline
 & 1 & 1 & & 1 & & & & \\
1 & & & 1 & & & & 1 & \\
 & & & & & 1 & 1 & & 1 \\
\hline
1 & & & & & & 1 & 1 & \\
1 & & & & 1 & & & & 1 \\
 & 1 & 1 & & 1 & & & & \\
\hline
1 & & 1 & & & & & & 1 \\
 & & & 1 & 1 & & 1 & & \\
1 & 1 & & & & 1 & & & \\
\hline
1 & 1 & & & & & 1 & & \\
 & 1 & & & & & & 1 & 1 \\
 & & 1 & 1 & & 1 & & &
\end{array}
\right).
$$

Figure 5.2: The incidence matrix of $AG_2(\mathbf{F}_3)$

so if we let h be the smallest integer with $\mathrm{w}_q(h) = \mu + 1$, then

$$
h = (q - s - 1)q^{m-r-1} + \sum_{i=0}^{m-r-2} (q-1)q^i = (q-s)q^{m-r-1} - 1.
$$

From the construction, every integer u with $0 \le u < h$ satisfies $\mathrm{w}_q(u) \le \mu$, and thus the elements $\omega^1, \omega^2, \ldots, \omega^{h-1}$ are all roots of the generator polynomial of the code. Thus $\mathcal{R}_{F_q}(\nu, m)^*$ is a subcode of a BCH code of designed distance $(q-s)q^{m-r-1} - 1$ as stated. $\square$

The designed distance is the true minimum distance, as the following theorem shows by an explicit construction of codewords of this weight.

Theorem 5.5.3 *For any ν such that $0 \le \nu < m(q-1)$, where we set $\nu = r(q-1) + s$ with $0 \le s < q-1$, $\mathcal{R}_{F_q}(\nu, m)$ has vectors of weight $(q-s)q^{m-r-1}$ that consist of the sum of multiples of the incidence vectors of $(q-s)$ parallel $(m-r-1)$-flats, all contained in an $(m-r)$-flat.*

Proof: Given arbitrary elements $w_i \in E$, for $i = 1, \ldots, r$, and s *distinct* elements w'_j in E, let $p(x_1, \ldots, x_m)$ be the polynomial

$$
\prod_{i=1}^{r} \left(1 - (x_i - w_i)^{q-1}\right) \prod_{j=1}^{s} (x_{r+1} - w'_j).
$$

Then $p(x_1,\ldots,x_m)$ has degree $r(q-1)+s=\nu$ and is zero in $E^m = V$ unless

$$x_i \;=\; w_i, \text{ for } i=1,\ldots,r, \tag{5.4}$$
$$x_{r+1} \;\neq\; w'_j \text{ for } j=1,\ldots,s. \tag{5.5}$$

There are $(q-s)q^{m-r-1}$ vectors in E^m satisfying both equations and the codeword corresponding to $p(x_1,\ldots,x_m)$ has this weight.

To establish the geometric nature of the codewords defined by such polynomials, consider the q^{m-r-1} points of E^m satisfying (5.4) and the additional equation $x_{r+1}=c$, where c is an element of E that is not amongst the w'_j. Then these points all belong to an $(m-r-1)$-flat and the corresponding coordinate positions in the codeword of $p(\mathbf{x})$ have the constant value

$$\prod_{j=1}^{s}(c-w'_j)$$

on these points. The same is true of each of the $(q-s)$ elements of E that are not amongst the w'_j, and hence we get a vector of the stated form. $\square$

Remark If $s=0$, then the polynomial $p(x_1,\ldots,x_m)$ is the incidence vector of an $(m-r)$-flat.

Corollary 5.5.4 *If* $\nu = r(q-1)+s < m(q-1)$ *with* $0 \le s < q-1$, *then* $\mathcal{R}_{F_q}(\nu,m)$ *has minimum weight* $(q-s)q^{m-r-1}$ *and* $\mathcal{R}_{F_q}(\nu,m)^*$ *has minimum weight* $(q-s)q^{m-r-1}-1$.

Proof: By taking the flat with $w_i = 0$ for $i=1,\ldots,r$, and the $w'_j \neq 0$ for $j=1,\ldots,s$, it follows that the coordinate at the point $\mathbf{0}$ of the corresponding polynomial is non-zero, so that the corresponding codeword in $\mathcal{R}_{F_q}(\nu,m)^*$ has weight $(q-s)q^{m-r-1}-1$. This is the minimum weight by Theorem 5.5.2. By translation invariance the minimum weight of $\mathcal{R}_{F_q}(\nu,m)$ must be $(q-s)q^{m-r-1}$ since it has vectors of that weight and, as we have just seen, the minimum weight of $\mathcal{R}_{F_q}(\nu,m)^*$ is $(q-s)q^{m-r-1}-1$. $\square$

For the special case of $m=1$, $r=0$ and $s=\nu$, and thus $\mathcal{R}_{F_q}(\nu,1)^*$ is a $[q-1,\nu+1,q-\nu-1]$ q-ary code. It is a Reed-Solomon MDS code, as defined in Section 2.8.

Corollary 5.5.5 *Let p be a prime. The code over $\mathbf{F}_p$ of the design of points and r-flats of the affine geometry $AG_m(\mathbf{F}_{p^t})$ has minimum weight p^{tr}.*

Proof: Apply Theorem 5.5.3 with $s = 0$. Since the code of the design is a subset of the generalized Reed-Muller code, it must have at least this minimum weight, and since it has vectors of this weight, this must be the minimum weight. $\square$

Corollary 5.5.6 *Let p be a prime. The code over $\mathbf{F}_p$ generated by the differences of the incidence vectors of two parallel r-flats of the affine geometry $AG_m(\mathbf{F}_{p^t})$ has minimum weight $2p^{tr}$.*

Proof: Set $q = p^t$ and take $\nu = (m - r - 1)(q - 1) + (q - 2)$. Each of the generating vectors of the code in question is in $\mathcal{R}_{F_q}(\nu, m)$ which, even as a code over $\mathbf{F}_q$, has minimum weight $2q^r$. Thus the code of the design, being a subset and having vectors of this weight, must also have minimum weight $2q^r$. $\square$

5.6 The geometric codes

In order to be able to state Delsarte's theorem [81] giving the precise relationship between the generalized Reed-Muller codes and the codes of the designs obtained from finite projective and affine geometries, we need to derive the *subfield* subcodes from $\mathcal{R}_{F_q}(\nu, m)$, and *non-primitive* (in the BCH sense) subfield codes from $\mathcal{R}_{F_q}(\nu, m)^*$. These codes are all most simply specified in terms of the single-variable approach of Mattson and Solomon [212]; they are described fully in Delsarte *et al.* [83] and also in Blake and Mullin [42].

Taking ω to be a primitive element of $K = \mathbf{F}_{q^m}$ and using the same notation as in the preceding section, set $v = q^m - 1$ and consider the vector space of polynomials in Z with coefficients in K and of degree less than v. Then Lagrange interpolation shows that any function from $K^\times$ to K is given uniquely by such a polynomial, viewed as a polynomial function in the single variable Z. In terms of the characteristic functions of the points ω^i of $V^* = V - \{\mathbf{0}\} \approx K^\times$, where $V = \mathbf{F}_q^m \approx K$, such a polynomial function can be written as

$$q(Z) = \sum_{i=0}^{v-1} q(\omega^i) g_i(Z), \tag{5.6}$$

where $g_i(Z)$ denotes the characteristic function of $\{\omega^i\}$, i.e.

$$v^{\omega^i} = g_i(Z) = -\omega^i \frac{(Z^v - 1)}{(Z - \omega^i)} \tag{5.7}$$

by Lagrange interpolation.

Obviously the polynomials Z^i, for $i = 0, 1, \ldots, v-1$, form an alternative basis, and the correspondence is given as follows: if

$$q(Z) = \sum_{j=0}^{v-1} c_j Z^j, \text{ where } c_j \in K,$$

then, using the discrete Fourier transform and noting that $1/v$ is -1 when viewed in K,

$$c_j = -\sum_{i=0}^{v-1} q(\omega^i)\omega^{-ji}. \tag{5.8}$$

We want to restrict to those functions taking values in $E = \mathbf{F}_q \subseteq K$ (i.e. we must consider E^{V^*}) so that we require that $q(\omega^i) \in E$ for all i. This is equivalent, from Equation (5.8), to $(c_j)^q = c_{qj}$, where the subscripts are taken modulo $v = q^m - 1$. These functions are, clearly, an E-subspace of K^{V^*}. Denoting this subspace by L and writing its vectors in terms of the basis of characteristic functions, we have

$$L = \left\{ (q(1), \ldots, q(\omega^{v-1}))|q(Z) = \sum_{j=0}^{v-1} c_j Z^j, c_j \in K, (c_j)^q = c_{qj} \right\}.$$

The space L corresponds with the polynomial ring $E[X]/(X^v - 1)$ via $a(X) = \sum_{i=0}^{v-1} q(\omega^i)X^i$. In fact a polynomial $b(X) = \sum_{i=0}^{v-1} b_i X^i$ corresponds to the function $q(Z)$ defined by $q(Z) = \sum_{j=0}^{v-1} c_j Z^j$, where $c_j = -b(\omega^{-j})$. If the polynomial $g(X)$ divides $X^v - 1$, then the cyclic code generated by $g(X)$ contains $b(X)$ if and only if $b(\omega^{-j}) = 0$ for all zeros ω^{-j} of $g(X)$. The corresponding $q(Z) \in L$ has the property that $c_j = 0$ if ω^{-j} is a root of $g(X)$. Thus the cyclic code with zeros $\{\omega^{-j}|j \in \mathcal{J}\}$, where $\mathcal{J} \subseteq \{0, 1, \ldots, v-1\}$, can be characterized as

$$\left\{ (q(1), \ldots, q(\omega^{v-1}))|q(Z) \in L, q(Z) = \sum_{j=0}^{v-1} c_j Z^j, c_j = 0 \text{ if } j \in \mathcal{J} \right\}.$$

To get the generalized Reed-Muller codes $\mathcal{R}_{F_q}(\nu, m)^*$, where $0 \leq \nu < m(q-1)$, we will see that the relevant subcode of L is given by

$$\mathcal{Q}_\nu = \left\{ q(Z)|q(Z) \in L, q(Z) = \sum_{j=0}^{v-1} c_j Z^j, \text{ and } c_j = 0 \text{ for } w_q(j) > \nu \right\}.$$

If we write, for any ν such that $0 \leq \nu \leq m(q-1)$,

$$U_\nu = \{j|0 \leq j \leq q^m - 1, w_q(j) \leq \nu\},$$

then

$$Q_\nu = \left\{ q(Z) | q(Z) = \sum_{u \in U_\nu} c_u Z^u, c_u \in K, c_{uq} = (c_u)^q \right\}. \quad (5.9)$$

That this is $\mathcal{R}_{F_q}(\nu, m)^*$ follows from the fact that it is a cyclic code with generator polynomial that has ω^{v-j} as a root if and only if $0 < \mathrm{w}_q(v - j) \leq \mu$, where $0 < (v - j) < q^m - 1$. Since $\mathrm{w}_q(v - j) = m(q - 1) - \mathrm{w}_q(j) \leq \mu$, this holds if and only if $\mathrm{w}_q(j) > \nu$. Thus

$$\mathcal{R}_{F_q}(\nu, m)^* = \{(q(1), q(\omega), \ldots, q(\omega^{v-1})) | q(Z) \in Q_\nu\}. \quad (5.10)$$

The reader should note that if a positive integer u has an orbit of length i under the map $j \mapsto jq$ modulo v, i.e. if $uq^i \equiv u \pmod{v}$ and i is the smallest integer satisfying the congruence, then the choice of the coefficient of Z^u must be in a field of degree i over E; this agrees, of course, with the dimensional requirements.

To get the extended code, $\mathcal{R}_{F_q}(\nu, m)$, adjoin the extra coordinate position corresponding to **0**, where the entry is $-\sum_{i=0}^{v-1} q(\omega^i)$. Since $q(0) = c_0$, (5.8) implies that

$$\mathcal{R}_{F_q}(\nu, m) = \{(q(0), q(1), q(\omega), \ldots, q(\omega^{v-1})) | q(Z) \in Q_\nu\}. \quad (5.11)$$

It follows from the above that every polynomial $q(Z)$ in $K[Z]$ of degree less than v has the property that $\sum_{z \in K} q(z) = 0$; thus a polynomial $r(Z) = \sum_{j=0}^{v} c_j Z^j$ satisfies $\sum_{z \in K} r(z) = 0$ if *and only if* it is of degree less than v — i.e. has $c_v = 0$ — since such polynomials form a subspace of codimension 1 (as a vector space over K).

We next wish to establish the correspondence of polynomials in $\mathbf{x} = (x_1, \ldots, x_m)$ with the polynomials $q(Z) \in L$. We merely sketch our development — which is different in one crucial respect from that given in [83]: here the Mattson-Solomon polynomials live in $K[Z]/(Z^{q^m} - Z)$ rather than $K[Z]/(Z^v - 1)$.

In order to explain the correspondence we first introduce the trace: let $\mathrm{Tr}_{K/E}$ denote the trace from K to E, i.e. for $z \in K$,

$$\mathrm{Tr}_{K/E}(z) = z + z^q + z^{q^2} + \cdots + z^{q^{m-1}}.$$

Since the trace furnishes a universal linear transformation from the vector space K over E to E (see, for example, Lidl and Niederreiter [185, Theorem 2.24]) given any basis $\{\alpha_1, \alpha_2, \ldots, \alpha_m\}$ for K over E, there is a unique *complementary basis* $\{\beta_1, \beta_2, \ldots, \beta_m\}$ for K over E such that

$$\mathrm{Tr}_{K/E}(\alpha_i \beta_j) = \delta_{ij},$$

where δ_{ij} denotes the Kronecker delta function. Moreover, it is clear from the definition of the complementary basis that the matrix

$$
A = \begin{pmatrix}
\alpha_1 & \alpha_1^q & \cdots & \alpha_1^{q^{m-1}} \\
\alpha_2 & \alpha_2^q & \cdots & \alpha_2^{q^{m-1}} \\
\vdots & \vdots & \vdots & \vdots \\
\alpha_m & \alpha_m^q & \cdots & \alpha_m^{q^{m-1}}
\end{pmatrix}, \tag{5.12}
$$

is invertible with its inverse giving the complementary basis, i.e.

$$
A^{-1} = \begin{pmatrix}
\beta_1 & \beta_2 & \cdots & \beta_m \\
\beta_1^q & \beta_2^q & \cdots & \beta_m^q \\
\vdots & \vdots & \vdots & \vdots \\
\beta_1^{q^{m-1}} & \beta_2^{q^{m-1}} & \cdots & \beta_m^{q^{m-1}}
\end{pmatrix}. \tag{5.13}
$$

Using the basis $\{1, \omega, \omega^2, \ldots, \omega^{m-1}\}$, where ω is a primitive element for K, let the complementary basis be $\{\beta_1, \beta_2, \ldots, \beta_m\}$. Then $z \in K$ satisfies $z = \sum_{i=1}^m x_i \omega^{i-1}$ if and only if $x_i = \mathrm{Tr}_{K/E}(\beta_i z)$.

We use the fundamental property of polynomial rings and the fact that $E \subseteq K$ to define a ring homomorphism

$$
\theta : E[x_1, \ldots, x_m] \to K[Z]
$$

by sending x_i to

$$
\beta_i Z + (\beta_i Z)^q + \ldots + (\beta_i Z)^{q^{m-1}} = \mathrm{Tr}_{K/E}(\beta_i Z).
$$

Following θ by the natural map

$$
K[Z] \to K[Z]/(Z^{q^m} - Z),
$$

using the standard representatives — namely polynomials in Z of degree less than and equal to v — and taking the usual liberty of viewing Z as $Z + (Z^{q^m} - Z)$, we see that $(\mathrm{Tr}_{K/E}(\beta_i Z))^q = \mathrm{Tr}_{K/E}(\beta_i Z)$; hence we get the induced ring homomorphism,

$$
\bar{\theta} : E[x_1, \ldots, x_m]/(x_1^q - x_1, \ldots, x_m^q - x_m) \to K[Z]/(Z^{q^m} - Z).
$$

We can thus convert any "reduced" polynomial in the variables x_i with coefficients in E into a polynomial in Z, of degree less than or equal to $v = q^m - 1$, with coefficients in K. It follows from Lemma 5.4.2 and the fact that

$$
(\mathrm{Tr}_{K/E}(\beta_1 z), \ldots, \mathrm{Tr}_{K/E}(\beta_m z))
$$

takes on every value in E^m as z varies over K, that the image of $p(x_1, \ldots, x_m)$ is of degree *less than v* provided $p(x_1, \ldots, x_m)$ is of degree less than $m(q-1)$. Moreover, the polynomial $q(Z) = \sum_{j=0}^{v-1} c_j Z^j$ has $c_j^q = c_{jq}$ — with subscripts computed modulo v — if and only if $q(Z)^q = q(Z)$ in the ring $K[Z]/(Z^{q^m} - Z)$ — since $qj \neq v$ for $j < v$ and therefore computing subscripts modulo v is the same as viewing the polynomial in $K[Z]/(Z^{q^m} - Z)$. Since $(p(x_1, \ldots, x_m))^q = p(x_1, \ldots, x_m)$ in the ring $E[x_1, \ldots, x_m]/(x_1^q - x_1, \ldots, x_m^q - x_m)$, the reduced polynomials of degree less than $m(q-1)$ have images, under $\bar{\theta}$, in L.

Conversely, we define a ring homomorphism,

$$K[Z] \to K[x_1, \ldots, x_m]/(x_1^q - x_1, \ldots, x_m^q - x_m),$$

by sending Z to $\sum_{i=1}^m x_i \omega^{i-1}$. Since $(\sum_{i=1}^m x_i \omega^{i-1})^{q^m} = \sum_{i=1}^m x_i \omega^{i-1}$, we obtain a ring homomorphism

$$K[Z]/(Z^{q^m} - Z) \to K[x_1, \ldots, x_m]/(x_1^q - x_1, \ldots, x_m^q - x_m).$$

Notice that if $q(Z)^q = q(Z)$, then the image of $q(Z)$ must lie in $E[x_1, \ldots, x_m]/(x_1^q - x_1, \ldots, x_m^q - x_m)$, since this is the subring of $K[x_1, \ldots, x_m]/(x_1^q - x_1, \ldots, x_m^q - x_m)$ left fixed by the Frobenius automorphism, $x \mapsto x^q$. Let R denote the subring of $K[Z]/(Z^{q^m} - Z)$ left pointwise fixed by the Frobenius map $x \mapsto x^q$; then we have a ring homomorphism

$$\psi : R \to E[x_1, \ldots x_m]/(x_1^q - x_1, \ldots, x_m^q - x_m)$$

and using the fact that $\mathrm{Tr}_{K/E}(x_i \beta) = x_i \mathrm{Tr}_{K/E}(\beta)$ one shows easily that $\psi \circ \bar{\theta}$ is the identity map. Moreover, since in $K[Z]/(Z^{q^m} - Z)$ one has $(Z^v)^k = Z^v$ for any positive integer k, $q(Z)^q = q(Z) = \sum_{j=0}^v c_j Z^j$ if and only if $c_v \in E$ and $\sum_{j=0}^{v-1} c_j Z^j \in L$. Both rings have dimension q^m as E-algebras and hence $\bar{\theta}$ is an isomorphism of rings — with ψ restricted to R its inverse. In addition, under this ring isomorphism the E-subspace L is isomorphic to $\mathcal{R}_{F_q}(m(q-1)-1, m)$.

A simple calculation shows that Z^j corresponds to a polynomial in the x_i of degree $\mathrm{w}_q(j)$; thus, if $q(Z) \in L$ is such that $c_j = 0$ for $\mathrm{w}_q(j) > \nu$, its image in $E[x_1, \ldots, x_m]/(x_1^q - x_1, \ldots, x_m^q - x_m)$ has degree less than or equal to ν. It now follows that the isomorphism above carries $\mathcal{Q}_\nu$ to $\mathcal{N}_E(\nu, m)$ for $0 \leq \nu < m(q-1)$.

Example 5.6.1 Let $m = 2$ and $q = 3$. Then $E = \mathbf{F}_3$ and $K = \mathbf{F}_9$. A primitive element for K is ω, a root of the primitive polynomial $X^2 + X - 1$. The basis complementary to $\{1, \omega\}$ can easily be computed; it is $\{-\omega, \omega - 1\}$, i.e. $\{\omega^5, \omega^6\}$. A polynomial $p(x_1, x_2)$ will correspond to $p(\mathrm{Tr}_{K/E}(\beta_1 Z), \mathrm{Tr}_{K/E}(\beta_2 Z)) = p(-\omega Z - \omega^3 Z^3, -\omega^2 Z - \omega^6 Z^3)$. For example, if $p(x_1, x_2) = x_1$, then $q(Z) = -\omega Z - \omega^3 Z^3 = \omega^5 Z + \omega^7 Z^3 = \sum_{j=0}^{7} c_j Z^j$. (Notice that $(c_j)^3 = c_{3j}$.) It can be verified that $c_j = 0$ when $\mathrm{w}_3(j) > \nu = 1$. The corresponding codeword in $\mathcal{R}_{F_3}(1, 2)$ is

$$(q(0), q(1), \ldots, q(\omega^7)) = (0, 1, 0, 1, -1, -1, 0, -1, 1),$$

which clearly checks with the codeword corresponding to the monomial $p(x_1, \ldots, x_m) = x_1$ when the vectors in $V = E^2$ are written in the order

$$0, 1, \omega, \omega^2, \omega^3, \omega^4, \omega^5, \omega^6, \omega^7$$

in the basis $\{1, \omega\}$, which is

$$(0,0), (1,0), (0,1), (1,-1), (-1,-1), (-1,0), (0,-1), (-1,1), (1,1).$$

To check that we obtain $p(x_1, \ldots, x_m) = x_1$ from $q(Z)$, put $Z = (\mathbf{x}, \mathbf{a}) = x_1 + x_2 \omega$ in $q(Z) = \omega^5 Z + \omega^7 Z^3$ and recall that the x_i are reduced modulo $x_i^q - x_i = x_i^3 - x_i$.

As we saw in Proposition 5.3.2, the punctured Reed-Muller codes were those that turned out to be the codes of designs from projective geometries over $\mathbf{F}_2$ and we would expect something analogous to be true for the general prime power. Here the punctured code has length $q^m - 1$ which is not, in general, the number of points of a projective geometry over $\mathbf{F}_q$. However, the observation that the punctured codes are subcodes of BCH codes and that we can thus consider associated *non-primitive* BCH codes of lengths dividing $q^m - 1$ will lead us to the required correspondence. We write $q^m - 1 = nb$, where our main application will be to the case $b = q - 1$ and

$$n = \frac{q^m - 1}{q - 1} = q^{m-1} + q^{m-2} + \cdots + q + 1.$$

Now define, for any ν such that $0 \le \nu \le m(q-1)$,

$$\mathcal{Q}_{\nu/b} = \{q(Z) | q(Z) \in \mathcal{Q}_\nu, q(\omega^i) = q(\omega^{i+n})\}, \text{ for } i = 0, 1, \ldots, \nu - 1\}. \quad (5.14)$$

For such a function $q(Z)$, it follows from (5.8) that $c_j = \omega^{jn} c_j$ for each j, so that $c_j = 0$ unless $j \equiv 0 \pmod{b}$. Thus

$$q(Z) = \sum_{i=0}^{n-1} c_{ib} Z^{ib},$$

and $\mathcal{Q}_{\nu/b}$ is the subset of $\mathcal{Q}_\nu$ with the property that $c_j = 0$ unless $j \equiv 0 \pmod{b}$. From what we already have for the dimension of the generalized Reed-Muller codes (see Corollary 5.5.3), the dimension of $\mathcal{Q}_{\nu/b}$ is

$$|\{j|0 \le j \le q^m - 1, b \text{ divides } j, \mathrm{w}_q(j) \le \nu\}|.$$

Now since $q(\omega^i) = q(\omega^{i+n})$, the usual vector of length $q^m - 1$ will consist of b repetitions of the vector

$$(q(1), q(\omega), q(\omega^2), \dots, q(\omega^{n-1})),$$

and we will take this to be the length of the non-primitive code.

Definition 5.6.1 *The* **non-primitive generalized Reed-Muller code** $\mathcal{R}^b_{F_q}(\nu, m)^*$ **of order** ν, *where* $0 \le \nu < m(q - 1)$ *and* b *divides* $q^m - 1$, *is the code of length* $n = (q^m - 1)/b$ *and dimension*

$$|\{j|0 \le j \le q^m - 1, b \text{ divides } j, \mathrm{w}_q(j) \le \nu\}|$$

given as the set of vectors

$$\{(q(1), q(\omega), \dots, q(\omega^{n-1}))|q(X) \in \mathcal{Q}_{\nu/b}\}.$$

We now assume that b divides $q - 1$ and hence that

$$j \equiv \mathrm{w}_q(j) \pmod{b}.$$

Then, setting

$$\mathcal{N}^b_E(\nu, m) = \left\langle x_1^{i_1} x_2^{i_2} \dots x_m^{i_m} \in \mathcal{N}_E(\nu, m) \Big| \sum_{k=1}^m i_k \equiv 0 \pmod{b} \right\rangle,$$

we have that $\mathcal{R}^b_{F_q}(\nu, m)^*$ is

$$\{(p(\mathbf{e}^t), p((S\mathbf{e})^t), \dots, p((S^{n-1}\mathbf{e})^t))|p(x_1, \dots, x_m) \in \mathcal{N}^b_E(\nu, m)\},$$

where $\mathbf{e}$ and S are as in the proof of Theorem 5.4.5.

Notice that when b divides $(q - 1)$ then we also have that $\mathrm{w}_q(jb)$ is divisible by b, and hence ν can be restricted to multiples of b, since other values will not give new codes: if $\nu = kb + r$, where $0 \le r < b$, then $\mathcal{R}^b_{F_q}(\nu, m)^* = \mathcal{R}^b_{F_q}(kb, m)^*$. Further, notice that the condition given in (5.14) that $q(\omega^i) = q(\omega^{i+n})$ implies immediately that with $c = \omega^n$, $p(\mathbf{w}) = p(\mathbf{w}c)$, so that $p(\mathbf{e}^t) = p((S^n \mathbf{e})^t)$ and the vector in terms of the basis of length $q^m - 1$ consists of b repetitions of the first n positions. In fact the factor group $\langle S \rangle / \langle S^n \rangle$ will preserve the code $\mathcal{R}^b_{F_q}(\nu, m)^*$.

Example 5.6.2 Construct the code $\mathcal{R}^2_{F_3}(2,3)^*$, of length 13 and dimension — from the above description — 7. The multi-variable formulation is the easiest to give, since the generating monomials are immediately seen to be $\{1, x_1x_2, x_1x_3, x_2x_3, x_1^2, x_2^2, x_3^2\}$. If the irreducible cubic $X^3 - X^2 + 1$, with root ω, is used to obtain $\mathbf{F}_{27}$, then the matrix S of (5.2) is

$$S = \begin{pmatrix} 0 & 0 & -1 \\ 1 & 0 & 0 \\ 0 & 1 & 1 \end{pmatrix}.$$

Now form vectors according to the method described above using the seven generating monomials, and we get the generator matrix

$$G = \begin{pmatrix} 1 & 1 & 1 & 1 & 1 & 1 & 1 & 1 & 1 & 1 & 1 & 1 & 1 \\ 0 & 0 & 0 & 0 & 1 & 1 & 0 & 0 & -1 & 1 & -1 & -1 & 0 \\ 0 & 0 & 0 & -1 & -1 & 0 & 0 & 1 & -1 & 1 & 1 & 0 & 0 \\ 0 & 0 & 0 & 0 & -1 & 0 & 1 & 0 & 1 & 1 & -1 & 0 & -1 \\ 1 & 0 & 0 & 1 & 1 & 1 & 0 & 1 & 1 & 1 & 1 & 1 & 0 \\ 0 & 1 & 0 & 0 & 1 & 1 & 1 & 0 & 1 & 1 & 1 & 1 & 1 \\ 0 & 0 & 1 & 1 & 1 & 0 & 1 & 1 & 1 & 1 & 1 & 0 & 1 \end{pmatrix}.$$

This code is the code over $\mathbf{F}_3$ of the projective plane of order 3, as will follow from Delsarte's theorem. The code vectors in G can be described geometrically: for example, labelling the columns 1 to 13, to represent the points, and using sets of these numbers to represent lines, then the penultimate row (corresponding to the monomial x_2^2) represents the complement of the line $\{1, 3, 4, 8\}$, i.e. the vector $\jmath - v^{\{1,3,4,8\}}$ in our usual notation for characteristic functions. The second row (corresponding to the monomial x_1x_2) is the vector $v^{\{3,5,6,10\}} - v^{\{3,9,11,12\}}$. In terms of homogeneous coordinates for the projective geometry, the line $\{1, 3, 4, 8\}$ represents the point set $\{(1,0,0), (0,0,1), (-1,0,1), (1,0,1)\}$, which is $(0,1,0)^t$ in homogeneous coordinates.

The non-primitive generalized Reed-Muller codes can also be viewed in the setting of BCH codes. With the usual notation, and ω a primitive $(q^m - 1)^{\text{th}}$ element for $\mathbf{F}_{q^m}$, Corollary 5.5.2 states that ω^j is a root of $(\mathcal{R}_{F_q}(\nu, m)^*)^\perp$ if and only if $w_q(j) \le \nu$. Thus

$$G_\nu = \begin{pmatrix} 1 & 1 & 1 & \cdots & 1 \\ 1 & \omega^{h_1} & \omega^{2h_1} & \cdots & \omega^{(q^m-1)h_1} \\ 1 & \omega^{h_2} & \omega^{2h_2} & \cdots & \omega^{(q^m-1)h_2} \\ \vdots & \vdots & \vdots & & \vdots \\ 1 & \omega^{h_i} & \omega^{2h_i} & \cdots & \omega^{(q^m-1)h_i} \end{pmatrix} \tag{5.15}$$

gives a parity-check matrix for $(\mathcal{R}_{F_q}(\nu, m)^*)^\perp$, i.e. it will give a generator matrix for $((\mathcal{R}_{F_q}(\nu, m)^*)^\perp)^\perp = \mathcal{R}_{F_q}(\nu, m)^*$, where $\{0, h_1, h_2, \ldots, h_i\}$ are the integers h such that $0 \le h \le q^m - 1$, and $w_q(h) \le \nu$.

Let c be a primitive n^{th} root of unity in $K = \mathbf{F}_{q^m}$, where $nb = (q^m - 1)$. Thus we may take $c = \omega^b$. Considering the rows of G_ν for which b divides h_j, write $t_j = h_j/b$ for these rows (where $1 \le j \le i$). Since

$$(\omega^{h_j})^{nl} = (\omega^{bn})^{t_j l} = 1$$

for $l = 1, 2, \ldots, b-1$, the rows being considered each consist of b replications of the first n column entries. Since $nb = q^m - 1$, the number of rows for which b divides h_j can be written as $n - k$. Now let H be the matrix consisting of these $n - k$ rows, taking only the first n columns. Since b divides h_j, this matrix can be written as

$$H = \begin{pmatrix} 1 & c^{t_1} & c^{2t_1} & \cdots & c^{(n-1)t_1} \\ 1 & c^{t_2} & c^{2t_2} & \cdots & c^{(n-1)t_2} \\ \vdots & \vdots & \vdots & & \vdots \\ 1 & c^{t_{n-k}} & c^{2t_{n-k}} & \cdots & c^{(n-1)t_{n-k}} \end{pmatrix}. \tag{5.16}$$

Define the non-primitive ν^{th}-order generalized Reed-Muller code to be the code over $\mathbf{F}_q$ obtained as the dual of the null space of H, i.e. it is the dual of the $[n, k]$ cyclic code over $\mathbf{F}_q$ with generator polynomial having ω^j as a root if and only if b divides j and $w_q(j) \le \nu$.

Example 5.6.3 Take $m = 3$, $q = 3$, $\nu = 2$, $b = q - 1 = 2$ as in our previous example, so that we are looking at $\mathcal{R}_{F_3}^2(2, 3)^*$. The integers j such that $0 \le j \le 26$ with 2 dividing j and $w_3(j) \le 2$ are $\{0, 2, 4, 6, 10, 12, 18\}$. Putting $c = \omega^2$, a primitive 13th root of unity, the generator polynomial for the orthogonal code is

$$h(X) = (X - 1)(X - c)(X - c^2)(X - c^3)(X - c^5)(X - c^6)(X - c^9),$$

and that for $\mathcal{R}_{F_3}^2(2, 3)^*$ is

$$g(X) = (X - c)(X - c^2)(X - c^3)(X - c^5)(X - c^6)(X - c^9).$$

Since $g(X)$ divides $h(X)$, the orthogonal code is in $\mathcal{R}_{F_3}^2(2, 3)^*$, which is a fact that also follows from Delsarte's theorem that we soon shall state; it implies that the $\mathcal{R}_{F_3}^2(2, 3)^* = C_3(PG_2(\mathbf{F}_3))$. Since the design of points and lines of a projective plane is a symmetric design of prime order, its code must contain the orthogonal: see Chapter 4. Computation yields that

$$g(X) = 1 - X - X^2 - X^3 + X^4 - X^5 + X^6$$

which gives a generator matrix in the usual way:

$$
\begin{pmatrix}
1 & -1 & -1 & -1 & 1 & -1 & 1 & 0 & 0 & 0 & 0 & 0 & 0 \\
0 & 1 & -1 & -1 & -1 & 1 & -1 & 1 & 0 & 0 & 0 & 0 & 0 \\
0 & 0 & 1 & -1 & -1 & -1 & 1 & -1 & 1 & 0 & 0 & 0 & 0 \\
0 & 0 & 0 & 1 & -1 & -1 & -1 & 1 & -1 & 1 & 0 & 0 & 0 \\
0 & 0 & 0 & 0 & 1 & -1 & -1 & -1 & 1 & -1 & 1 & 0 & 0 \\
0 & 0 & 0 & 0 & 0 & 1 & -1 & -1 & -1 & 1 & -1 & 1 & 0 \\
0 & 0 & 0 & 0 & 0 & 0 & 1 & -1 & -1 & -1 & 1 & -1 & 1
\end{pmatrix}
$$

which we denote by G. Labelling the columns $1, 2, \ldots, 13$ and adding the first two rows of G, gives a weight-4 vector which must be the incidence vector of the line $\{1, 3, 4, 8\}$ of the plane of order 3. Cycling this incidence vector under the cycle $(1, 2, \ldots, 13)$ gives all the 13 lines of $PG_2(\mathbf{F}_3)$. This is, of course, exactly the same plane we obtained in our previous example.

The codes orthogonal to $\mathcal{R}^b_{F_q}(\nu, m)^*$, where without loss of generality we take b dividing ν, are obtained in a manner similar to those of the primitive codes: see Corollary 5.5.2.

Theorem 5.6.1 *If* $0 \leq \nu < m(q - 1)$ *and* b *divides* ν, *then with* $\mu' = m(q - 1) - \nu - b$,

$$
(\mathcal{R}^b_{F_q}(\nu, m)^*)^{\perp} = \mathcal{R}^b_{F_q}(\mu', m)^* \cap (\mathbf{F}_q \jmath)^{\perp}.
$$

Proof: The code $\mathcal{Q}_{\nu/b}$ consists of b repetitions of $\mathcal{R}^b_{F_q}(\nu, m)^*$, and has for its *non-zeros*, ω^{-kb} where $\mathrm{w}_q(kb) \leq \nu$, and these are also non-zeros for $\mathcal{R}^b_{F_q}(\nu, m)^*$. Since ω^{-kb} is a root of $X^n - 1 = 0$, these are the non-zeros of $\mathcal{R}^b_{F_q}(\nu, m)^*$, and hence the zeros of the orthogonal code are ω^{kb} where $0 \leq \mathrm{w}_q(kb) \leq \nu$. Setting $ib = v - kb = q^m - 1 - kb$, we have $\nu \geq \mathrm{w}_q(kb) = \mathrm{w}_q(v - ib) = m(q-1) - \mathrm{w}_q(ib)$, so that $\mathrm{w}_q(ib) \geq m(q-1) - \nu > m(q-1) - \nu - b = \mu'$. Thus for the orthogonal code, $c_{ib} = 0$ for $\mathrm{w}_q(ib) > \mu'$ and $c_0 = 0$ since 1 is a zero. This completes the proof. $\square$

Corollary 5.6.1 *If* $b = (q-1)$ *and* $\nu = r(q-1)$ *then* $\mu' = (m-r-1)(q-1)$ *and*

$$
(\mathcal{R}^{q-1}_{F_q}(r(q - 1), m)^*)^{\perp} = \mathcal{R}^{q-1}_{F_q}((m - r - 1)(q - 1), m)^* \cap (\mathbf{F}_q \jmath)^{\perp}.
$$

Remark In Example 5.6.3, $\nu = 2$ gives $\mu' = 6 - 2 - 2 = 2$, and the orthogonal is $\mathcal{R}^2_{F_3}(2, 3)^* \cap (\mathbf{F}_3 \jmath)^{\perp}$, as expected.

Theorem 5.6.2 *Suppose b divides $q - 1$. Then the minimum weight of $\mathcal{R}^b_{F_q}(r(q-1), m)^*$ is*

$$(q^{m-r} - 1)/b.$$

Proof: Since, by Corollary 5.5.4, the minimum weight of $\mathcal{R}_{F_q}(r(q-1), m)^*$ is $q^{m-r} - 1$, the minimum weight is at least $(q^{m-r}-1)/b$. But the polynomial

$$p(x_1, \ldots, x_m) = \prod_{i=1}^{r} \left(1 - (x_i)^{q-1}\right)$$

is such that each of its monomials has degree divisible by b and is, hence, in $\mathcal{N}^b_E(r(q-1), m)$; since it takes the value 1 at $\mathbf{0}$, it obviously yields a vector of weight $(q^{m-r} - 1)/b$ in $\mathcal{R}^b_{F_q}(r(q-1), m)^*$. $\square$

Corollary 5.6.2 *The non-primitive code $\mathcal{R}^{q-1}_{F_q}(r(q-1), m)^*$ has minimum weight*

$$(q^{m-r} - 1)/(q - 1) = q^{m-r-1} + q^{m-r-2} + \cdots + q + 1.$$

Observe that the polynomial above that yields a minimum-weight vector is, in fact, the incidence vector of an $(m - r)$-dimensional subspace of $\mathbf{F}_q^m$. Thus projectively it is an $(m - r - 1)$-dimensional subspace of $PG_{m-1}(\mathbf{F}_q)$.

5.7 The subfield subcodes

So far the codes have all been over the field $E = \mathbf{F}_q$, which is not necessarily a prime field. In case $q = p$, a prime, then the primitive generalized Reed-Muller codes $\mathcal{R}_{F_p}((m - r)(p - 1), m)$ are, as we will see, the codes of the affine geometry designs given by the r-flats; similarly, the non-primitive generalized Reed-Muller codes $\mathcal{R}^{p-1}_{F_p}((m - r)(p - 1), m + 1)^*$ are the codes of the projective-geometry designs given by the r-dimensional subspaces. To obtain the codes of the designs in general, we need to restrict the codes, $\mathcal{R}_{F_q}(r(q-1), m)$ and $\mathcal{R}^{q-1}_{F_q}((m-r)(q-1), m+1)^*$, to **subfield subcodes**.

Definition 5.7.1 *Let C be a linear code over a field E and let F be a subfield of E. The set C' of vectors in C, all of whose coordinates lie in F, is called the **subfield subcode** of C over F.*

It is easy to verify that C' is a linear code over F and that any permutation of the coordinate positions preserving C also preserves C'. We are, of course, interested only in the case where $F = \mathbf{F}_p$, the prime subfield of E.

From now on we take $E = \mathbf{F}_q$, as before, with $q = p^t$ where p is a prime and we set $F = \mathbf{F}_p$. Further, denote by $\mathcal{A}_{F_q/F_p}(\nu, m)$ the subfield subcode of the generalized Reed-Muller code $\mathcal{R}_{F_q}(\nu, m)$; if q and p are clear from the context we will frequently omit the subscript and write $\mathcal{A}(\nu, m) = \mathcal{A}_{F_q/F_p}(\nu, m)$. Taking first the single-variable approach, $q(Z) \in \mathcal{A}(\nu, m)$ if $q(\omega^j) \in F$ for all j, and $q(0) \in F$, which is equivalent to $c_{jp} = c_j^p$ for all j, the subscripts being read modulo v as usual. Writing

$$V_\nu = \{u | 0 \le u \le q^m - 1, w_q(up^j) \le \nu \text{ for } j = 0, 1, \ldots, t - 1\}$$

(where up^j is taken reduced modulo $q^m - 1$, for the same reasons as before), we have that $\mathcal{A}_{F_q/F_p}(\nu, m)$ is

$$\left\{ (q(0), \ldots, q(\omega^{\nu-1})) | q(Z) = \sum_{u \in V_\nu} c_u Z^u, c_{up} = (c_u)^p \right\} \qquad (5.17)$$

and that

$$\dim(\mathcal{A}_{F_q/F_p}(\nu, m)) = |V_\nu|.$$

Clearly, by Theorem 5.4.4, $\mathcal{A}_{F_q/F_p}(\nu, m)$ will contain the incidence vector of any $(m - r)$-flat when $\nu \ge r(q - 1)$. Its minimum weight d_ν is bounded by

$$(q - s)q^{m-r-1} \le d_\nu \le q^{m-r},$$

where $\nu = r(q - 1) + s$ and $0 \le s < q - 1$, and, from Theorem 5.5.3, attains the lower bound if and only if there are s distinct w_j' in E such that the c_k, as defined above, are in F. In particular, if $s = 0$, then this holds and $d_\nu = q^{m-r}$. Further, in the case $s = q - 2$ this is also the case: the vector obtained is the difference of the incidence vectors of two parallel r-flats, which is clearly a vector of the subfield subcode. See also Theorem 5.7.5.

For the orthogonal code, $\mathcal{A}(\nu, m)^\perp$, clearly we have

$$\mathcal{A}(\mu, m) \subseteq \mathcal{A}(\nu, m)^\perp,$$

where $\mu = m(q - 1) - \nu - 1 = (m - r - 1)(q - 1) + (q - 2 - s)$ (and $\nu = r(q - 1) + s$, and $0 \le s < q - 1$, as above). Its minimum weight $d_\nu^\perp$ thus certainly satisfies

$$d_\nu^\perp \le d_\mu \le q^{r+1}, \qquad (5.18)$$

from the above discussion. A lower bound for $d_\nu^\perp$ follows from the BCH bound, and some evaluations of these are quoted in [83, Theorem 4.3.1]. In particular, for $\nu = r(q - 1)$ this gives

$$d_{r(q-1)}^\perp \ge (p + q)q^{r-1},$$

and for $\nu = r(q-1) + (q-2) = (r+1)(q-1) - 1$, it gives

$$d_{r(q-1)+(q-2)}^{\perp} \geq q^{r+1},$$

which, from (5.18), yields

$$d_{r(q-1)+(q-2)}^{\perp} = q^{r+1}.$$

For the codes of designs from projective geometries, we must take the subfield subcodes of the non-primitive codes

$$\mathcal{R}_{F_q}^{q-1}(\nu, m)^*$$

when $\nu = r(q-1)$. Here is the notation we will use and the relevant facts about this important code:

Proposition 5.7.1 *The subfield subcode of $\mathcal{R}_{F_q}^{q-1}(r(q-1), m)^*$, which we denote by $\mathcal{P}_{F_q/F_p}(r, m)$, has minimum weight*

$$(q^{m-r} - 1)/(q-1)$$

and among the minimum-weight vectors are the multiples of the incidence vectors of the $(m - r - 1)$-dimensional subspaces. Further, the p-rank is given by the cardinality of the set of integers u satisfying

- $0 \leq u \leq q^m - 1$,

- $q - 1$ *divides* u,

- $w_q(up^j) \leq r(q-1)$ *for* $j = 0, 1, \ldots, t-1$

where up^j is reduced modulo $q^m - 1$.

Once again, if the context is clear we will omit the subscript and write $\mathcal{P}(r, m)$ for $\mathcal{P}_{F_q/F_p}(r, m)$. Notice also that we have dropped the factor $q-1$ since all our ν's will be multiples of $q - 1$ when using this notation. We will also use the obvious notation, $\mathcal{A}_{F_q/F_p}^b(\nu, m)^*$ in the general case, i.e. for the subfield subcode of $\mathcal{R}_{F_q}^b(\nu, m)^*$.

Our interest is in the codes given by the designs of r-flats of the affine spaces and r-dimensional subspaces of the projective spaces. Just as in the binary case we must first analyse the codimension 1 case — in the projective case the design of points and hyperplanes of a projective space. This case was, historically, the one given the most attention and was introduced for projective *planes* by Prange — as we remarked in Chapter 2 — with Rudolph considerably enriching the subject and making serious conjectures.

The first systematic treatment in the case of planes was given by Graham and MacWilliams [109]. These results were generalized to higher dimensions by Goethals and Delsarte [105] and MacWilliams and Mann [202]; in particular, these authors computed the dimension of the code of the design of points and hyperplanes of an arbitrary projective space. Another, somewhat cleaner, proof of this result was given by Smith [267]. Here is the result:

Theorem 5.7.1 *If $q = p^t$, the p-rank of the design of points and hyperplanes of $PG_m(\mathbf{F}_q)$ is*

$$\binom{m+p-1}{m}^t + 1.$$

Moreover, this code is precisely $\mathcal{P}(1, m+1)$.

Proof: Smith's proof is elementary and easy to follow and we will not reproduce it here. But, let C denote the code (over $\mathbf{F}_p$) of the theorem. It is a subset of $\mathcal{R}_{F_q}^{q-1}(q-1, m)^*$ and, in fact, a subcode of $\mathcal{P}(1, m+1)$. Smith's argument in fact shows that the dimension is that of $\mathcal{P}(1, m+1)$ and hence we have the final equality. In other words $C = \mathcal{P}(1, m+1)$. $\square$

Consider the code C of the above theorem. It has minimum weight $q^{m-1} + \cdots + 1$ and the incidence vectors of the hyperplanes are amongst the minimum-weight vectors. Given any hyperplane and viewing such a hyperplane as the points at infinity, the code C projects onto the code of the design of points and $(m-1)$-flats of $AG_m(\mathbf{F}_q)$. Clearly, the dimension drops by precisely one, since the incidence vector of the hyperplane generates the kernel of the projection. Hence we have:

Corollary 5.7.1 *If $q = p^t$, the p-rank of the design of points and $(m-1)$-flats of the affine geometry $AG_m(\mathbf{F}_q)$ is*

$$\binom{m+p-1}{m}^t.$$

Moreover, the code over $\mathbf{F}_p$ of this design is $\mathcal{A}(q-1, m)$.

Proof: Only the "moreover" statement has not already been proved but, once again, we have an inclusion and a dimension argument suffices. $\square$

This discussion also shows that one of the short exact sequences of Theorem 5.3.2 holds, properly interpreted, for *any* finite field — at least when $r = m - 1$. In order to discuss the entire matter we introduce some temporary notation: let $\mathcal{A}_{r,m}$ denote the design of points and r-flats of

$AG_m(\mathbf{F}_q)$ and $\mathcal{P}_{r,m}$ denote the design of points and r-dimensional subspaces of $PG_m(\mathbf{F}_q)$. Note that we are deliberately suppressing the field in the notation: it will always be $\mathbf{F}_q$ where $q = p^t$ and p is a prime. Then the discussion above shows that the following short exact sequence arises from any embedding of $PG_{m-1}(\mathbf{F}_q)$ into $PG_m(\mathbf{F}_q)$:

$$0 \to C_p(\mathcal{P}_{m-1,m-1}) \to C_p(\mathcal{P}_{m-1,m}) \to C_p(\mathcal{A}_{m-1,m}) \to 0,$$

the kernel being simply $\mathbf{F}_p\jmath$.

Consider next the subcode $E_{r,m}$ of $C_p(\mathcal{A}_{r,m})$ generated by the differences of incidence vectors of parallel r-flats. Just as in the binary case $E_{r,m}$ is in the kernel of the projection of $C_p(\mathcal{P}_{r,m})$ onto the coordinates corresponding to the embedded $(m-1)$-dimensional projective space, the image of the projection being, just as in the binary case, $C_p(\mathcal{P}_{r-1,m-1})$. Observe that by using the $(q-1)$-to-1 map of $V - \{\mathbf{0}\}$ onto $PG_{m-1}(\mathbf{F}_q)$, where V is the m-dimensional vector space over $\mathbf{F}_q$ defining the projective space, we can pull the code $C_p(\mathcal{P}_{r-1,m-1})$ back to $\mathbf{F}_p^{V^*}$, where we are writing V^* for $V - \{\mathbf{0}\}$; this simply amounts to repeating each column $(q-1)$ times. By adjoining an overall parity check to this pull-back we get the code in $\mathbf{F}_p^V$ that is generated by the incidence vectors of the r-dimensional *subspaces* of V. Call this code $P_{r,m}$. Viewing $C_p(\mathcal{A}_{r,m})$ and $E_{r,m}$ in this same ambient space we have, clearly, that

$$E_{r,m} + P_{r,m} = C_p(\mathcal{A}_{r,m}).$$

This equation points to the reason why the binary case is so easy: when $q = 2$, $E_{r,m} \subseteq P_{r,m}$ and thus we need analyse only the projective geometry codes.

For the same reason as in the binary case, $P_{r+1,m} \subseteq P_{r,m}$ and, furthermore, $P_{r+1,m} \subseteq E_{r,m}$ since, if T is any $(r+1)$-dimensional subspace, S any r-dimensional subspace contained in it, and $\mathbf{v}$ is in T but not in S, then

$$-v^T = \sum_{a \in \mathbf{F}_q, a \neq 0} (v^S - v^{a\mathbf{v} + S}).$$

Letting $a_{r,m}$ be the p-rank of $\mathcal{A}_{r,m}$, $p_{r,m}$ be the p-rank of $\mathcal{P}_{r,m}$ and setting $e_{r,m} = \dim(E_{r,m})$, we have that $\dim(P_{r,m}) = p_{r-1,m-1}$ and that

$$p_{r-1,m-1} + e_{r,m} \geq a_{r,m} + p_{r,m-1} \tag{5.19}$$

since the intersection, $P_{r,m} \cap E_{r,m}$, contains $P_{r+1,m}$. Further, we have the following:

Lemma 5.7.1 *Given an embedding of $PG_{m-1}(\mathbf{F}_q)$ in $PG_m(\mathbf{F}_q)$, if the first of the following sequences,*

$$0 \to C_p(\mathcal{P}_{r,m-1}) \to C_p(\mathcal{P}_{r,m}) \to C_p(\mathcal{A}_{r,m}) \to 0$$

and

$$0 \to E_{r,m} \to C_p(\mathcal{P}_{r,m}) \to C_p(\mathcal{P}_{r-1,m-1}) \to 0,$$

that arise from the embedding is exact, then so is the second and, moreover, in that case we have

$$p_{r,m} = a_{r,m} + p_{r,m-1} \quad and \quad e_{r,m} + p_{r-1,m-1} = p_{r,m}.$$

Proof: Clearly, just as in the binary case, the sequences follow easily from the embedding, and we need only check that the kernels are as described. That the codes are contained in the kernels is obvious; thus in order to prove that they are the kernels we must check the dimensions. From the discussion preceding the lemma, in particular (5.19), and the second sequence, we have that

$$p_{r,m-1} + a_{r,m} \le p_{r-1,m-1} + e_{r,m} \le p_{r,m}$$

and the result follows since, if the first sequence is exact, $p_{r,m-1} + a_{r,m} = p_{r,m}$. $\square$

Applying the lemma to the case $r = m - 1$ we have the following:

Proposition 5.7.2 *The sequences*

$$0 \to C_p(\mathcal{P}_{m-1,m-1}) \to C_p(\mathcal{P}_{m-1,m}) \to C_p(\mathcal{A}_{m-1,m}) \to 0$$

and

$$0 \to E_{m-1,m} \to C_p(\mathcal{P}_{m-1,m}) \to C_p(\mathcal{P}_{m-2,m-1}) \to 0$$

are exact. In particular, the dimension of the code generated by the difference of the incidence vectors of parallel $(m-1)$-dimensional flats of $AG_m(\mathbf{F}_q)$ is

$$\binom{m+p-1}{m}^t - \binom{m+p-2}{m-1}^t.$$

Remark For the desarguesian affine plane, $AG_2(\mathbf{F}_q)$, the proposition says that the subcode of the code of the plane generated by the differences of the incidence vectors of parallel lines has codimension q in the code of the plane. This is true for arbitrary affine planes: see Theorem 6.3.2.

We next begin an investigation of the nature of the minimum-weight vectors in the codes of the projective designs. We first use the short exact sequences currently at our disposal to prove the following:

Proposition 5.7.3 *The minimum-weight vectors of the code over* $\mathbf{F}_p$ *of the design of points and hyperplanes of* $PG_m(\mathbf{F}_q)$, *where* q *is a power of the prime* p, *are the scalar multiples of the incidence vectors of these hyperplanes.*

Proof: The proof is by induction on m, but we first note that the result is classical for arbitrary projective planes of any order (see Chapter 6) and we could start the induction at $m = 2$. We prefer to start at $m = 1$ where the result is trivial since the code is the entire ambient space. The inductive step uses the exact sequences just as in the binary case.

Given a minimum-weight vector of the code of the design of points and hyperplanes of $PG_m(\mathbf{F}_q)$, if it is in the kernel of the projection onto the affine space it must be a scalar multiple of the incidence vector of the embedded projective space $PG_{m-1}(\mathbf{F}_q)$ and we are done. If it is in the kernel of the projection onto the embedded projective space, it is in $E_{m-1,m}$ and has weight at least $2q^{m-1}$, an impossibility since its weight is $q^{m-1} + \cdots + 1$ and, for any $q > 1$, an easy induction on n shows that $q^n > q^{n-1} + \cdots + 1$ for $n > 0$. Thus its projection on the embedded projective space has weight $q^{m-2} + \cdots + 1$ while its projection onto the affine space (the points not at infinity) has weight q^{m-1}. Further, by induction the projection onto the embedded projective space is a scalar multiple of a hyperplane of that space. Normalizing so that this scalar multiple is 1, construct a minimum-weight vector of the large code that is the incidence vector of a hyperplane whose projection on the embedded projective space is equal to the projection of the vector in question and that has its support intersecting the support of the projection of the minimum-weight vector on the affine space non-trivially. Subtracting these two vectors gives a vector in $E_{m-1,m}$ with weight less than $2q^{m-1}$, a contradiction unless it is the zero vector. $\square$

We next want to describe the minimum-weight vectors of the other projective designs and, in order to do that, we first establish the existence of the short exact sequences in the general case. We first remark once again that $C_p(\mathcal{P}_{r,m}) \subseteq \mathcal{P}(m-r, m+1)$ and that $C_p(\mathcal{A}_{r,m}) \subseteq \mathcal{A}((m-r)(q-1), m)$. We begin with a lemma.

Lemma 5.7.2 *We have the following recursion for the dimensions of the relevant generalized Reed-Muller codes:*

$$\dim \mathcal{P}(m - r - 1, m) + \dim \mathcal{A}((m - r)(q - 1), m) = \dim \mathcal{P}(m - r, m + 1).$$

Proof: Let Q, A and P be the sets of integers whose cardinalities give the dimensions of $\mathcal{P}(m - r - 1, m)$, $\mathcal{A}((m - r)(q - 1), m)$ and $\mathcal{P}(m - r, m + 1)$, respectively. Then Q is the set of integers satisfying

- $0 \le u \le q^m - 1$

- $(q-1)$ divides u

- $w_q(up^j) \le (m-r-1)(q-1)$ for $j = 0, 1, \ldots, t-1$

where up^j is computed modulo $q^m - 1$ and $q = p^t$. Similarly, A is the set of integers satisfying

- $0 \le u \le q^m - 1$

- $w_q(up^j) \le (m-r)(q-1)$ for $j = 0, 1, \ldots, t-1$

and P the set of integers satisfying

- $0 \le u \le q^{m+1} - 1$

- $q-1$ divides u

- $w_q(up^j) \le (m-r)(q-1)$ for $j = 0, \ldots, t-1$

with up^t computed modulo $q^{m+1} - 1$ for P. Divide P into the following two disjoint sets: Q', the set of those integers in P whose q-ary expansion has $u_m = q - 1$, and A', those integers in P whose q-ary expansion has $u_m < q - 1$. For $u \in Q'$ set $f(u) = u - (q-1)q^m$ and for $u = u_0 + \cdots + u_m q^m \in A'$ set $f(u) = u - u_m q^m$. The reader will have no difficulty seeing that were it not for the condition involving up^j we would have, via f, a one-to-one correspondence between Q' and Q and between A' and A — and thus have proved the result in the case of prime q. We next indicate how to deal with the complication posed by the added condition when q is not prime.

Recall that the up^j for P are computed modulo $q^{m+1} - 1$ and those for Q and A modulo $q^m - 1$. It is trivial to verify that, modulo $q^{m+1} - 1$, $p^j(q-1)q^m \equiv p^j - 1 + (q - p^j)q^m$ and hence that both q-weights are $q - 1$. Were it not for the "carries" one easily now sees that Q' and Q are in correspondence via f. The carries are dealt with by the fact that, for $0 \le u \le q - 1$, $p^j u q^i = r q^i + b q^{i+1}$ with $0 \le r \le q - p^j$ and $0 \le b < p^j$, a fact easily verified by writing, using the division algorithm, $u = bp^{t-j} + r'$ with $0 \le r' \le p^{t-j} - 1$ and noting that $b < p^j$ since $u < q$. The reader can now complete the proof as an exercise. $\square$

Lemma 5.7.2 is the analogue of the Pascal triangle equality in the binary case and the result plays a role in the following theorem analogous to that played by that equality in the binary case. Here then is the promised result and its inductive proof:

Theorem 5.7.2 *Let q be a power of the prime p. Let $\mathcal{K}$ be the design of points and r-dimensional subspaces of $PG_{m-1}(\mathbf{F}_q)$, $\mathcal{L}$ the design of points and $(r-1)$-dimensional subspaces of $PG_{m-1}(\mathbf{F}_q)$, $\mathcal{D}$ the design of points and r-dimensional subspaces of $PG_m(\mathbf{F}_q)$, $\mathcal{A}$ the design of points and r-flats of $AG_m(\mathbf{F}_q)$, and let E be the code over $\mathbf{F}_p$ generated by the differences of the incidence vectors of parallel r-flats of $AG_m(\mathbf{F}_q)$, chosen so that $E \subseteq C_p(\mathcal{A})$.*

Then any embedding of an $(m-1)$-dimensional projective space in the m-dimensional projective space yields the following short exact sequences:

$$0 \to C_p(\mathcal{K}) \to C_p(\mathcal{D}) \to C_p(\mathcal{A}) \to 0$$

and

$$0 \to E \to C_p(\mathcal{D}) \to C_p(\mathcal{L}) \to 0.$$

Moreover, $C_p(\mathcal{D}) = \mathcal{P}(m-r, m+1)$ and $C_p(\mathcal{A}) = \mathcal{A}((m-r)(q-1), m)$.

Proof: The proof will be by induction on m and will mirror the binary case. Observe first that for any m we have the result for $r = m-1$ by Theorem 5.7.1, Corollary 5.7.1 and Proposition 5.7.2. Since we must have $0 < r < m$, the induction can start at $m = 2$ where r must be 1 and we have the result. We assume that for all r we have that $C_p(\mathcal{A}) = \mathcal{A}((m-r)(q-1), m)$ and that for all $0 < r < m$ we have that $C_p(\mathcal{K}) = \mathcal{P}(m-r-1, m)$ and that the sequences are exact in projective dimensions less than m. Our task is to establish the result for projective dimension m and affine dimension $m+1$. Because of Lemma 5.7.1 we need only check that the first sequence is exact. Let W be the $(m+1)$-dimensional vector space over $\mathbf{F}_q$ defining $PG_m(\mathbf{F}_q)$ and let H be a hyperplane of W. Then H defines the projective space $PG_{m-1}(\mathbf{F}_q)$. Let $\overline{H}$ be any coset of H in W not equal to H. Now $\overline{H}$ is an affine space isomorphic to $AG_m(\mathbf{F}_q)$. Any $(r+1)$-dimensional subspace of W is either entirely in H or else intersects $\overline{H}$ in an r-flat; further, any r-flat of $\overline{H}$ is such an intersection. Thus we have a natural projection of $C_p(\mathcal{D})$ onto $C_p(\mathcal{A})$ and we need only show that the kernel is a subcode of $C_p(\mathcal{D})$ isomorphic to $C_p(\mathcal{K})$. As in the binary case this code, viewed via H, is clearly in the kernel and we only need check that the dimensions are correct.

We have, of course, that $r > 0$ and if $r = m-1$ we know the sequence is exact and that the theorem is true. Thus we may assume that $r < m-1$. The induction assumptions together with Lemma 5.7.2 show that the dimension of $C_p(\mathcal{D})$ is at least that of $\mathcal{P}(m-r, m+1)$ and hence we have equality and $C_p(\mathcal{D}) = \mathcal{P}(m-r, m+1)$. Thus the first sequence must now be exact. We still must show that the affine design in dimension $m+1$ has its code as announced. To do this we pull the code of the corresponding projective design back to $\mathbf{F}_p^W$ and note that taking the code over $\mathbf{F}_p$ generated by it and its translates gives on the one hand the code of the affine design and, using the polynomials, gives, on the other hand, $\mathcal{A}((m-r)(q-1), m+1)$. $\square$

Corollary 5.7.2 (Delsarte) *For $0 \le r \le m$, the code $\mathcal{A}((m-r)(q-1), m)$ is the code over $\mathbf{F}_p$ of the design of points and r-flats of the affine geometry $AG_m(\mathbf{F}_q)$.*

Corollary 5.7.3 *Let $p_{r,m}$ denote the p-rank of the design of points and r-dimensional subspaces of $PG_m(\mathbf{F}_q)$ and let $a_{r,m}$ denote the p-rank of the design of points and r-flats of $AG_m(\mathbf{F}_q)$ where, of course, q is a power of the prime p. Then*

$$p_{r,m} = p_{r,m-1} + a_{r,m}.$$

Corollary 5.7.4 *With the notation of the above corollary we have that the dimension of the code generated by the differences of incidence vectors of parallel r-flats of $AG_m(\mathbf{F}_q)$ has dimension $p_{r,m} - p_{r-1,m-1}$. In fact, the intersection of the code generated by the differences and the code generated by the incidence vectors of the r-dimensional subspaces of the affine space is the code generated by the incidence vectors of the $(r+1)$-dimensional subspaces.*

Proof: Since the sequences are exact we have by Lemma 5.7.1 that the dimension of the code E generated by the differences of the incidence vectors of parallel r-flats is $p_{r,m} - p_{r-1,m-1}$ and pulling everthing back to $\mathbf{F}_p^W$ as we did in the proof of Theorem 5.7.2 then gives the assertion about the intersection. $\square$

The nature of the minimum-weight vectors of the *projective* designs now follows easily by induction, just as in the binary case, but one must make use of the inequality $q^n > q^{n-1} + \cdots + 1$ just as we did in the proof of Proposition 5.7.3. Thus, we have also the following:

Corollary 5.7.5 *The minimum-weight vectors of the code over $\mathbf{F}_p$ generated by the r-dimensional subspaces of $PG_m(\mathbf{F}_q)$ are precisely the scalar multiples of these subspaces.*

Proof: Exercise for the reader. $\square$

We have not been able to see how to extract the same result for the affine case (which is rather trivial when $q = 2$). However, the result is true in the affine case and, moreover, the code generated by the difference of the incidence vectors of parallel r-flats has precisely these vectors and their scalar multiples as minimum-weight vectors. These results are due to Delsarte. In fact the minimum-weight vectors for *all* the generalized Reed-Muller codes are known. We shall shortly state that result — without giving the somewhat technical proof. We do want to remark, however, that were it not for our inability to extract the affine results from the projective we would have been able to imitate the easy binary results for the geometric designs.

In order to actually construct the subfield subcodes, the m-variable approach is once again the most straightforward. The subfield subcode case is, however, more complicated; it is described fully in [83]. Before describing the construction, we need some notation: if k satisfies $0 \leq k \leq q-1$, and $k = \sum_{i=0}^{t-1} k_i p^i$, where $0 \leq k_i \leq p-1$, then write

$$[pk] = k_{t-1} + k_0 p + \cdots + k_{t-2} p^{t-1} = pk - k_{t-1}(q-1), \qquad (5.20)$$

i.e.

$$[pk] = \begin{cases} pk \bmod (q-1) & \text{if } k < q-1 \\ q-1 & \text{if } k = q-1. \end{cases}$$

Further, write $[k] = k$.

Theorem 5.7.3 *For any ν such that $0 \leq \nu \leq m(q-1)$, the code*

$$\mathcal{A}_{F_q/F_p}(\nu, m)$$

consists of the following polynomial functions, in terms of the usual basis of characteristic functions on $\mathbf{F}_q^m$:

$$p(x_1, \ldots, x_m) = \sum_{l_1, \ldots, l_m} d(l_1, l_2, \ldots, l_m) x_1^{l_1} x_2^{l_2} \ldots x_m^{l_m},$$

where $0 \leq l_i \leq q-1$, $d(l_1, l_2, \ldots, l_m) \in \mathbf{F}_q$, and

(1) $\sum_{i=1}^m [p^j l_i] \leq \nu$, *for* $j = 0, 1, \ldots, t-1$;

(2) $d([p^j l_1], \ldots, [p^j l_m]) = (d(l_1, \ldots, l_m))^{p^j}$, *for* $j = 0, 1, \ldots, t-1$.

Example 5.7.1 Take $m = 2$ and $q = 4 = 2^2$. Thus $t = 2$ and $0 \leq \nu \leq 6$. Taking $\nu = 3$ will give, by Theorem 5.7.2, $\mathcal{A}(3, 2) = C_2(AG_2(\mathbf{F}_4))$. Then $V_3 = \{0, 1, 2, 3, 4, 6, 8, 9, 12\}$ and so $\dim(\mathcal{A}(3, 2)) = 9$. If ω is a primitive element for $E = \mathbf{F}_4$, a root of $X^2 + X + 1 = 0$, then polynomials that generate the code, according to Theorem 5.7.3, are $\{1, x_1^3, x_2^3, x_1 + x_1^2, x_2 + x_2^2, \omega x_1 + \omega^2 x_1^2, \omega x_2 + \omega^2 x_2^2, x_1 x_2^2 + x_1^2 x_2, \omega x_1 x_2^2 + \omega^2 x_1^2 x_2\}$. A generator matrix from these polynomials can be constructed, and the entries are all, of course, in $\mathbf{F}_2$. For example, if $K = \mathbf{F}_{16}$ is constructed from E using the primitive polynomial $X^2 + \omega X + \omega$ and a is a primitive root of this, then ordering the vectors of E^2 in the usual way, i.e. $\mathbf{0}, 1, a, a^2, \ldots, a^{14}$ (corresponding, of course, to the action of all the powers of the companion matrix

$$S = \begin{pmatrix} 0 & \omega \\ 1 & \omega \end{pmatrix}$$

on $\mathbf{e}$) then the codeword obtained from the polynomial $\omega x_2 + \omega^2 x_2^2$ is

$$(0, 0, 1, 1, 1, 1, 0, 1, 0, 1, 1, 0, 0, 1, 0, 0).$$

Example 5.7.2 Let $\mathcal{D}_r$ be the design of points and r-dimensional subspaces and let $\mathcal{D}_{m-r}$ be the design of points and $(m-r)$-dimensional subspaces of $PG_m(\mathbf{F}_q)$. Then, as we have seen,

$$\begin{aligned} \mathcal{P}(m-r, m+1) &= C_p(\mathcal{D}_r), \text{ and} \\ \mathcal{P}(r, m+1) &= C_p(\mathcal{D}_{m-r}). \end{aligned}$$

By the dimension theorem for subspaces, an r-space will meet an $(m-r)$-space in at least a point, and thus the number of points of the intersection will always be congruent to 1 modulo p. Thus for any two blocks S and T of $\mathcal{D}_{m-r}$, $v^S - v^T \in C_p(\mathcal{D}_r)^\perp$. Hence (see Definition 2.4.4), $E_p(\mathcal{D}_{m-r}) \subseteq C_p(\mathcal{D}_r)^\perp$ and $E_p(\mathcal{D}_r) \subseteq C_p(\mathcal{D}_{m-r})^\perp$. Since $\jmath \in C_p(\mathcal{D}_r)$ for any r, it follows that

$$E_p(\mathcal{D}_r) = C_p(\mathcal{D}_r) \cap \langle \jmath \rangle^\perp \supseteq H_p(\mathcal{D}_r) = C_p(\mathcal{D}_r) \cap C_p(\mathcal{D}_r)^\perp.$$

If, further, $r \geq m - r$, then $C_p(\mathcal{D}_r) \subseteq C_p(\mathcal{D}_{m-r})$, and it follows that $E_p(\mathcal{D}_r) = H_p(\mathcal{D}_r) \subseteq E_p(\mathcal{D}_{m-r})$. In particular, if $r = m-1$, then $\mathcal{D}_{m-1}$ is the symmetric design of points and hyperplanes and $\mathcal{D}_1$ is the design of points and lines.

The codes $\mathcal{P}(r, m)$ can be constructed in a manner analogous to the primitive case as in Theorem 5.7.3. With the added condition that $(q-1)$ divides $\sum_i l_i$, the codewords are given by the first $n = (q^m - 1)/(q-1)$ coordinates, as in Definition 5.6.1.

Example 5.7.3 For $m = 3$, $q = 4$, $n = (4^3 - 1)/(4-1) = 21$, and $r = 1$, i.e. $\nu = 3$, will produce $\mathcal{P}(1,3)$ as the binary code of the projective plane of order 4, $PG_2(\mathbf{F}_4)$. If ω is a primitive element for $E = \mathbf{F}_4$, then the polynomials that generate (over $\mathbf{F}_2$) $\mathcal{P}(1,3)$ are $\{1, x_1^3, x_2^3, x_3^3, x_1 x_2^2 + x_1^2 x_2, \omega x_1 x_2^2 + \omega^2 x_1^2 x_2, x_2 x_3^2 + x_2^2 x_3, \omega x_2 x_3^2 + \omega^2 x_2^2 x_3, x_3 x_1^2 + x_3^2 x_1, \omega x_3 x_1^2 + \omega^2 x_3^2 x_1\}$. The dimension is thus 10 and a generator matrix is given by the codewords corresponding to each of these ten polynomials $p(x_1, \ldots, x_m)$, i.e.

$$(p(e^t), p((Se)^t), \ldots, p((S^{20}e)^t)),$$

where $\mathbf{e} = (1, 0, 0)^t$ and

$$S = \begin{pmatrix} 0 & 0 & \omega \\ 1 & 0 & \omega \\ 0 & 1 & \omega^2 \end{pmatrix},$$

taking $X^3 + \omega^2 X^2 + \omega X + \omega$ for the generating polynomial of $\mathbf{F}_{64}$ over $\mathbf{F}_4$. For example, the codeword corresponding to $p(x_1, \ldots, x_m) = \omega x_1 x_2^2 + \omega^2 x_1^2 x_2$ is

$$(0, 0, 0, 1, 1, 0, 1, 0, 0, 1, 1, 0, 0, 1, 0, 0, 1, 0, 1, 0, 0).$$

This is the vector $v^L - v^M$ where L and M are lines, $L = \{3, 4, 7, 17, 19\} = (1, 1, 0)^t$, and $M = \{3, 5, 10, 11, 14\} = (1, \omega, 0)^t$, where the points are labelled $1, 2, \ldots, 21$ in the order given.

Here, finally, is the most general result, special cases of which we have shown to be true. This important result of Delsarte, Goethals and Mac-Williams — proved in [83, Theorem 2.6.3, Appendix] — yields *all* the minimum-weight vectors via the group in question and those of Theorem 5.5.3. It has a technical proof; we will merely state it here.

Theorem 5.7.4 *All minimum-weight vectors of $\mathcal{R}_{F_q}(\nu, m)$ can be obtained, under the action of $AGL_m(\mathbf{F}_q)$, from scalar multiples of those described in Theorem 5.5.3.*

An important special case arises when $\nu = r(q - 1)$, for then a theorem of Delsarte [81] asserts that the code $\mathcal{A}(\nu, m)$ is spanned by the incidence vectors of the $(m - r)$-flats. Thus, by the theorem above, they give the minimum-weight vectors. We are now able to say something about the nature of the minimum-weight vectors of the orthogonal code.

Theorem 5.7.5 *If $\nu = r(q - 1)$ then the minimum weight of $\mathcal{R}_{F_q}(\nu, m)^{\perp}$ is $2q^r$ and the minimum-weight vectors are the scalar multiples of the differences of the incidence vectors of two parallel r-flats, i.e. two r-flats that are distinct cosets of the same subspace.*

Proof: If $\nu = r(q - 1)$, then $\mu = m(q - 1) - \nu - 1 = (m - r)(q - 1) - 1 = (m - r - 1)(q - 1) + (q - 2)$, expressing it in the form required to use Theorem 5.5.3. Thus $\mathcal{R}_{F_q}(\mu, m)$ contains the incidence vector of every $(r + 1)$-flat and has minimum weight $2q^r$.

The minimum-weight vectors are made up of a linear combination of two parallel r-flats and it is a simple matter to show that, in this case, if the coefficient of one of the flats is c, then the coefficient of the other is $-c$. The quoted Theorem 5.7.4 then completes the proof. $\square$

We also wish to state another result which we will need but have not proved. It is crucial to the development in Chapter 6 and is a consequence of the main result above.

Theorem 5.7.6 *For $0 \le r \le (m-1)$, and $\nu = r(q-1) + (q-2) = (r+1)(q-1)-1$, the subfield subcode $\mathcal{A}(\nu, m)$ has minimum weight $2q^{m-r-1}$ and the minimum-weight vectors can be obtained from the differences of the incidence vectors of two parallel $(m-r-1)$-flats, i.e. two distinct cosets of the same subspace. Further,*

$$\mathcal{A}((r+1)(q-1)-1, m) = \left\langle v^M - v^N | M, N \text{ parallel } (m-r-1)\text{-flats in } V \right\rangle.$$

The minimum weight of the orthogonal code is q^{r+1}.

Exercise 5.7.1 Show that the supports of the minimum-weight vectors of $\mathcal{R}_{F_3}(2(m-1), m)$ form a Steiner triple system, i.e. a design with parameters 2-$(3^m, 3, 1)$, and that the design has a doubly-transitive automorphism group.

The interpretation of the generalized Reed-Muller codes as the codes of designs from geometries raises the question of the appearance of codewords corresponding to the incidence vectors of geometrical configurations in the geometry. In particular, hermitian varieties appear in this way, which is a question we will look at again in Chapter 6 with respect to projective planes. The following result is in Key [158]:

Theorem 5.7.7 *Let $\mathcal{D}$ denote the symmetric design of points and hyperplanes of the projective geometry $PG_{m-1}(\mathbf{F}_{q^2})$, where $q = p^t$, and p is a prime. Then $C_p(\mathcal{D})$ contains the incidence vector of any hermitian variety in the geometry.*

Proof: From Section 3.7, (3.3), a hermitian variety H is given as the set of points P with homogeneous coordinates x_i satisfying the equation

$$\sum_{i=1}^{m} X_i^{q+1} = 0.$$

If $v^H \in C = C_p(\mathcal{D})$ then the same is true for *any* hermitian variety, since the collineation group $P\Gamma L_m(\mathbf{F}_{q^2})$ is transitive on the varieties (see Chapter 3 or, for a more detailed discussion, [85, Section 1.4]). By Delsarte's results, we need only show that $v^H \in \mathcal{P}(1, m)$. Thus, we need to find a polynomial to represent v^H. Consider

$$p(x_1, \ldots, x_m) = 1 - \left(\sum_{i=1}^{m} x_i^{q+1} \right)^{q-1}. \tag{5.21}$$

Then $p(x_1, \ldots, x_m) = 1$ precisely when $\mathbf{x} = (x_1, x_2, \ldots, x_m)$ gives the homogeneous coordinates of an absolute point of the unitary polarity, and $p(x_1, \ldots, x_m) = 0$ elsewhere. Further, $p(x_1, \ldots, x_m)$ is given in terms of monomial functions in the non-primitive generalized Reed-Muller code $\mathcal{R}_{F_{q^2}}^{q^2-1}(q^2 - 1, m)$, since each monomial has degree precisely $q^2 - 1$, and, since it gives a vector in the code with all entries in the subfield $\mathbf{F}_p$ (the entries being 0 and 1 only), the incidence vector of length $(q^{2m} - 1)/(q^2 - 1)$ that it defines is in the subfield code, $\mathcal{P}(1, m) = C$. $\square$

Remark (1) The number of points in the hermitian variety H, i.e. the weight of the codeword v^H in $C_p(\mathcal{D})$ of the theorem, is

$$\frac{(q^m + (-1)^{m-1})(q^{m-1} - (-1)^{m-1})}{(q^2 - 1)}.$$

See, for example, Hirschfeld [136, Theorem 5.1.8, p.102] for a proof of this.

(2) For any r such that $1 \le r \le m$, the polynomial

$$p(x_1, \ldots, x_m) = 1 - \left(\sum_{i=1}^{r} x_i^{q+1} \right)^{q-1}$$

will also be in the code of the design, for the same reason.

In order to do the following exercises the reader may wish to employ Lucas's theorem which, for the convenience of the reader, we now state.

Theorem 5.7.8 (Lucas) *Let p be a prime. Suppose the p-ary representations of n and k are given by $n = \sum_{i=0}^{t} n_i p^i$ and $k = \sum_{i=0}^{t} k_i p^i$, where $0 \le n_i \le p - 1$ and $0 \le k_i \le p - 1$, then*

$$\binom{n}{k} \equiv \prod_{i=0}^{t} \binom{n_i}{k_i} \bmod p.$$

Exercise 5.7.2 (1) Show that all the monomials that occur in the expansion of the equation (5.21) above do satisfy the requirements of Theorem 5.7.3. For this note that the coefficient of the term $x_1^{r_1(q+1)} x_2^{r_2(q+1)} \ldots x_m^{r_m(q+1)}$ in the expansion is

$$\binom{q-1}{r_1} \binom{q-1-r_1}{r_2} \ldots \binom{q-1-r_1-r_2\cdots-r_{m-1}}{r_m},$$

where $\sum_{i=1}^m r_i = q-1$ and then use Lucas's theorem.

(2) Show that the incidence vector of any quadric (see Section 3.7) in $PG_m(\mathbf{F}_q)$ is in the code over $\mathbf{F}_p$ (where q is a power of p) of the design of points and $(m-2)$-dimensional subspaces of $PG_m(\mathbf{F}_q)$. In particular, the incidence vector of an ovoid defined by an elliptic quadric will be in the code of the design of points and lines of $PG_3(\mathbf{F}_q)$.

Note: Bagchi and Sastry [27] use their result Corollary 4.6.1 applied to certain classical generalized quadrangles to show also that the incidence vector of a Tits ovoid is in the binary code of the design of points and lines of $PG_3(\mathbf{F}_q)$ where $q = 2^{2m+1}$ with $m \geq 1$.

Finally, we summarize the results identifying the generalized Reed-Muller codes with the codes of the designs from geometries, in analogy with Theorem 5.3.3. Notice that everything stated below is true also for $p = 2$, but that Theorem 5.3.3 gives more precise information in that case. In the statement below $w_q(u)$ denotes the q-weight of an integer u, and is defined in Definition 5.5.1, page 160.

Theorem 5.7.9 *Let m be any positive integer, $q = p^t$ where p is a prime, and let $0 \leq r \leq m$.*

(1) *The code over $\mathbf{F}_p$ of the design of points and r-flats in the affine geometry $AG_m(\mathbf{F}_q)$, is $\mathcal{A}((m-r)(q-1),m)$. It has minimum weight q^r and the minimum-weight vectors are the multiples of the incidence vectors of the r-flats. The p-rank is given by the cardinality of the set of integers u satisfying*

- $0 \leq u \leq q^m - 1$
- $w_q(up^j) \leq (m-r)(q-1), j = 0, 1, \ldots, t-1$

where up^j is reduced modulo $q^m - 1$. The orthogonal code satisfies

$$\mathcal{A}((m-r)(q-1),m)^\perp \supseteq \mathcal{A}(r(q-1)-1,m)$$

and

$$\mathcal{A}(r(q-1)-1,m) = \langle v^M - v^N | M, N \text{ parallel } (m-r)\text{-flats in } V\rangle.$$

This latter code has minimum weight $2q^{m-r}$ with minimum-weight vectors multiples of the difference of the incidence vectors of two parallel $(m-r)$-flats . The minimum weight, $d^{\perp}_{(m-r)(q-1)}$, of the orthogonal code satisfies

$$(q+p)q^{m-r-1} \leq d^{\perp}_{(m-r)(q-1)} \leq 2q^{m-r}.$$

When $q = p$, $\mathcal{A}((m-r)(q-1),m)^{\perp} = \mathcal{A}(r(q-1)-1,m)$.

(2) The code over $\mathbf{F}_p$ of the design of points and r-dimensional subspaces of the projective geometry $PG_m(\mathbf{F}_q)$ is $\mathcal{P}(m-r,m+1)$. It has minimum weight $(q^{r+1}-1)/(q-1)$ and the minimum-weight vectors are the multiples of the incidence vectors of the blocks. The p-rank is given by the cardinality of the set of integers u satisfying

- $0 \leq u \leq q^{m+1}-1$
- $(q-1)$ *divides u*
- $w_q(up^j) \leq (m-r)(q-1), j = 0,1,\ldots,t-1$

where up^j is reduced modulo $q^{m+1}-1$. The orthogonal code satisfies

$$\mathcal{P}(m-r,m+1)^{\perp} \supseteq \mathcal{P}(r,m+1) \cap \langle \jmath \rangle^{\perp}$$

and has minimum weight at least $(q^{m-r+1}-1)/(q-1)+1$. If $q = p$ then $\mathcal{P}(m-r,m+1)^{\perp} = \mathcal{P}(r,m+1) \cap \langle \jmath \rangle^{\perp}$.

In particular, the code over $\mathbf{F}_p$ of the design of points and hyperplanes in the affine geometry $AG_m(\mathbf{F}_q)$ is $\mathcal{A}(q-1,m)$ and the code over $\mathbf{F}_p$ of the design of points and hyperplanes of the projective geometry $PG_m(\mathbf{F}_q)$ is $\mathcal{P}(1,m+1)$.

Note: The designs formed from affine or projective geometries may happen to have orders divisible by primes other than the characteristic prime for the geometry. The codes for such primes will not be of any interest — a result that follows from work of Mortimer [219] on the modular representations of doubly-transitive groups.

5.8 Summation formulas for the p-rank

The dimensions of the generalized Reed-Muller codes are given in the preceding section in terms of the number of integers with q-weight satisfying certain properties. From now on we are concerned only with the subfield codes over the prime field, and for these, with $q = p^t$ and all the notation of the preceding sections, we have that $\dim(\mathcal{A}(\nu, m))$ is

$$|\{u | 0 \leq u \leq q^m - 1; \mathrm{w}_q(up^j) \leq \nu, j = 0, 1, \ldots, t - 1\}|,$$

and, for b dividing $(q^m - 1)$, $\dim(\mathcal{A}^b(\nu, m)^*)$ is

$$|\{u | 0 \leq u \leq q^m - 1; b \text{ divides } u; \mathrm{w}_q(up^j) \leq \nu, j = 0, 1, \ldots, t - 1\}|,$$

where up^j is reduced modulo $q^m - 1$ in both cases. There seems to be no simple summation formula that will cover all possibilities (except in the case of the Reed-Muller codes, where $q = p = 2$) even if attention is restricted to codes from geometries, i.e. $\nu = r(q - 1)$ and $b = q - 1$. For these latter cases, Hamada [124, 125] found a multiple summation formula:

Theorem 5.8.1 (Hamada) *Let $q = p^t$ and let $\mathcal{D}$ denote the design of points and r-dimensional subspaces of the projective geometry $PG_m(\mathbf{F}_q)$, where $0 \leq r \leq m$. Then the p-rank of $\mathcal{D}$ is given by*

$$p_{r,m} = \sum_{s_0} \cdots \sum_{s_{t-1}} \prod_{j=0}^{t-1} \sum_{i=0}^{L(s_{j+1}, s_j)} (-1)^i \binom{m+1}{i} \binom{m + s_{j+1}p - s_j - ip}{m},$$

where $s_t = s_0$ and summations are taken over all integers s_j (for $j = 0, 1, \ldots, t - 1$) such that

$$r + 1 \leq s_j \leq m + 1, \text{ and } 0 \leq s_{j+1}p - s_j \leq (m + 1)(p - 1),$$

and

$$L(s_{j+1}, s_j) = \lfloor \frac{s_{j+1}p - s_j}{p} \rfloor,$$

i.e. the greatest integer not exceeding $(s_{j+1}p - s_j)/p$.

This formula is deduced in [124, 125]. It simplifies in certain cases, in particular in the case of designs of points and hyperplanes, when the formula becomes that found earlier by Graham and MacWilliams [109] for planes, and by Goethals and Delsarte [105], MacWilliams and Mann [202], and Smith [267], for general m. It becomes in that case, as we have already remarked in Theorem 5.7.1,

$$\dim(\mathcal{P}(1, m + 1)) = \binom{m + p - 1}{m}^t + 1.$$

Another particular case that will turn out to be useful in results on translation planes (Chapter 6) is the following from Key and Mackenzie [160]:

Theorem 5.8.2 *If $\mathcal{D}$ is the design of points and m-flats in $AG_{2m}(\mathbf{F}_p)$, where p is a prime, then the p-rank of $\mathcal{D}$ is given by*

$$\dim\left(\mathcal{A}(m(p-1), 2m)\right) = \text{rank}_p(\mathcal{D}) = \sum_{i=0}^{m-1}(-1)^i\binom{2m}{i}\binom{m+(m-i)p}{2m}.$$

Corollary 5.8.1 *For any $m \geq 1$,*

$$2^{2m-1} + \frac{1}{2}\binom{2m}{m} = \sum_{i=0}^{m-1}(-1)^i\binom{2m}{i}\binom{3m-2i}{2m}.$$

Proof: This follows immediately from the theorem, putting $p = 2$, and noting that in this case we simply have a Reed-Muller code, $\mathcal{R}(2m, m)$, whose dimension is given by the left-hand side of the identity. $\square$

Note: The identity deduced in the corollary has subsequently been checked by Zeilberger and Wilf using the "snake oil" method described in Wilf [298].

There are also certain particular cases amongst these classes of designs where simpler arguments might give the p-rank or even a basis in terms of incidence vectors for the p-ary code. In [27], Bagchi and Sastry have produced a very simple proof of the result giving the dimension of the binary code of the design of points and planes in $PG_3(\mathbf{F}_{2^t})$, by finding a set of planes whose incidence vectors form a basis:

Theorem 5.8.3 (Bagchi and Sastry) *Let $\mathcal{D}$ be the design of points and planes in $PG_3(\mathbf{F}_{2^t})$ and let $\mathcal{O}$ be an ovoid in $PG_3(\mathbf{F}_{2^t})$. Then the incident vectors of the tangent planes to the ovoid form a basis for $C_2(\mathcal{D})$. It follows that $\dim(C_2(\mathcal{D})) = 2^{2t} + 1$.*

Proof: Let $\mathcal{O}$ be any ovoid in $PG_3(\mathbf{F}_{2^t})$, not necessarily a quadric. For any $X \in \mathcal{O}$, let H_X denote the tangent plane at X. The claim is that $\mathcal{H} = \{v^{H_X} | X \in \mathcal{O}\}$ is a basis for $C_2(\mathcal{D})$. It is immediate that the set $\mathcal{H}$ is linearly independent and so all that is needed is to show that $\mathcal{H}$ spans $C_2(\mathcal{D})$. We prove that, if H is any plane, then

$$v^H = \sum_{X \in \mathcal{O} \cap H} v^{H_X}. \tag{5.22}$$

Any plane in $PG_3(\mathbf{F}_q)$ is either a tangent plane or meets the ovoid in a $(q+1)$-arc (see, for example, Hirschfeld [137, Section 16.1]), the latter type being called a secant plane. For any point $P \notin \mathcal{O}$ of $PG_3(\mathbf{F}_{2^t})$, denote by $\text{Star}(P)$ the union of all tangent lines to the ovoid that pass through P. Then $\text{Star}(P)$ is a plane (again see [137]) and $\text{Star}(X) = H_X$ for $X \in \mathcal{O}$.

Now let H be a secant plane. Set $\mathcal{C} = H \cap \mathcal{O}$ and let K be the nucleus of the $(q+1)$-arc, $\mathcal{C}$. Then $H = \text{Star}(K)$. Notice that for $X \in \mathcal{O}$, a point Q is on $H_X = \text{Star}(X)$ if and only if $X \in \text{Star}(Q)$. Considering the right-hand side of equation (5.22), a non-zero entry will appear at the coordinate position of a point P if P is on an odd number of the planes H_X for $X \in \mathcal{C}$. Since $P \in H_X$ if and only if $X \in \text{Star}(P)$, and $\text{Star}(P)$ meets $\mathcal{C}$ in $0, 1, 2$ or $2^t + 1$ points, the result will follow if the 1 or $2^t + 1$ occur if and only if $P \in H$. Now $\text{Star}(P)$ meets $\mathcal{C}$ in $0, 1, 2$ or $2^t + 1$ points, according to whether (i) $P \notin H$ and the line joining P and K is a secant to $\mathcal{O}$; (ii) $P \in H$, $P \neq K$, and PK is a tangent to $\mathcal{C}$; (iii) $P \notin H$ and the line joining P and K is exterior to $\mathcal{O}$; or (iv) $P = K$. This proves the assertion. $\square$

Another result which describes an explicit basis of incidence vectors of lines in the affine plane $AG_2(\mathbf{F}_p)$, where p is a prime, has been obtained by Moorhouse [218], using k-*nets*. In this case the dimension of the code over $\mathbf{F}_p$ is

$$\binom{p+1}{2} = \sum_{i=1}^{p} i$$

and a basis can be had by ordering, arbitrarily, any p of the $p+1$ parallel classes and taking one line from the first, two from the second, etc., even the choices of the lines being made arbitrarily; the basis consists of the incidence vectors of the selected lines.

It is rather easy to see, for an affine plane of any order and for any prime dividing that order, that, for any line ℓ, its incidence vector, v^ℓ is

$$\jmath - \sum_{P \in m, m \neq \ell} v^m,$$

where P is any fixed point of ℓ, and this explains why at least one parallel class can be discarded when choosing a basis from the incidence vectors of the lines.

Below is a dictionary that will allow the reader to translate the notation of [83] into that used here:

$$
\begin{aligned}
C_\nu(m,q) &= \mathcal{R}_{F_q}(\nu,m) \\
HC_\nu(m,q) &= \mathcal{R}_{F_q}(\nu,m)^* \\
HC_\nu^b(m,q) &= \mathcal{R}_{F_q}^b(\nu,m)^* \\
B_\nu(m,q) &= \mathcal{A}_{F_q/F_p}(\nu,m) \\
HB_\nu^b(m,q) &= \mathcal{A}_{F_q/F_p}^b(\nu,m)^*.
\end{aligned}
$$

When there is no possibility of confusion we will omit the subscript, F_q/F_p, and simply write $\mathcal{A}(\nu,m)$ and $\mathcal{A}^b(\nu,m)^*$. Moreover, when ν is a multiple of $q-1$ and $b = q-1$, we set

$$
\mathcal{P}(r,m) = \mathcal{A}^{q-1}(r(q-1),m)^*.
$$

Chapter 6

Codes from planes

6.1 Introduction

We have already met some projective planes in their roles as symmetric designs with $\lambda = 1$, i.e. designs with parameters $(n^2 + n + 1, n + 1, 1)$. Moreover, we have constructed the classical projective planes, $PG_2(F)$, F being a field; in fact, F could have been a division ring. But, historically, they arose synthetically or, in other words, via an axiomatic treatment. The undefined terms were "point", "line" and "incident" and, besides an axiom designed to eliminate trivial cases[1], the axioms were simply that *any two distinct points are incident with a unique line* and that *any two distinct lines are incident with a unique point*. These axioms are duals of one another in the sense that interchanging "point" and "line" interchanges the two axioms. The **principle of duality** then states that in any theorem concerning projective planes, the statement obtained by interchanging the words "point" and "line" is also a theorem concerning projective planes.

Likewise, we have met some affine planes, i.e. designs with parameters $2\text{-}(n^2, n, 1)$. These also arose synthetically, the undefined terms being the same, but the axioms less elegant. Again leaving aside the axiom eliminating trivial cases, the axioms stated that *any two distinct points are incident with a unique line*, but the dual axiom was replaced by an axiom which we will specify shortly. It was the inelegance that arose from the lack of duality (among other things) that led geometers to treat projective planes with more respect and to go (usually) directly to the projective completion when confronted with an affine plane. One notable exception to the preference

---

[1] The axiom requires the existence of a quadrangle in a projective plane, and of a triangle in an affine plane.

for projective over affine was amongst those mathematicians interested in translation planes. The coding-theoretical approach we adopt here makes it quite clear why this was so.

In an affine plane two distinct lines are incident with *at most* one point; two distinct lines are said to be **parallel** if there is no point incident with both. The famous **parallel postulate** that *given a line and a point not incident with it there is a unique line incident with the given point and parallel to the given line* replaces the other projective axiom. Allowing a line to be parallel to itself, parallelism is then an equivalence relation on the set of all lines, defining distinct parallel classes.

To complete a given affine plane to a projective plane, "points at infinity" must be adjoined, and this is done by constructing a new point for each parallel class. One new line, the "line at infinity", is added having only the new points incident with it. Each new point is defined to be incident not only with the new line, but also with each of the lines in the parallel class that defines it. Conversely, removing a line together with its points from a projective plane yields an affine plane; in fact, this process gave rise to the notion of a residual design.

The synthetic treatment, when cast in set-theoretical language, yields the incidence structures of projective and affine planes and a *finite projective plane* or a *finite affine plane* is one whose point set is finite. Elementary arguments then show that the designs obtained are those that we have already met.

We will have no need here for an extensive synthetic development and we will view a **projective plane of order** n simply as a symmetric (v, k, λ) design with $\lambda = 1$, and $n = k - 1$. It is thus any 2-$(n^2 + n + 1, n + 1, 1)$ design where $n \geq 2$. The trivial cases with $n = 0$ or $n = 1$ are excluded from this definition, but we can extend the definition in a natural way to include these *degenerate planes*. Thus, the case $n = 0$ would give a $(1, 1, 1)$ design: we take the complement of this design to be the projective plane of order 0, i.e. the plane with one point, one line, and no incidences. A projective plane of order $n = 1$ is a $(3, 2, 1)$ design, i.e. a triangle.

Similarly, a 2-$(n^2, n, 1)$ design (with $n > 1$) will be an **affine plane of order** n. Should we want to consider affine planes of orders $n = 0$ and $n = 1$, and preserve the passage from a projective plane to an affine plane by discarding a line and the points on it, an affine plane of order 0 will consist of one point and no lines while an affine plane of order 1 will consist of one point and two lines, each containing the solitary point. The reader may find it amusing to check that we are forced, by duality and the felt need to preserve the passages from projective to affine and back, to insist on the above definitions at orders 0 and 1. See Figure 6.1 for a pictorial representation of these degenerate planes. Having said this, we

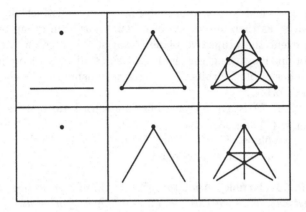

Figure 6.1: Planes of orders 0, 1 and 2

will, however, from now on assume that we are dealing with *non-trivial* designs, i.e. we will assume our planes have order at least 2.

6.2   Projective to affine and back

Given a projective plane Π and a line L of Π, the residual design $\pi = \Pi^L$ (see Definition 1.3.4) is easily seen to be an affine plane whose parallel classes are in one-to-one correspondence with the points of the line L. We call π an **affine part** of Π. A projective plane has as many affine parts as it has lines. Given an affine part π, the projective plane is easily reconstructed by the process indicated in Section 6.1 and described precisely below; it is uniquely determined by the affine part. This implies immediately that if two affine parts, $\pi = \Pi^L$ and $\sigma = \Pi^M$, are isomorphic designs, then the projective plane Π has an automorphism that carries the line L to the line M and "extends" the given isomorphism of the two affine parts. Conversely, if there is an automorphism carrying the line L to the line M, then the respective affine parts are isomorphic affine planes.

The projective plane obtained from an affine plane π is called the **projective completion** of π. To be precise: given $\pi = (\mathcal{P}, \mathcal{L}, \mathcal{I})$, the set $\mathcal{E}$ of parallel classes is taken to be the set of new points, and one new line, L_∞, is introduced. The projective completion $\Pi = (\tilde{\mathcal{P}}, \tilde{\mathcal{L}}, \tilde{\mathcal{I}})$, has $\tilde{\mathcal{P}} = \mathcal{P} \cup \mathcal{E}$, $\tilde{\mathcal{L}} = \mathcal{L} \cup \{L_\infty\}$ and $\mathcal{I}$ is augmented by including all pairs (E, L_∞) for $E \in \mathcal{E}$ and all pairs (E, ℓ) for $\ell \in E \in \mathcal{E}$.

Thus every affine plane yields a unique projective completion and a projective plane yields, up to isomorphism, as many distinct affine planes (as affine parts) as there are orbits of its automorphism group on its line set. In the event the projective plane is *not* self-dual twice this number could be obtained by using the dual also, since Block's theorem ensures that the number of block orbits of the automorphism group is equal to the number of line orbits.

If $\Pi = (\mathcal{P}, \mathcal{L}, \mathcal{I})$ is a projective plane, a **subplane** of Π is any substructure $\Pi_0 = (\mathcal{P}_0, \mathcal{L}_0, \mathcal{I}_0)$, where $\mathcal{P}_0 \subseteq \mathcal{P}$, $\mathcal{L}_0 \subseteq \mathcal{L}$ and $\mathcal{I}_0 \subseteq \mathcal{I}$, that is also a projective plane. The following theorem, due to Bruck, restricts the possibilities for the order of a subplane.

Theorem 6.2.1 (Bruck) *A proper subplane Π_0 of a projective plane Π of order n must have order m satisfying $n = m^2$ or $n \geq m^2 + m$.*

Proof: If L is a line of Π_0, then L has $(m+1)$ points of Π_0 and $(n-m)$ points outside the subplane. Since any two lines of Π_0 intersect in the subplane, and Π_0 has $m^2 + m + 1$ lines, there are $(m^2 + m + 1)(n-m)$ points outside of Π_0 that are incident with a line of Π_0. Thus

$$n^2 + n + 1 \geq m^2 + m + 1 + (m^2 + m + 1)(n - m)$$

and hence

$$0 \geq (m^2 - n)(n - m).$$

Since $n > m$, this implies that $m^2 \leq n$. If equality holds here, all the inequalities above are equalities and thus every point of Π is incident with a line of Π_0.

Suppose $m^2 < n$. Then there is a point P of Π that is not on a line of the subplane. Thus every line through P contains at most one point of Π_0. Since every point of Π_0 is on some line through P, the total number of lines through P is at least as great as the number of points of Π_0, i.e. $n + 1 \geq m^2 + m + 1$, or $n \geq m^2 + m$. $\square$

In fact no instances of a subplane with $n = m^2 + m$ are known. In the case $n = m^2$, it was noted in the proof that every point of Π is incident with a line of Π_0, and, arguing dually, every line of Π is incident with a point of Π_0: such a subset of elements of a plane is called a **Baer subset** and, in case it is also a subplane, it is called a **Baer subplane**. Passing to the affine plane $\pi = \Pi^L$, the point set of a subplane might meet the line L in a line of the subplane, a point, or not at all. In the first case the restriction of the subplane to π is an **affine subplane**, and in the last, a **projective subplane**. An affine subplane must then be a 2-$(m^2, m, 1)$ design, where $m^2 = n$ in the case of an affine Baer subplane.

Example 6.2.1 If $K = \mathbf{F}_q$, where $q = p^t$, then, if s divides t and $F = \mathbf{F}_{p^s}$, the plane $PG_2(F)$ of order p^s is embedded as a subplane inside the plane $PG_2(K)$ of order p^t in the natural way. If $t \neq 2s$, then $p^t > p^s(p^s + 1)$.

Not only is it true that no instances of a subplane of order m in a plane of order $m^2 + m$ are known to exist, but there are no known planes of orders that are not powers of a prime. Because of the recent work of Lam *et al.* the first undecided case of "composite order" is, at present, $n = 12$ and, indeed, a projective plane of order 12 might contain a subplane of order 3. It is quite easy to show that the planes of orders 0 through 5 are unique (i.e. they are the degenerate planes and the classical planes coming from the appropriate finite fields). It is harder to show that there is but one plane of order 7, i.e. $PG_2(\mathbf{F}_7)$, and there is only a computer proof showing that there is only one plane of order 8, i.e. $PG_2(\mathbf{F}_8)$. Four planes of order 9 (which the reader will encounter in this chapter) were discovered early in the century, and recently Lam, Kolesova and Thiel [177] have shown, by computer, that there are only these four. As remarked, there has been shown to be no plane of order 10: see Lam [176], and Lam, Thiel and Swiercz [179]. It is not known whether or not the desarguesian plane of order 11 is unique. It is known, however, that there are at least two nonisomorphic planes of order p^2 for every odd prime p and, in fact, the number of nonisomorphic planes of order p^2 tends to infinity with p: see Ebert [100].

6.3 The codes

In the late 1950s and early 1960s coding theorists, notably Prange and Rudolph, recognized that projective planes could be used to produce error-correcting codes. The main interest then was in binary codes and Prange [243] made an exhaustive study of the codes associated with the classical projective plane of order 8. Rudolph [249] was the first to sense that the classical planes should yield codes with the least redundancy and he conjectured the formula for the rank of the design of points and hyperplanes of a projective space. It was much later, however, that Hamada and then, independently, Sachar made the precise conjecture that we treat in Section 6.9. In the meantime there was an enormous amount of work done by coding theorists on so-called *geometric* codes; the most important work, for our purposes, was done by Delsarte *et al.* We have described that work fully in Chapter 5.

The most transparent example of the relationship of geometry to algebraic coding theory is that given by the $[7, 4, 3]$ binary Hamming code and the projective plane of order 2; pictorially it can be seen immediately by comparing the Venn diagram description of this code (Figure 2.2) with the diagram for the Fano plane (Figure 1.1). The reader may check, using the Venn diagram, that the vectors of weight 3 of the code correspond precisely to the lines of the plane. Put another way, an incidence matrix of the plane yields a row space over the two-element field that is the $[7, 4, 3]$ binary Hamming code.

More generally, given any projective plane Π and any field F, the code $C_F(\Pi)$ is, essentially, the row space over F of an incidence matrix of Π. As we have seen, this code will be of interest only when the characteristic of the field divides the order of the plane; throughout this chapter the base field will always be the field with p elements where p is a prime dividing n, the order of the plane. If the prime in question is important, we denote the code by $C_p(\Pi)$ and otherwise simply by $C(\Pi)$. We will see that this code is an $[n^2 + n + 1, k, n + 1]$ code and that the minimum-weight vectors are simply the scalar multiples of the incidence vectors of the lines. This latter fact came as something of a surprise, since usually the minimum weight and the minimum-weight vectors of a code are difficult to determine and the dimension easy, whereas in this case the reverse is true: the dimension k will, in general, depend heavily on the structure of the projective plane. The situation with respect to the hull is more interesting and here neither the minimum weight nor the minimum-weight vectors are known in general. Moreover, it is rather difficult to prove even for the classical planes (see Chapter 5) that both the minimum weight and the minimum-weight vectors are as expected: the minimum weight is $2n$ and the minimum-weight vectors are the scalar multiples of $v^L - v^M$, where L and M are distinct lines of the projective plane.

Some appreciation of the significance of this assertion can be gleaned from the fact that, were it true in general, then, for a projective plane of order 10, the non-existence of ovals in the plane and vectors of weight 16 in the code would follow immediately. These results about a putative plane of order 10 were uncovered relatively recently (see Lam *et al.* [180]) and, as already remarked, Lam, Thiel and Swiercz have shown by extensive computer work that there is no plane of order 10. An historical account of these results can be found in Lam [176].

In fact, the non-existence of ovals in *any* plane of order n greater than 2 and congruent to 2 modulo 4 would follow simply if one knew that the minimum weight of the hull of the plane were $2n$, the codes taken to be binary.

We first summarize these and other now "classical" facts concerning the codes given by a projective plane. The development of this material — with the new results to come — is largely taken from Assmus and Key [11, 12, 13]. For an account of the classical facts from the group-theoretical perspective the reader may wish to consult Hering [133].

Recall that the hull, H, of a design is the intersection of its code and its orthogonal. For a given code C, we always have the following tower:

$$H = C \cap C^\perp \subseteq C \subseteq C + C^\perp = H^\perp$$

and, as we shall prove, for projective planes all codes between C and $H^\perp$ have the same minimum-weight vectors. It is only for H that questions arise and where, it seems, the nature of the plane makes itself strongly felt as far as the minimum-weight vectors are concerned.

The ambient space for the codes we are here considering will, as usual, be $F^\mathcal{P}$ where $\mathcal{P}$ is the point set. For a vector v in that space and a point Q we will usually denote the value of v at Q by v_Q, but sometimes by $v(Q)$. Moreover, Supp(v) will denote the *support* of the vector v.

Theorem 6.3.1 *Let Π be a projective plane of order n and let p be a prime dividing n. Set $H = \mathrm{Hull}_p(\Pi)$ and $B = H^\perp$. If k is the dimension of $C_p(\Pi)$, then B is an $[n^2 + n + 1, n^2 + n + 2 - k, n + 1]$ code. The minimum-weight vectors of B are precisely the vectors of the form av^L, where a is a non-zero scalar and L is a line of Π. Moreover,*

$$\mathrm{Hull}_p(\Pi) = \left\{ c \in C_p(\Pi) \mid \sum_Q c_Q = 0 \right\}$$

$$= \langle v^L - v^M \mid L \text{ and } M \text{ lines of } \Pi \rangle.$$

Proof: We first prove the last assertion. Since, for any three lines L, M, and N of Π, $(v^N, v^L - v^M) = 0$ we have that $v^L - v^M$ is in $C_p(\Pi)^\perp$ for any two lines, L and M; these vectors are clearly in $C_p(\Pi)$ and hence in H. Since $n + 1 \equiv 1 \pmod{p}$, $\sum_L v^L = \jmath$, the all-one vector. Thus $\jmath \in C_p(\Pi)$. Further, $(v^L - \jmath, v^M) = 0$ for all L and M, so $v^L - \jmath \in H$ for all L, and $C_p(\Pi) = H \oplus F_p \jmath$, since $\jmath$ is not in H. Thus the space spanned by the vectors $v^L - \jmath$ is a subspace of H which is a subspace of $\{c \in C_p(\Pi) \mid \sum_Q c_Q = 0\}$, and as all these have codimension 1 in $C_p(\Pi)$, they must be equal. Further, since for a line L and a fixed point Q not on L, $v^L - \jmath = \sum_{M \ni Q}(v^L - v^M)$, the set $\{v^L - v^M \mid L \text{ and } M \text{ lines of } \Pi\}$ also spans H.

It follows that $v \in B$ if and only if $(v, v^L - v^M) = 0$ for any two lines, L and M, of Π and hence that $a = (v, v^L)$ is independent of L. If $a = 0$ (i.e. if $v \in C_p(\Pi)^\perp$), it follows easily that $\mathrm{wt}(v) \geq n + 2$. Since $C_p(\Pi) \subseteq B$, the minimum weight of B is at most $n + 1$. Thus for a minimum-weight vector v and all L, $a = (v, v^L) \neq 0$. For such a vector v, let L_0 be a line of Π meeting Supp(v) at least twice. If $L_0 \subseteq \mathrm{Supp}(v)$, then $L_0 = \mathrm{Supp}(v)$ and, choosing a point Q not on L_0, we have for all lines L through Q, $(v, v^L) = a = v_{L_0 \cap L}$; i.e. $v = a v^{L_0}$. If $L_0 \not\subseteq \mathrm{Supp}(v)$, let Q be a point of L_0 not in $\mathrm{Supp}(v)$. Every line through Q meets $\mathrm{Supp}(v)$ since $(v, v^L) \neq 0$ and thus the $n + 1$ lines through Q force v to have weight at least $n + 2$. To get the dimension of B, we use the fact that $\dim H = k - 1$ since $C_p(\Pi) = H \oplus \mathbf{F}_p \jmath$. Thus, B has dimension $n^2 + n + 1 - (k - 1)$, as required. $\square$

We can extract a bit more from the proof just given when the order n of the plane is even and we take $p = 2$: then the argument shows that $C_2(\Pi)^\perp$ has minimum weight at least $n + 2$ and, if the minimum weight of this orthogonal *is* $n + 2$, then the minimum-weight vectors are all of the form v^X where X is an oval of Π. We shall have more to say about these vectors shortly, but we record these facts now in a slightly more general form, the proof of which should be clear to the reader: see also Lemma 2.4.2.

Corollary 6.3.1 *If Π is a projective plane of order n and p is a prime dividing n, then the minimum weight of $C_p(\Pi)^\perp$ is at least $n + 2$. If the minimum weight is $n + 2$ then, necessarily, $p = 2$ and n is even; moreover, the minimum-weight vectors are all of the form v^X where X is an oval of Π.*

We emphasize that for a projective plane Π of order n and a prime p dividing n we have $\mathrm{Hull}_p(\Pi) \subset C_p(\Pi) \subseteq \mathrm{Hull}_p(\Pi)^\perp$ and the code with the least redundancy (since the minimum weight and the minimum-weight vectors remain the same between the code of the plane and the orthogonal of the hull) and minimum weight $n + 1$ will be obtained from the plane whose code has the least dimension. Moreover, for *any* code B with $C_p(\Pi) \subseteq B \subseteq \mathrm{Hull}_p(\Pi)^\perp$ the minimum weight of B is $n + 1$ and the minimum-weight vectors are precisely the scalar multiples of the incidence vectors of the lines of Π. Hence any such B uniquely determines Π. As we shall see, the situation is entirely different in the affine case, to which we now turn, keeping the notation defined above.

In order to get an analogous result for affine planes we will first analyse, in coding-theoretical terms, the passage from the projective to the affine. Let L be any line of the projective plane Π and set $\pi = \Pi^L$. As designs, both Π and π have the same order n and for any prime p dividing n there is a natural projection of $\mathbf{F}_p^{\mathcal{P}}$ onto $\mathbf{F}_p^{\mathcal{A}}$ obtained by deleting the coordinates corresponding to the points of L. (Here we are denoting by $\mathcal{P}$ the point set of the projective plane and by $\mathcal{A}$ the point set of the affine plane.) The restriction to B of the linear transformation from $\mathbf{F}_p^{\mathcal{P}}$ to $\mathbf{F}_p^{\mathcal{A}}$ that effects this natural projection we denote by T. Since B has minimum weight $n + 1$ the kernel of T is 1-dimensional and consists precisely of the scalar multiples of v^L. Thus,

$$\dim C_p(\pi) = \dim C_p(\Pi) - 1;$$

also

$$\dim C_p(\pi)^\perp = n^2 - \dim C_p(\pi).$$

Clearly, if $v \in C_p(\Pi)^\perp$ and $v_Q = 0$ for Q on L, $T(v) \in C_p(\pi)^\perp$. Setting

$$D = \{v \in C_p(\Pi)^\perp | v_Q = 0 \text{ for } Q \text{ on } L\}$$

we have that $T(D) \subseteq C_p(\pi)^\perp$.

The dimension of D is easy to compute: if $v \in C_p(\Pi)^\perp$ and $v_Q = 0$ for all but one point Q_0 on L, then $v_{Q_0} = 0$ since $(v^L, v) = \sum_{Q \in L} v_Q$, and requiring that $v_Q = 0$ for one point Q on L lowers the dimension by 1 the first n times. Hence, $\dim D = n^2 + n + 1 - \dim C_p(\Pi) - n = n^2 - \dim C_p(\pi)$ and therefore $T(D) = C_p(\pi)^\perp$ since $D \cap \ker T = \{0\}$ and $\dim T(D) = \dim(D)$.

Observe next that $B = \mathrm{Hull}_p(\Pi)^\perp = C_p(\Pi) + C_p(\Pi)^\perp = C_p(\Pi) + D$ since any $v \in C_p(\Pi)^\perp$ can be written as the sum of an element of D and an appropriate linear combination of the vectors $v^M \in C_p(\Pi)$, where M runs through the lines of Π except L. (More precisely, take for each P on L a line L_P through P with $L_P \neq L$ and, if $v \in C_p(\Pi)^\perp$, form $v - \sum_{P \in L} v_P v^{L_P} = w$; clearly $w_P = 0$ for $P \in L$ and hence $w \in D$ since $\sum_{P \in L} v_P v^{L_P} \in C(\Pi)^\perp$ in view of the fact that $\sum_{P \in L} v_P = (v, v^L) = 0$.) Hence

$$T(C_p(\Pi) + C_p(\Pi)^\perp) = C_p(\pi) + C_p(\pi)^\perp$$

or, in other words,

$$T(\mathrm{Hull}_p(\Pi)^\perp) = \mathrm{Hull}_p(\pi)^\perp.$$

Thus, $\dim \mathrm{Hull}_p(\pi)^\perp = \dim \mathrm{Hull}_p(\Pi)^\perp - 1 = n^2 + n + 1 - (\dim C_p(\Pi) - 1) - 1 = n^2 + n - \dim C_p(\pi)$ and $\dim \mathrm{Hull}_p(\pi) = \dim C_p(\pi) - n$.

We are now in a position to state and finish the proof of the following result. In what follows the line L above will be the line at infinity and will be denoted by L_∞.

Theorem 6.3.2 *Let π be an affine plane of order n and let Π be its projective completion with L_∞ the line at infinity. Then, for a prime p dividing n,*

$$\dim \mathrm{Hull}_p(\pi) = \dim C_p(\pi) - n,$$

both $C_p(\pi)$ and $\mathrm{Hull}_p(\pi)^\perp$ are the images of the natural projection of $C_p(\Pi)$ and $\mathrm{Hull}_p(\Pi)^\perp$ respectively into $\mathbf{F}_p^\mathcal{A}$, $\mathrm{Hull}_p(\pi)$ is the projection of the sub-code $\{c \in \mathrm{Hull}_p(\Pi) | c_Q = 0 \text{ for } Q \text{ on } L_\infty\}$, and

$$\mathrm{Hull}_p(\pi) = \langle \{v^\ell - v^m | \text{ where } \ell \text{ and } m \text{ are parallel lines of } \pi\} \rangle.$$

Proof: Only the last two assertions need to be verified. If ℓ and m are two parallel lines of π, their extensions, L and M, to Π meet at a point of L_∞. Thus, $v^L - v^M$ is not only in $C_p(\Pi)^\perp$ but also in D (defined above), i.e. $T(v^L - v^M) = v^\ell - v^m \in C_p(\pi)^\perp$. Since $v^\ell - v^m$ is clearly in $C_p(\pi)$ we need only show that these vectors form a generating set of $\mathrm{Hull}_p(\pi)$. Let E be the subspace of $\mathbf{F}_p^\mathcal{A}$ generated by the set $\{v^\ell - v^m | \ell$ and m parallel lines of $\pi\}$. Then $\jmath = -\sum_{m\|\ell}(v^\ell - v^m)$ for any fixed ℓ since there are n lines in a parallel class and p divides n. Thus $\jmath \in E \subseteq \mathrm{Hull}_p(\pi)$. Fix a point Q of π and choose n lines through Q, $\ell_1, \ldots, \ell_n$ say. The subspace generated by E together with $v^{\ell_1}, \ldots, v^{\ell_n}$ clearly contains the $(n+1)^{\text{th}}$ line through Q, namely $\jmath - \sum_{i=1}^n v^{\ell_i}$. Since every $v^\ell - v^m$, $\ell \| m$, is in E this subspace must be $C_p(\pi)$. Thus $\dim E \geq \dim C_p(\pi) - n$; i.e. $E = \mathrm{Hull}_p(\pi)$. Clearly $\mathrm{Hull}_p(\pi)$ is the projection of $\{c \in \mathrm{Hull}_p(\Pi) | c_Q = 0$ for Q on $L_\infty\}$. $\square$

Notice that the hull of the affine plane is essentially that subcode of the hull of its projective completion consisting of those vectors with zeros at the points at infinity.

We now turn to the minimum weight of B and the minimum-weight vectors, where, as in the projective case, we have $B = \mathrm{Hull}_p(\pi)^\perp \supseteq C_p(\pi) \supset \mathrm{Hull}_p(\pi)$. For the sake of economy of language we call a vector in $\mathbf{F}^\mathcal{A}$ a *constant* vector if it is a scalar multiple of a vector all of whose entries are 0 or 1. As we shall see, the minimum weight is as expected, but the minimum-weight vectors cannot be described so easily. Here is the result.

Theorem 6.3.3 *If π is an affine plane of order n and p is a prime dividing n, then the minimum weight of $B = \mathrm{Hull}_p(\pi)^\perp$ is n and all minimum-weight vectors are constant.*

Proof: Clearly the minimum weight of B is at most n since $C_p(\pi)$ is contained in B. Let v be a minimum-weight vector in B. Then $(v, v^\ell - v^m) = 0$ for every pair of parallel lines, ℓ and m, of π. Thus $a = (v, v^\ell)$ depends only on the parallel class in which ℓ lies. Since $\mathrm{wt}(v) \leq n$, if $Q \in \mathrm{Supp}(v)$, not all of the $n+1$ lines through Q can meet $\mathrm{Supp}(v)$ again. Let ℓ be a line with $|\ell \cap \mathrm{Supp}(v)| = 1$. Then $(v, v^\ell) = a \neq 0$ where $a = v_Q$. Since $(v, v^m) = a$ for m parallel to ℓ, $\mathrm{wt}(v) \geq n$ and hence $\mathrm{wt}(v) = n$ and v is a constant vector, $\mathrm{Supp}(v)$ being a transversal for that parallel class. $\square$

We will soon return to the ideas in this proof, but let us now remark that we also have the conclusion of the theorem for $C_p(\pi)$ since $C_p(\pi) \subseteq B$. An important point to note here is that it can very well happen that B will contain minimum-weight vectors that are not scalar multiples of lines. If we take $\Pi = PG_2(q^2)$ where q is an arbitrary prime power and let S be the point set of a Baer subplane of Π, then $|S| = q^2 + q + 1$ and for every line

L of Π, $|L \cap S|$ is 1 or $q+1$. Hence, taking p to be the characteristic of the field $\mathbf{F}_q$ we have that $|S \cap L| \equiv 1 \pmod{p}$ for all L. It follows that $\jmath - v^S$ is in $C_p(\Pi)^\perp$ and hence

$$v^S \in \mathrm{Hull}_p(\Pi)^\perp = C_p(\Pi)^\perp \oplus \mathbf{F}_p \jmath.$$

If L is a line of Π meeting S in $q+1$ points and $\pi = \Pi^L$, then the projection of $\mathrm{Hull}_p(\Pi)^\perp$ onto $\mathrm{Hull}_p(\pi)^\perp$ takes v^S to a constant vector of weight $q^2 = n$ of $\mathrm{Hull}_p(\pi)^\perp$.

For $q = 2$, for example, $\mathrm{Hull}_2(AG_2(4))$ is the $[16, 5, 8]$ first-order Reed-Muller code $\mathcal{R}(1, 4)$ with weight enumerator

$$1 + 30X^8 + X^{16}$$

and $\mathrm{Hull}_2(AG_2(4))^\perp = \mathcal{R}(2, 4)$ is the $[16, 11, 4]$ binary extended Hamming code with 140 weight-4 vectors. These vectors actually form a 3-design which can be partitioned into seven affine planes of order 4, and, as a 2-design, they contain 112 subdesigns which are affine planes of order 4, all of which have the same hull. If 20 vectors forming an affine plane of order 4 are chosen, the remaining 120 produce Baer subplanes of that plane. As we shall see, this phenomenon concerning the Baer subplanes is not peculiar to order 4 and it will occur whenever the order is the square of a prime. See Lemma 6.4.2 below.

6.4 The minimum-weight vectors

Throughout the following discussion, π will be an affine plane of order n and Π its projective completion with L_∞ the line at infinity. We fix a prime p dividing n and set $H = \mathrm{Hull}_p(\pi)$ and $B = H^\perp$. We know that B has minimum weight n and that all minimum-weight vectors are constant; further, amongst the minimum-weight vectors are the scalar multiples of the incidence vectors of the $n^2 + n$ lines of π. Our purpose here is to investigate the nature of those minimum-weight vectors of B which are *not* obtained from the lines of π. In fact we want to relate these vectors to their pre-images, under the natural projection, in $\mathrm{Hull}_p(\Pi)^\perp$. Set $\overline{B} = \mathrm{Hull}_p(\Pi)^\perp$ and, as before, let $T : \overline{B} \to B$ be the natural projection.

We begin with a technical result.

Lemma 6.4.1 *Let v be a minimum-weight vector in B. Then, for p odd, there is a unique constant vector v' in $\overline{B}$ with $T(v') = v$. Moreover $\mathrm{wt}(v') = n + r$ with $r \equiv 1 \pmod{p}$, $r < n$; r is the number of parallel classes in π for which $\mathrm{Supp}(v)$ is not a transversal. When $p = 2$ there are precisely* *two vectors—both constant—in the preimage of v; uniqueness is achieved by requiring that v' be not in $C_2(\Pi)^\perp$.*

Proof: Since $\overline{B} = C_p(\Pi)^\perp \oplus \mathbf{F}_p \jmath$ and $v^{L\infty}$ is not in $C_p(\Pi)^\perp$, $\overline{B} = C_p(\Pi)^\perp \oplus \mathbf{F}_p v^{L\infty}$. Thus if c is any vector in $\overline{B}$ with $T(c) = v$, all such are of the form $c + av^{L\infty}$, $a \in \mathbf{F}_p$. Without loss of generality we assume that the non-zero coordinates of v are 1's. By a proper choice of a we can find a $c' \in C_p(\Pi)^\perp$ with $T(c') = v$.

If ϑ is a parallel class of π for which $\mathrm{Supp}(v)$ is a transversal, then each extension of an ℓ in ϑ to Π goes through a fixed point Q on L_∞ with, necessarily, $c'_Q = -1$. On the other hand, if ϑ is a parallel class of π for which $\ell \cap \mathrm{Supp}(v)$ is empty for some ℓ in ϑ, then for the corresponding Q on L_∞, $c'_Q = 0$. Clearly, for p odd, the only constant vector amongst $c' + av^{L\infty}$, $a \in \mathbf{F}_p$, is $v' = c' + v^{L\infty}$, with r as described. Here $r \equiv 1(\mathrm{mod}\ p)$ since now $(v', v^L) = 1$ for all $L \neq L_\infty$ of Π, which forces $(v', \jmath) = 1 \equiv r(\mathrm{mod}\ p)$ (and $r = n + 1$ is impossible since then $\mathrm{wt}(v' - v^{L\infty}) = n$). The amendment for $p = 2$ should be obvious. $\square$

The constant vector v' of the lemma has a support that is a **blocking set** of the projective plane, i.e. a set of points in the plane that meets every line and the complement of every line. Thus, if v is not a line of the affine plane (i.e. if $r \neq 1$), we have the inequalities

$$1 + \sqrt{n} \leq r < n$$

with $r \equiv 1(\mathrm{mod}\ p)$. For a proof of the lower bound see Bruen [54]

The next lemma will, among other things, completely characterize the minimum-weight vectors in B in case $n = p$ or $n = p^2$.

Lemma 6.4.2 *Suppose $n = sp$ where p is a prime and $1 \leq s \leq p$. Then for an affine plane π of order n, and $B = \mathrm{Hull}_p(\pi)^\perp$, the minimum-weight vectors of B are scalar multiples of the incidence vectors of the lines, unless $s = p$ where they are scalar multiples of the incidence vectors of either lines or Baer subplanes of π.*

Proof: Let v' be the constant vector of $\overline{B}$ described above. If $r = 1$, v' is a scalar multiple of the incidence vector of a line of Π and hence v is a scalar multiple of the incidence vector of a line of π. Suppose v is a weight-n vector of B, normalized so that its non-zero coordinates are 1, and suppose its support is not a line of π. Then, for some parallel class ϑ of π, we have $\ell \in \vartheta$ with $1 < |\ell \cap \mathrm{Supp}(v)| < sp$ and, in fact, for each ℓ in ϑ, either $\ell \cap \mathrm{Supp}(v)$ is empty or $|\ell \cap \mathrm{Supp}(v)|$ is a multiple of p. There must be at least two lines ℓ_1 and ℓ_2 in ϑ with $\ell_i \cap \mathrm{Supp}(v)$ non-empty. Now if ℓ is a line through a point of $\ell_1 \cap \mathrm{Supp}(v)$ and a point of $\ell_2 \cap \mathrm{Supp}(v)$, then $|\ell \cap \mathrm{Supp}(v)| \geq 2$ and is hence a multiple of p and thus $|\ell \cap \mathrm{Supp}(v)| \geq p$. Clearly, then there are at least p lines in ϑ meeting $\mathrm{Supp}(v)$ in a subset of cardinality at least p. This is impossible unless $s = p$ and, in this case, every line of π meets $\mathrm{Supp}(v)$ in 0,1, or p points. Thus $\mathrm{Supp}(v)$ is a Baer subplane of π. $\Box$

Corollary 6.4.1 *If π is an affine plane of prime order p, then $\mathrm{Hull}_p(\pi)^\perp = C_p(\pi)$. Moreover, the minimum-weight vectors of $C_p(\pi)$ are precisely the scalar multiples of the incidence vectors of the lines of π and $\mathrm{Hull}_p(\pi)$ uniquely determines π.*

Proof: For a projective plane Π of prime order p, we have $\dim C_p(\Pi) = \frac{1}{2}(p^2 + p + 2)$ and $C_p(\Pi)^\perp \subseteq C_p(\Pi)$: see Theorem 4.6.2. In this case, therefore, $\mathrm{Hull}_p(\Pi) = C_p(\Pi)^\perp$ and $\mathrm{Hull}_p(\Pi)^\perp = C_p(\Pi)$. Going to the affine plane we have $\dim(C_p(\pi)) = \frac{1}{2}(p^2 + p), \dim(C_p(\pi)^\perp) = \frac{1}{2}(p^2 - p)$ with, again, $\mathrm{Hull}_p(\pi) = C_p(\pi)^\perp$ and $\mathrm{Hull}_p(\pi)^\perp = C_p(\pi)$. $\Box$

Dougherty [97] has recently generalized the fact that the hull determines the plane in the case of prime order to *nets* of prime order: if their hulls are isomorphic, either they are themselves isomorphic or they have a common extension to a larger net.

Corollary 6.4.2 *If π is an affine plane of order p^2 where p is a prime, the minimum-weight vectors of $\mathrm{Hull}_p(\pi)^\perp$ are scalar multiples of the incidence vectors of either lines or Baer subplanes of π.*

The classification of projective or affine planes of prime order p is tantamount to a classification of certain codes over $\mathbf{F}_p$ of block length $p^2 + p + 1$ and minimum weight $p+1$ (the projective case) or block length p^2 and minimum weight p (the affine case). The only known planes of prime order are the desarguesian planes and it is widely believed that these planes are the only ones of prime order. There have been serious attempts to construct non-desarguesian planes of prime order (see, for example, Mendelsohn and Wolk [213] where planes of orders 13 and 17 were sought) but none have been uncovered; the question is open for all primes greater than 7, the planes of orders 2, 3, 5 and 7 being unique.

Example 6.4.1 (1) For $p = 2$ the affine hull is a binary $[4, 1, 4]$ code, viz. $\mathbf{F}_2 \mathbf{J}$.

(2) For $p = 3$ the affine hull is a $[9, 3, 6]$ ternary code. It is unique and is generated by the vectors:

$$(1, 1, 1, -1, -1, -1, 0, 0, 0),$$

$$(1, 1, 1, 0, 0, 0, -1, -1, -1),$$

and

$$(1, -1, 0, 1, -1, 0, 1, -1, 0).$$

The above two examples yield trivial proofs of the rather trivial facts that the affine and projective planes of orders 2 and 3 are unique.

Corollary 6.4.3 *If* Π *is a projective plane of order* n *and* p *is a prime dividing* n *with the minimum weight of* $\mathrm{Hull}_p(\Pi) = 2n$, *then every affine part* π *of* Π *has the property that* $C_p(\pi)$ *has, as minimum-weight vectors, only the scalar multiples of the incidence vectors of the lines of* π.

Proof: Let $v \in C_p(\pi)$ have weight n. Then the v' of Lemma 6.4.1 can be chosen in $C_p(\Pi)$. If $r = 1$ then v', and hence v, is a line. Assume $1 < r$; then $r < n$ and, normalizing so that the non-zero coordinates of v' are 1, we have $\sum_{Q \in \Pi} v'_Q = 1$ since $r \equiv 1 \pmod{p}$. Thus, $\sum_Q (v' - v^{L_\infty})_Q = 0$ and $v' - v^{L_\infty} \in C_p(\Pi)^\perp$. But $\mathrm{wt}(v' - v^{L_\infty}) = n + (n + 1 - r) = 2n + 1 - r < 2n$, a contradiction, since $v' - v^{L_\infty}$ is clearly in $C_p(\Pi)$, and hence in $\mathrm{Hull}_p(\Pi)$, which has minimum weight $2n$. $\square$

For the desarguesian planes, the results of Chapter 5 tell us everything about the minimum-weight vectors of the code and the hull:

Theorem 6.4.1 *If* $q = p^t$, *where* p *is a prime, then the minimum-weight vectors of* $C_p(AG_2(\mathbf{F}_q))$ *are the scalar multiples of the incidence vectors of the lines of* $AG_2(\mathbf{F}_q)$, *and the minimum-weight vectors of* $\mathrm{Hull}_p(AG_2(\mathbf{F}_q))$ *are the scalar multiples of the differences of the incidence vectors of any two distinct parallel lines.*

Proof: $C_p(AG_2(\mathbf{F}_q))$ is the generalized Reed-Muller code $\mathcal{A}_{F_q/F_p}(q - 1, 2)$, which, by Theorem 5.7.9, has precisely (the scalar multiples of) the incidence vectors of the lines as minimum-weight vectors. By Theorem 5.7.5 the generalized Reed-Muller code $\mathcal{A}_{F_q/F_p}(q - 2, 2)$ is spanned by its minimum-weight vectors which are all obtained from two parallel lines. Thus, by Theorem 6.3.2, $\mathrm{Hull}_p(AG_2(\mathbf{F}_q)) = \mathcal{A}_{F_q/F_p}(q - 2, 2)$. $\square$

Corollary 6.4.4 *If* $q = p^t$, *where* p *is a prime, then* $\mathrm{Hull}_p(PG_2(\mathbf{F}_q))$ *has minimum weight* $2q$ *and the minimum-weight vectors are the scalar multiples of the differences of the incidence vectors of any two distinct lines.*

Proof: We first show that the minimum weight is $2q$. Suppose this is not the case; then letting c be a vector in $\mathrm{Hull}_p(PG_2(\mathbf{F}_q))$ with $0 < \mathrm{wt}(c) < 2q$, there is a line L of $PG_2(\mathbf{F}_q)$ with $L \cap \mathrm{Supp}(c)$ empty: such a line exists, for otherwise $|L \cap \mathrm{Supp}(c)| \geq 2$ for all lines L and it follows that $\mathrm{wt}(c) \geq 2q + 1$. Now the affine plane determined by taking L to be the line at infinity is, of course, desarguesian and has a vector $T(c)$ (where T is the natural projection) in $\mathrm{Hull}_p(AG_2(\mathbf{F}_q))$ of weight less than $2q$, which is a contradiction. A similar argument shows that the minimum-weight vectors of $\mathrm{Hull}_p(PG_2(\mathbf{F}_q))$ must be scalar multiples of vectors of the form $v^L - v^M$, where L, M are lines. $\square$

We do not have an example of an affine plane whose code has weight-n vectors whose supports are not lines. The only general result we know is the one we have given for planes of prime order. It would be very interesting to have a more general result since, for example, ovals in projective planes of even order congruent to 2 modulo 4 give rise to weight-n vectors in the binary codes (of certain affine parts) which are not the characteristic functions of lines — when $n \neq 2$.

For the convenience of the reader we now summarize in a single theorem some of the results of Chapter 5, taken together with the facts we have just established, for the desarguesian planes.

Theorem 6.4.2 *Let p be any prime, $q = p^t$, and $\Pi = PG_2(\mathbf{F}_q)$. Then $C_p(\Pi) = \mathcal{A}_{F_q/F_p}^{q-1}(q-1,3)^* = \mathcal{P}(1,3)$ and has dimension $\binom{p+1}{2}^t + 1$. The minimum-weight vectors of $C_p(\Pi)$ and of $C_p(\Pi) + C_p(\Pi)^\perp$ are the scalar multiples of the incidence vectors of the lines. The minimum weight of $\mathrm{Hull}_p(\Pi)$ is $2q$ and its minimum-weight vectors are the scalar multiples of the differences of the incidence vectors of distinct lines of Π. The minimum weight $d^\perp$ of $C_p(\Pi)^\perp$ satisfies*

$$q + p \leq d^\perp \leq 2q,$$

with equality at the lower bound if $p = 2$.

If $\pi = AG_2(\mathbf{F}_q)$, then $C_p(\pi) = \mathcal{A}_{F_q/F_p}(q-1,2) = \mathcal{A}(q-1,2)$ and has dimension $\binom{p+1}{2}^t$. The minimum weight of both $C_p(\pi)$ and $C_p(\pi) + C_p(\pi)^\perp$ is q, and the minimum-weight vectors of $C_p(\pi)$ are the scalar multiples of the incidence vectors of the lines of π. The minimum weight of $\mathrm{Hull}_p(\pi) = \mathcal{A}(q-2,2)$ is $2q$, and the minimum-weight vectors are the scalar multiples of the differences of the incidence vectors of distinct parallel lines in π. The minimum weight $d^\perp$ of $C_p(\pi)^\perp$ satisfies

$$q + p \leq d^\perp \leq 2q,$$

with equality at the lower bound when $p = 2$.

This theorem incorporates the results of Theorems 5.7.1 and 5.7.9 and Corollary 5.7.1 together with the results established for general planes in this chapter. The salient feature concerns the hull; as we shall see, the nature of the minimum-weight vectors in the non-desarguesian case is different.

Before beginning a discussion of the non-desarguesian case in Section 6.8 we shall spend some time embellishing the classical case; to this end we pause, in the next section, to recall some of the central group-theoretical results that we will need.

The reader wishing to continue with the structural account of planes via algebraic coding theory may wish to go directly to Section 6.8 where the vein we have opened begins to be explored.

Exercise 6.4.1   (1) Prove that any two ovals in a projective plane of even order n have at most $(n+2)/2$ points in common.

(2) If $\Pi = PG_2(\mathbf{F}_{2^m})$, where $m \geq 3$ and $C = C_2(\Pi)$, show that if $c \in C^\perp$ satisfies $\mathrm{wt}(c) > 2^m + 2$, then $\mathrm{wt}(c) \geq 2^m + 8$.

6.5   Central collineations

The 50 years of effort that have been directed toward the understanding and classification of projective planes have been dominated by coordinate methods combined with group-theoretical methods. The modern attack began with Hall's classic paper [115] and the results were generalized, organized and codified a quarter of a century later by Dembowski [85] and again, more recently, by Beth *et al.* [37].

As with any class of designs, one of the main methods of classifying planes is through the nature of their symmetries, i.e. through the type of collineations they admit. For example, a classic theorem of Ostrom and Wagner [225] states that if a finite projective plane has a doubly-transitive collineation group, then it must be desarguesian and, for a more recent example, a theorem of Gluck [104] shows that an affine plane of prime order with a regular, abelian group of automorphisms must be desarguesian.

The existence of sufficient *central collineations* ensures that a plane is desarguesian, and the number and nature of such collineations of a plane has been used in the main classification of planes which is due to Lenz and Barlotti: see Dembowski [85, Section 3.1]. The natural analogues of these types of collineations in higher-dimensional geometries from vector spaces are the transvections, which arise from elementary matrices. The book by Biggs and White [39] has some examples of this. We have already noted some facts about collineations of projective planes as symmetric designs — in particular, Theorem 4.3.1. Here we need to know something about central collineations. These are defined through their fixed elements, so we first examine the possibilities. We will give some proofs, mostly to illustrate the geometric arguments that are involved. A fuller treatment of the material to follow in this section is given in Hughes and Piper [143, Chapter IV].

For any projective plane $\Pi = (\mathcal{P}, \mathcal{L})$ and collineation $\sigma \in \text{Aut}(\Pi)$, let $\text{Fix}(\sigma)$ denote the set of fixed elements of σ; thus $\text{Fix}(\sigma)$ consists of the union of the set of fixed points and fixed lines; by Brauer's theorem there are as many points as lines in $\text{Fix}(\sigma)$. Clearly, $\text{Fix}(\sigma)$ is a *closed* configuration, i.e. if any two points P and Q are in $\text{Fix}(\sigma)$, then so is the line PQ, and dually. From this it follows that, if $\text{Fix}(\sigma)$ contains a quadrangle, then $\text{Fix}(\sigma)$ is a subplane of Π.

Lemma 6.5.1 *If σ is an involution of the plane Π then $\text{Fix}(\sigma)$ is either a Baer subplane or consists of a line with all its points, together with a point and all the lines through it.*

Proof: We first show that $\text{Fix}(\sigma)$ is a **Baer subset**, i.e. that every element of Π is incident with an element of $\text{Fix}(\sigma)$. Clearly $\text{Fix}(\sigma) \neq \Pi$, since we are assuming that σ is an involution. By Theorem 4.3.1 we know that the number of points in $\text{Fix}(\sigma)$ is the same as the number of lines, so we can assume there is a point $A \notin \text{Fix}(\sigma)$. Thus AA^σ is a line, and clearly $AA^\sigma \in \text{Fix}(\sigma)$. A dual argument shows that any line $L \notin \text{Fix}(\sigma)$ is incident with a point of $\text{Fix}(\sigma)$, *viz.* $L \cap L^\sigma$. Thus every element *not* in $\text{Fix}(\sigma)$ is incident with an element of $\text{Fix}(\sigma)$.

Next let B be a point in $\text{Fix}(\sigma)$; if there is another point C in $\text{Fix}(\sigma)$ then B is incident with BC which is in $\text{Fix}(\sigma)$ and, by duality, we have the result. Thus, $\text{Fix}(\sigma)$ is a Baer subset unless it consists of a single point B and a single line L. But this is impossible since, if C is a point other than B and not on L, then CC^σ is a fixed line containing C which is clearly not L.

To complete the proof of the lemma, notice that if $\text{Fix}(\sigma)$ contains a quadrangle then it is a subplane, and hence a Baer subplane. Suppose it does not contain a quadrangle. If it contains a triangle with vertices A, B and C all other fixed points must be on one of the sides, BC say, and then all other fixed lines must pass through A. Since every other point not on one of the sides must lie on a fixed line it must be on a line through A and it follows that all points of BC are fixed and all lines through A are. If it does not contain a triangle the fixed points are collinear and on a fixed line L. Now another line M must be fixed and arguments similar to those already used show that the fixed lines are those through the intersection of L and M and that all the points of L are fixed. $\square$

In case an involution σ fixes a Baer subplane, σ will be called a **Baer**

involution. Any other involution will be a central collineation, according to the following definition:

Definition 6.5.1 *Let* Π *be a projective plane and let* $\sigma \in \mathrm{Aut}(\Pi)$. *Then* σ *is a* **central collineation** *of* Π *if the set of fixed points of* σ *includes all the points incident with a line of* Π.

Lemma 6.5.2 *If* Π *is a projective plane and* σ *a non-trivial central collineation that fixes all the points of a line* L, *then* σ *fixes at most one further point of* Π. *Further, there is a point* C *such that* σ *fixes every line through* C.

Proof: Suppose P and Q are distinct points not on L that are fixed by σ. If $P \in M$, and $M \cap L = R$, then $M = PR$ and $M^\sigma = P^\sigma R^\sigma = PR = M$, so every line through P (or Q) is fixed by σ. Let T be any other point of Π not on L. Then PT and QT are fixed, and thus $T^\sigma \in PT \cap QT$. If $T \notin PQ$ then this implies that $T^\sigma = T$, and if $T \in PQ$ then pick another point S off PQ, show that $S^\sigma = S$ and now use the fact that T cannot be on both PS and QS. Thus $\sigma = 1$, contradicting our assumption. Thus σ fixes at most one more point.

Suppose that σ fixes a point $C \notin L$. Then σ fixes every line through C, as required. Suppose that σ does not fix any point not on L. Let $P \notin L$; then $P \neq P^\sigma$, so PP^σ is a line and meets L at a point C, say. The line PC is fixed by σ. Let R be any other point, not on L nor PC. Again, RR^σ is fixed by σ. If $S = RR^\sigma \cap PP^\sigma$, then $S^\sigma = S$, and so $S \in L$, and thus $S = C$. This shows that every line through C is fixed by σ. $\square$

Thus if σ is a non-trivial central collineation then there is a unique line L that is fixed pointwise, called the **axis** of σ, and there is a unique point C fixed linewise, called the **centre** of σ. Further, σ is an **elation** if the centre is on the axis; σ is a **homology** if the centre is off the axis. Either way, σ is referred to as a (C, L)-**central collineation**.

Exercise 6.5.1   (1) For $\Pi = PG_2(\mathbf{F}_q)$, show that the collineations induced by the linear transformations

$$
\begin{pmatrix} 1 & 0 & a \\ 0 & 1 & 0 \\ 0 & 0 & 1 \end{pmatrix} \text{ and } \begin{pmatrix} a & 0 & 0 \\ 0 & 1 & 0 \\ 0 & 0 & 1 \end{pmatrix},
$$

where $a \in \mathbf{F}_q$, are central collineations; find the centre and axis in each case.

(2) Prove that an elation of Π is completely determined by its axis L and the image of one point not on L and that a homology is completely determined by its axis L, its centre C, and the image of one point not on L and distinct from C.

(3) If σ is a (C, L)-central collineation of Π, and $\tau \in \mathrm{Aut}(\Pi)$, show that $\sigma^\tau = \tau^{-1}\sigma\tau$ is a (C^τ, L^τ)-central collineation.

(4) If σ is a non-trivial central collineation of Π, where Π has order n, and σ has order k, show that either k divides n and σ is an elation, or k divides $n - 1$ and σ is a homology.

For any point C and line L, the set $G(C, L)$ of all (C, L)-central collineations clearly forms a subgroup of $\mathrm{Aut}(\Pi)$. Also, all the elations with axis L, together with the identity, form a subgroup, $G(L)$; in fact we have the following:

Theorem 6.5.1 *The set $G(L)$, the identity together with the elations of Π with axis L, forms a subgroup of $\mathrm{Aut}(\Pi)$. If Π has at least two elations that have distinct centres on L, then $G(L)$ is elementary abelian.*

Proof: If all elations in $G(L)$ have the same centre they clearly form a group. Suppose $G(L)$ contains an elation σ with centre A and an elation τ with centre B, where $A \neq B$. If $P \notin L$, then $P \neq P^{\sigma\tau}$ and so $PP^{\sigma\tau}$ is a line that meets L at a point C that is neither A nor B, and which must be the centre of the central collineation $\sigma\tau$. (For if not, $\sigma\tau$ is a homology with centre D on $PP^{\sigma\tau}$; since $D^\sigma = D^{\tau^{-1}}$, this point lies on both AD and BD, i.e. on L.) Thus $G(L)$ is a group.

Consider the commutator $[\sigma, \tau] = \sigma^{-1}\tau^{-1}\sigma\tau$; then σ^τ is a $(A^\tau, L^\tau) = (A, L)$-elation, and $(\tau^{-1})^\sigma$ is a $(B^\sigma, L^\sigma) = (B, L)$-elation, by Exercise 6.5.1 above. So $[\sigma, \tau] = \sigma^{-1}\sigma^\tau \in G(A, L)$, and $[\sigma, \tau] = (\tau^{-1})^\sigma\tau \in G(B, L)$, and since $A \neq B$, this implies that $[\sigma, \tau] = 1$.

Now let $\kappa \in G(A, L)$. We need to show that $\sigma\kappa = \kappa\sigma$. We know that $\sigma\tau$ has centre distinct from A and thus commutes with κ, and also τ commutes with both σ and κ. Thus $(\sigma\tau)\kappa = \kappa(\sigma\tau)$, implying $(\tau\sigma)\kappa = \tau(\kappa\sigma)$ and $\sigma\kappa = \kappa\sigma$. So $G(L)$ is abelian. Now suppose σ has order t. Replacing τ by a power of τ allows us to assume that τ has prime order, p say. Then $(\sigma\tau)^p = \sigma^p$ has centre distinct from A or B, which is a contradiction, unless it is 1. It follows that $G(L)$ is elementary abelian. $\square$

As we have seen, if $\sigma \in G(C, L)$, then the image of any point $R \neq C$, $R \notin L$, must lie on the line RC. If every possible (C, L)-central collineation exists, the plane is said to be (C, L)-**transitive**. If, for a fixed line L, Π is

(P, L)-transitive for all $P \in L$, then L is called a **translation line** for Π and $G(L)$ is called the **translation group** of L.

Definition 6.5.2 *A projective plane Π that has a translation line is called a (projective)* **translation plane**. *If the translation line is L, then the affine plane $\pi = \Pi^L$ is an (affine) translation plane.*

Note: Usually the term "translation plane" is understood to mean an *affine* translation plane, since elations of a plane with axis taken to be the line at infinity of an affine plane become translations of the affine plane (the homologies become dilatations: see Artin [5] for a nice account). The translation line of a projective translation plane is called the "translation point" when viewed in the dual plane. Finite translation planes must have prime-power order, by Theorem 6.5.1.

Example 6.5.1 The desarguesian plane $PG_2(\mathbf{F}_q)$ has every line a translation line. Its translation group is elementary abelian of order q^2, acting regularly on the points off the translation line.

Finally we mention without proof the relevance of the existence of central collineations to Desargues' theorem. Baer [24] introduced the terminology of (C, L)-**desarguesian**, meaning that Desargues' Theorem holds for any two triangles in perspective from C with two sets of corresponding sides meeting on L (and dually). He then proved the following:

Theorem 6.5.2 (Baer) *A projective plane Π is (C, L)-transitive if and only if it is (C, L)-desarguesian, where (C, L) is a point-line pair.*

A proof of this theorem can be found in Chapter IV of [143].

6.6   Other geometric codewords

Besides the lines there are other geometric configurations in projective planes that give rise to constant vectors. The first case, historically, concerned ovals in projective planes of even order but, as we have already seen, Baer subplanes also play a role and, as we shall soon see, so do unitals. In this section we try to give an up-to-date account of the known results, but there are some interesting open questions which we have not been able to settle—even for classical projective planes.

We look at the designs formed by taking as blocks the set of all the configurations of a given type and the relationship of the codes of these designs to the code associated with the plane. In particular we will prove the following for desarguesian planes:

Theorem 6.6.1 *Let* $\Pi = PG_2(\mathbf{F}_{q^2})$, $q = p^r$, *and let* $\mathcal{S}$, $\mathcal{U}$ *and* $\mathcal{H}$ *be the designs consisting of the points of* Π *and the Baer subplanes, unitals and hermitian unitals, respectively, as blocks. Then* $C_p(\mathcal{S})$, $C_p(\mathcal{U})$ *and* $C_p(\mathcal{H})$ *all contain* $C_p(\Pi)$ *and are in* $\mathrm{Hull}_p(\Pi)^\perp$. *In fact,* $C_p(\Pi) = C_p(\mathcal{H})$.

If $\Pi = PG_2(\mathbf{F}_q)$, $q = p^r$, *and if* $\mathcal{O}$ *is the design whose blocks are the ovals of* Π, *then, when* p *is odd,* $C_p(\mathcal{O}) = \mathbf{F}_p^{\mathcal{P}}$. *When* $p = 2$, *we have that* $\mathrm{Hull}_2(\Pi) \subseteq C_2(\mathcal{O}) \subseteq C_2(\Pi)^\perp$ *with* $C_2(\mathcal{O}) = C_2(\Pi)^\perp$ *for* $q = 2, 4, 8$, *or* 16.

The proof of the theorem will emerge in the course of this section. We make a few comments on the result before beginning the technical development.

That $C_2(\mathcal{O}) = C_2(PG_2(\mathbf{F}_{2^r}))^\perp$ for $r = 1$, i.e. the Fano plane, is trivial and, for $r = 2$, i.e. the plane of order 4, easy to verify. For the planes of orders 8 and 16 electronic computation was necessary: see the remark in Assmus and Key [12]. In fact, for $PG_2(\mathbf{F}_{16})$, taking the orbit of any oval under a Singer cycle will give a generating set for $C_2(PG_2(\mathbf{F}_{16}))^\perp$. We suspect that we have the equality for all powers of 2 but we have not attempted to prove this. Recall that, for an *arbitrary* projective plane Π of even order n, the minimum weight of $C_2(\Pi)^\perp$ is at least $n + 2$ and, if it is $n + 2$, the minimum-weight vectors *are* the incidence vectors of the ovals: see Lemma 2.4.2. What we suspect, therefore, is that for the desarguesian planes of even order the minimum-weight vectors of the orthogonal generate the orthogonal.

As far as we have been able to determine, every *known* projective plane of even order has ovals. Certainly all the translation planes of order 16 do and, in fact, these ovals have been classified by Cherowitzo [71, 72]. It should be an easy matter to check whether or not the ovals in these planes generate the orthogonal.

The fact that $C_p(\mathcal{H}) \subseteq C_p(PG_2(\mathbf{F}_{p^{2s}}))$ follows from Theorem 5.7.7 but the special case $p = 2$ was obtained earlier by Bagchi and Sastry (see Corollary 4.6.1) for all dimensions, and for planes and any prime p by Blokhuis, Brouwer and Wilbrink [44]. In fact they showed that for any unital U embedded in $PG_2(\mathbf{F}_{q^2})$, U is hermitian if and only if the incidence vector of U is in $C(PG_2(\mathbf{F}_{q^2}))$. This characterization of hermitian unitals was first conjectured in Assmus and Key [12] and it is closely related to intersection properties of unitals with unitals, and unitals with Baer subplanes; these questions had been addressed by Baker and Ebert [29], Bruen and Hirschfeld [55], and Kestenband [156].

We know of no characterization of the ovals in desarguesian projective planes of even order that would distinguish, in coding-theoretic terms, the ovals that consist of a conic together with its nucleus from the "sporadic" ovals (see the discussion below). It would, of course, be very interesting were there such a characterization. See also the remarks in Sections 7.12 and 8.4.

Recall that a Baer subplane Σ of a projective plane Π of order $n = m^2$ is a subset of $m^2 + m + 1$ points of Π and $m^2 + m + 1$ lines of Π with the property that Σ, with incidence induced by the incidence of Π, is a projective plane of order m. Elementary counting arguments then show that every point of Π is incident with 1 or $m + 1$ lines of Σ and, dually, every line of Π is incident with 1 or $m + 1$ points of Σ.

A unital embedded in a projective plane Π of order $n = m^2$ is a subset U of $m^3 + 1$ points with the property that every line of Π meets U in 1 or $m + 1$ points. Then the structure with point set U and block set the set of all lines of Π that contain $m + 1$ points of U — the incidence once again being that induced by Π — forms a *unitary* design, i.e. a design with parameters 2-$(m^3 + 1, m + 1, 1)$. If $\Pi = PG_2(\mathbf{F}_{q^2})$, then the absolute points and non-absolute lines of a unitary polarity form a hermitian unital in Π (see Section 3.7). For $q > 2$ there are non-hermitian unitals in $PG_2(\mathbf{F}_{q^2})$: see Metz [215]. Thus, just as there are (in general) sporadic ovals in classical planes of even order, there are sporadic unitals in classical planes of square order.

An oval of a projective plane Π of odd order n is an $(n + 1)$-arc. Every line of Π is either an exterior line, a tangent or a secant of an oval and thus the cardinality of the intersection of a line with an oval can be 0, 1 or 2. An oval of a projective plane of even order n is an $(n + 2)$-arc. Every line of the plane is either an exterior line or a secant and thus the cardinality of the intersection of a line with an oval is either 0 or 2. For $\Pi = PG_2(\mathbf{F}_q)$ with q odd, Segre's theorem (Theorem 3.7.1) shows that the ovals are just the conics, i.e. the absolute points of an orthogonal polarity. For q even, a conic together with its nucleus is an oval. For $q < 16$ these are the only ovals but for all $q \geq 16$ sporadic ovals exist. For $q = 16$ there is, up to projective equivalence, only one oval that is not a conic together with its nucleus, the Hall oval — discovered by Lunelli and Sce [198] and thoroughly investigated by Hall [120]. For $q \geq 32$ they have not been classified.

On the point set $\mathcal{P}$ of Π we define the structures $\mathcal{S}, \mathcal{U}, \mathcal{H}$ and $\mathcal{O}$ where the blocks are given as follows:

- $\mathcal{S}$: the Baer subplanes of Π, when $n = m^2$;

- $\mathcal{U}$: the unitals in Π, when $n = m^2$;

- $\mathcal{H}$: the hermitian unitals in Π, when $\Pi = PG_2(\mathbf{F}_{q^2})$;

- $\mathcal{O}$: the ovals in Π, of size $n + 1$ for n odd, $n + 2$ for n even.

Recall that a Singer group is a collineation group generated by a Singer cycle, which is a collineation of order $n^2 + n + 1$ that acts regularly on the points (and hence on the lines) of Π. For $\Pi = PG_2(\mathbf{F}_q)$, Singer cycles always exist (see Section 3.2), but no other projective plane is known with a regular cyclic collineation group. (If a plane has more than one Singer group, then from Ott [226] the plane is desarguesian.)

For the remainder of this section Π will denote a projective plane of order n with point set $\mathcal{P}$ and line set $\mathcal{L}$. The codes $C_p(\Pi)$, $H_p(\Pi) = \text{Hull}_p(\Pi)$, $C_p(\Pi)^\perp$, and $B_p(\Pi) = H_p(\Pi)^\perp$ are defined over the field $\mathbf{F}_p$, where p, as usual, divides n. We will normally omit the subscript p. Recall that we have the tower

$$H(\Pi) \subset C(\Pi) \subset B(\Pi)$$

and that $H(\Pi)$ is generated by vectors of the form $v^L - v^M$ where $L, M \in \mathcal{L}$.

Lemma 6.6.1 *A constant vector $v \in \mathbf{F}_p^{\mathcal{P}}$ is contained in $B(\Pi)$ if and only if $|\text{Supp}(v) \cap L|$, taken modulo p, is independent of the line L of Π.*

Proof: Let $X = \text{Supp}(v)$, so that $v = av^X$, where $a \in \mathbf{F}_p^\times$, the multiplicative group of $\mathbf{F}_p$. Suppose $v \in B(\Pi)$. Then $(v, v^L - v^M) = 0$ and (v, v^L) is independent of $L \in \mathcal{L}$. Thus $(v, v^L) = (av^X, v^L) = a|X \cap L|$, and hence $|X \cap L|$ is also independent of L modulo p.

Conversely, suppose that $|X \cap L|$ is constant modulo p. Then $(v^X, v^L) \equiv |X \cap L| \pmod{p}$ is independent of L and $(v^X, v^L - v^M) = 0$ for all $L, M \in \mathcal{L}$; hence $v^X \in B(\Pi)$. Thus, since v is a constant vector, $v = av^X \in B(\Pi)$. □

This lemma gives immediate information about other possible geometric vectors in $\mathbf{F}_p^{\mathcal{P}}$. We collect this information in the following two propositions:

Proposition 6.6.1 *Let Π be an arbitrary projective plane of order n and X a subset of points of Π. Then*

(1) $v^X \in B(\Pi)$ whenever X is the set of points of a Baer subplane of Π,

(2) $v^X \in B(\Pi)$ whenever X is the set of points of a unital in Π,

(3) $v^X \in C_2(\Pi)^\perp \subset B_2(\Pi)$ whenever n is even and X is an oval of Π,

(4) $v^X \notin B(\Pi)$ whenever n is odd and X is an oval of Π.

Proof: The assertions are immediate from the lemma and the definitions of the geometric configurations. □

Proposition 6.6.2 *If* $\Pi = PG_2(\mathbf{F}_q)$, *where* $q = p^s$, *and* $s = ht$, *then, if* S *is the set of points of* $\Sigma = PG_2(\mathbf{F}_{p^t})$ *in its natural embedding in* Π, $v^S \in B(\Pi)$ *if and only if* $h \in \{1, 2\}$.

Proof: For $h = 1$ the subspace Σ is merely a line and for $h = 2$ it is a Baer subplane; hence we need only prove the only-if part of the proposition.

Let $h \geq 3$. Then $K = \mathbf{F}_{p^s}$ is a vector space of dimension h over $F = \mathbf{F}_{p^t}$; let a, b and c be three elements of K that are linearly independent over F and, using homogeneous coordinates for Π, consider the line $L = \langle (a, b, c) \rangle$. Now $S = \{\langle (x, y, z)^t \rangle | x, y, z \in F\}$ and, since the equation

$$(a, b, c) \begin{pmatrix} x \\ y \\ z \end{pmatrix} = ax + by + cz = 0$$

has no solutions in F, the line L is disjoint from the set S. However, some lines of Π will clearly meet Σ in $p^t + 1$ points. The lemma shows that $v^S \notin B(\Pi)$. $\square$

Lemma 6.6.2 *Let* $X \subseteq \mathcal{P}$ *with* $v^X \in C(\Pi)$. *If* $Y \subseteq \mathcal{P}$ *with* $v^Y \in B(\Pi)$ *and* $|Y \cap L| \equiv |X \cap L| \pmod{p}$ *for all* L, *then* $|X \cap Y| \equiv |X| \pmod{p}$.

Proof: Since $v^X \in C(\Pi) \subseteq B(\Pi)$, $|X \cap L|$ is independent of L, modulo p. Now $(v^X - v^Y, v^L) = (v^X, v^L) - (v^Y, v^L) \equiv |X \cap L| - |Y \cap L| \equiv 0 \pmod{p}$, for all $L \in \mathcal{L}$, so $v^X - v^Y \in C(\Pi)^\perp$. Hence $(v^X - v^Y, v^X) \equiv |X| - |X \cap Y| \equiv 0 \pmod{p}$, since $v^X \in C(\Pi)$. Hence $|X \cap Y| \equiv |X| \pmod{p}$. $\square$

Proposition 6.6.3 *If* X *is the set of points of a Baer subplane of* $\Pi = PG_2(\mathbf{F}_{q^2})$, *then* $v^X \notin C(\Pi)$.

Proof: Since $|X| = q^2 + q + 1$, $|X| \equiv 1 \pmod{p}$, where $q = p^r$. It is not difficult to see that there is a Baer subplane Y that meets X in a line segment of size $q + 1$, and in one other point off the line. Then $|X \cap L| \equiv 1 \equiv |Y \cap L| \pmod{p}$ and $|X \cap Y| = q + 2 \equiv 2 \pmod{p}$. But $v^Y \in B(\Pi)$ from Proposition 6.6.1, so, by the lemma, $v^X \notin C(\Pi)$. $\square$

Proposition 6.6.4 *If* X *is a unital in a projective plane* Π *and* $v^X \in C(\Pi)$ *then the intersection of every unital and Baer subplane of* Π *with* X *must be of cardinality congruent to 1 modulo* p.

Proof: This follows directly from the lemma and Proposition 6.6.1. $\square$

In connection with this proposition we note once again that intersection properties of hermitian unitals, and of hermitian unitals with Baer subplanes, have been examined by Kestenband [156] and by Bruen and Hirschfeld [55], respectively.

Lemma 6.6.3 *Let X be an arbitrary subset of $\mathcal{P}$. Suppose there exists a flag (P, L) with $P \notin X$ such that Π is (P, L)-transitive. Set $E = \langle \sigma \mid \sigma$ is a (P, L)-elation of $\Pi \rangle$ and $\mathcal{B} = \{X^\sigma | \sigma \in E\}$; let $\mathcal{D}$ be the incidence structure $(\mathcal{P}, \mathcal{B})$. If for those lines $M \in \mathcal{L}$ with $P \in M$ and $M \neq L$, $|X \cap M|$ is independent of M modulo p and not divisible by p, then $\jmath - v^L \in C_p(\mathcal{D})$. If $\jmath \in C_p(\mathcal{D})$, then $v^L \in C_p(\mathcal{D})$.*

Proof: Let $u = \sum_{\sigma \in E} v^{X^\sigma}$. The subgroup E, which has order n, acts regularly on the points of each of the lines $M \neq L$ through P. Thus $u_Q = n \equiv 0 \pmod{p}$ for $Q \in L \cap X$; $u_Q = 0$ for $Q \in L$, $Q \notin X$; $u_Q \equiv |M \cap X| \pmod{p}$ for $Q \notin L$. Thus $u = a(\jmath - v^L) \in C_p(\mathcal{D})$, where $a \equiv |X \cap M| \pmod{p}$. Hence $\jmath - v^L \in C_p(\mathcal{D})$ since $a \neq 0$. The last assertion is obvious. $\square$

Lemma 6.6.4 *Let X be an arbitrary subset of $\mathcal{P}$ with $|X| \not\equiv 0 \pmod{p}$. Suppose that Π has a Singer cycle S. Then, for $\mathcal{C} = \{X^\sigma | \sigma \in \langle S \rangle\}$ and $\mathcal{E} = (\mathcal{P}, \mathcal{C})$, $\jmath \in C_p(\mathcal{E})$.*

Proof: Set $a \equiv |X| \pmod{p}$ and let $w = \sum_{\sigma \in \langle S \rangle} v^{X^\sigma}$. Then $w_Q = a$ for all $Q \in \mathcal{P}$, so $w = a\jmath \in C_p(\mathcal{E})$. But $a \neq 0$ and hence $\jmath \in C_p(\mathcal{E})$. $\square$

Proposition 6.6.5 $C(\mathcal{S}) \supset C(PG_2(\mathbf{F}_{q^2}))$ *and* $C(\mathcal{H}) = C(PG_2(\mathbf{F}_{q^2}))$.

Proof: By Proposition 6.6.3 $C(\mathcal{S}) \neq C(PG_2(\mathbf{F}_{q^2}))$. Let X be a Baer subplane and let L be any line of $PG_2(\mathbf{F}_{q^2})$. Choose $P \in L$ with $P \notin X$. Now $|X| \equiv 1 \pmod{p}$ and $|X \cap L| \equiv 1 \pmod{p}$ for all lines L, where $q = p^r$. Now $PG_2(\mathbf{F}_{q^2})$ is (P, L)-transitive for every flag (P, L) and, moreover, has a Singer cycle. The lemmas ensure that $v^L \in C(\mathcal{S})$. Since L was an arbitrary line we have that $C(\mathcal{S}) \supset C(PG_2(\mathbf{F}_{q^2}))$.

If X is a unital, we have that $|X| \equiv 1 \pmod{p}$ and $|X \cap L| \equiv 1 \pmod{p}$ for all lines L of Π. Again, all the conditions are satisfied and the lemmas give that $C(\mathcal{H}) \supseteq C(PG_2(\mathbf{F}_{q^2}))$. Now Theorem 5.7.7 gives the reverse inequality, and the result follows. $\square$

Note that the lemmas actually prove a bit more than the propositions indicate and are, therefore, of some interest in their own right.

Proposition 6.6.6 $C(PG_2(\mathbf{F}_{2^s}))^\perp \supseteq C(\mathcal{O}) \supseteq \mathrm{Hull}(PG_2(\mathbf{F}_{2^s}))$; *the second containment is strict when $s > 1$.*

Proof: The first containment is obvious since the intersection of a line with an oval is either empty or of cardinality 2. Set $q = 2^s$. Since the cardinality of an oval is $q + 2$ and all vectors in $H(PG_2(\mathbf{F}_q))$ have weight congruent to 0 mod 4, the strict containment assertion is obvious once we prove the second containment. But if Y is an oval and L is a secant of Y, we apply the first lemma with $X = Y - L$ and P one of the two points of $X \cap L$. Then $\jmath + v^L$ is in $C(\mathcal{O})$ for every L and hence $v^L + v^M$ for every L and M. We remark that $\mathcal{O}$ could have been taken to be the orbit of any oval under $PGL_3(\mathbf{F}_q)$ and, in particular, we could have taken $\mathcal{O}$ simply to be those ovals obtained as a conic together with its nucleus. $\square$

Remark (1) We have now shown that

$$B(PG_2(\mathbf{F}_{q^2}) \supseteq C(\mathcal{S}) \supset C(PG_2(\mathbf{F}_{q^2})).$$

We conjecture that, for q prime, $C(\mathcal{S}) = B(PG_2(\mathbf{F}_{q^2}))$. This is true for $q^2 = 4$, 9 and 25.

(2) The result of Bagchi and Sastry (see Corollary 4.6.1) says that if $\mathcal{D}$ is any symmetric incidence structure of even order with a polarity, and if X is the set of absolute points of the polarity, then $v^X \in C_2(\mathcal{D})$. Thus for X a hermitian unital of $PG_2(\mathbf{F}_{2^s})$, we have that $v^X \in C_2(PG_2(\mathbf{F}_{2^s}))$, without the use of Theorem 5.7.7, and hence, by what we have just proved, $C_2(PG_2(\mathbf{F}_{2^s})) = C_2(\mathcal{H})$. (The inclusion $C_2(\mathcal{H}) \subseteq C_2(PG_2(\mathbf{F}_{2^s}))$ can also be proved from the results of Fisher, Hirschfeld and Thas [101].)

(3) The result of Bagchi and Sastry is false for p odd, i.e. if X is the set of absolute points of a polarity σ of a symmetric structure $\mathcal{D}$ of odd order, then v^X need not be in $C_p(\mathcal{D})$. For example, if $\mathcal{D} = PG_2(\mathbf{F}_q)$ with q odd, and if σ is an orthogonal polarity, then X is an oval, and $v^X \notin B(PG_2(\mathbf{F}_q)) \supseteq C(PG_2(\mathbf{F}_q))$.

(4) We have already noted that we suspect that $C(PG_2(\mathbf{F}_{2^s}))^\perp = C(\mathcal{O})$ for all s; we have verified it for $s < 5$.

Lemma 6.6.5 *Let Π be a translation plane and L its translation line. Suppose X is a subset of points that is disjoint from L with $|X| \not\equiv 0 \pmod{p}$. With T the translation group of Π, set $\mathcal{B} = \{X^\sigma | \sigma \in T\}$ and $\mathcal{D} = (\mathcal{P}, \mathcal{B})$. Then $\jmath - v^L \in C(\mathcal{D})$.*

Proof: Let $a \equiv |X| \pmod{p}$ with $a \in \mathbf{F}_p^\times$ and set $w = \sum_{\sigma \in T} v^{X^\sigma}$. Then $w_Q = 0$ for $Q \in L$ and $w_Q = a$ for $Q \notin L$, since T acts regularly on $\mathcal{P} - L$. Thus $w = a(\jmath - v^L) \in C(\mathcal{D})$ and hence $\jmath - v^L \in C(\mathcal{D})$. $\square$

Proposition 6.6.7 *For desarguesian projective planes of odd order,*

$$C(\mathcal{O}) = \mathbf{F}_p^{\mathcal{P}}.$$

Proof: Set $\Pi = PG_2(\mathbf{F}_q)$, where $q = p^r$ and p is odd. We first show that $C(\mathcal{O}) \supset C(\Pi)$. An oval X has $q + 1$ points, so $|X| \equiv 1 \pmod{p}$. Each line L of Π is exterior to some oval, and is a translation line for Π. So $\jmath - v^L \in C(\mathcal{O})$ for all lines L. We need only show that $\jmath \in C(\mathcal{O})$, and this follows, as usual, by taking a Singer cycle s for Π, and then setting $w = \sum_{g \in \langle s \rangle} v^{X^g} = (q + 1)\jmath = \jmath \in C(\mathcal{O})$. Since $v^X \notin C(\Pi)$ for every oval X, $C(\mathcal{O}) \supset C(\Pi)$.

For q odd, the ovals are precisely the conics, and through any five points, no three collinear, there passes a unique oval: see Hirschfeld [136]. Let Δ be a triangle of points $\{A, B, C\}$, with edges the lines K, L and M. A simple count yields that Δ is on $(q - 1)^2$ ovals. Call this set of ovals $\mathcal{B}$. Every point of Π not on K, L or M is on x members of $\mathcal{B}$, where $x(q^2 + q + 1 - 3q) = x(q - 1)^2 = (q - 1)^2(q - 2)$, i.e. $x = (q - 2)$. Let $v = \sum_{X \in \mathcal{B}} v^X$. Then $v_Q = 0$ for $Q \in K, L, M$, and $Q \notin \{A, B, C\}$; $v_Q = 1$ for $Q \in \{A, B, C\}$; $v_Q = -2$ for $Q \notin K, L, M$. Let $w = 2(v^K + v^L + v^M) - (v + 2\jmath)$. Then $w = v^\Delta$ and $w \in C(\mathcal{O})$, since $C(\mathcal{O}) \supset C(\Pi)$. This holds for any triangle, so we clearly have that $\langle \jmath \rangle^\perp \subseteq C(\mathcal{O})$, since any generator of $\langle \jmath \rangle^\perp$ can be obtained from the vector $v^{\Delta_1} - v^{\Delta_2}$, where Δ_1 and Δ_2 are suitably chosen distinct triangles with an edge in common. Since $\jmath \in C(\mathcal{O})$, $\jmath \notin \langle \jmath \rangle^\perp$, it follows that $C(\mathcal{O}) = \mathbf{F}_p^{\mathcal{P}}$. $\square$

Theorem 6.6.1 now follows from the propositions above.

It should be noted that the design of points and conics in *any* classical projective plane is an example of a design with a doubly-transitive automorphism group that is not amenable to a straightforward treatment (but see [22]) via algebraic coding theory, since the code of the design is uninteresting for *every* choice of p. Of course, the conics, when q is even, are not quite the correct objects to look at and the ovals do play a major coding-theoretical role in this case—as we have seen. Curiously, for q odd Segre's theorem characterizes the ovals.

6.7 Hermitian unitals

We give now the characterization of hermitian unitals that was conjectured in Assmus and Key [12] and proved by Blokhuis *et al.* [44]. The result characterizes the hermitian unitals as those unitals of $PG_2(\mathbf{F}_{q^2})$ whose incidence vectors appear as codewords in the plane's code. We already know from Theorem 5.7.7 (also Theorem 6.6.1) that the incidence vector of a hermitian unital is in the code of the plane over the field $\mathbf{F}_p$. We have also proved this independently above in Proposition 6.6.5; this result first appeared in [12]. Thus we will not describe the possibly somewhat more general situation developed by Blokhuis *et al.* [44] for this part of the proof; in that paper, projective planes given by abelian difference sets are discussed.

The theorem is as follows:

Theorem 6.7.1 (Blokhuis, Brouwer and Wilbrink) *Let q be a power of the prime p and let $\mathcal{U}$ be a unital embedded in $\Pi = PG_2(\mathbf{F}_{q^2})$ with point set U. Then $\mathcal{U}$ is hermitian if and only if $v^U \in C_p(\Pi)$*

We have the sufficiency of the condition from either Theorem 5.7.7 or Proposition 6.6.5; for the necessity we need a lemma.

Lemma 6.7.1 *For any prime power $q = p^m$, let X be a set of points of $\Pi = PG_2(\mathbf{F}_q)$ with $v^X \in C_p(\Pi)$. If P is a point that is not in X, then the points $Q \in X$ for which the line PQ is tangent to X (i.e. meets X only in Q) are collinear.*

Proof: The case of $PG_2(\mathbf{F}_2)$ can be done by inspection. Taking $q > 2$, suppose there are three points Q_i, for $i = 1, 2, 3$, in X for which PQ_i is a tangent line to X for each i. Using homogeneous coordinates, we may take $P = (1, 0, 0)$ and $Q_i = (x_i, y_i, 1)$, for $i = 1, 2, 3$, where $y_i \neq y_j$ for $i \neq j$, since P, Q_i and Q_j are not collinear. Thus $(1, 1, 1) \neq (y_1, y_2, y_3)$, and there is a unique line, $(w_1, w_2, w_3)^t$ say, through them, where $w_i \in \mathbf{F}_q$ for $i = 1, 2, 3$. Construct the vector $w \in \mathbf{F}_q^{\mathcal{P}}$ by defining $w(Q) = 0$ if $Q = P$ or if Q is not on any of the lines PQ_i, for $i = 1, 2, 3$, and $w(Q) = w_i x$ if $Q = (x, y_i, 1)$. Then $w \in C_{\mathbf{F}_q}(\Pi)^\perp$, since any line through P will have inner product 0 (if not one of PQ_i) or $w_i \sum_{x \in \mathbf{F}_q} x = 0$ if PQ_i; otherwise the line has coordinates $(1, a, b)^t$, and inner product $\sum_{i=1}^3 w_i(ay_i + b) = 0$. Since $v^X \in C_p(\Pi) \subseteq C_{\mathbf{F}_q}(\Pi)$, $(v^X, w) = 0 = w_1 x_1 + w_2 x_2 + w_3 x_3$, which finally shows that $Q_1 Q_2 Q_3$ is a line. $\square$

We are now in a position to prove the other half of the theorem.

Proof: Suppose $\mathcal{U}$ is a unital embedded in Π with $v^U \in C_p(\Pi)$, where U is the point set of the unital and q is a power of p.

Since $v^U \in C_p(\Pi)$, Lemma 6.7.1 implies that for $P \notin U$, the $q+1$ points Q_i of U for which PQ_i is a tangent are all on a line of Π, which we denote by $P^\perp$. If $P \in U$, then there is a unique tangent to U at P, which we will denote by $P^\perp$. The aim is to show that $\perp$ defines a polarity since then U will be the set of absolute points of $\perp$ and hence hermitian; for this it is enough to show that $Q \in P^\perp$ implies that $P \in Q^\perp$. The only case that is not immediate is if both P and Q are not on U; thus take this to be the case, with $Q \in P^\perp$. We can choose coordinates in such a way that $P = (1, 0, 0)$

and $P^\perp$ is the line $(1,0,0)^t$. Let $Q_i = (0, y_i, 1)$, for $i = 1, 2, \ldots, q+1$, be the points of U on $P^\perp$, and let $Q = (0, y_0, 1) \in P^\perp$. Then $PQ = (0, 1, -y_0)$.

Elementary linear algebra tells us that there exist $w_i \in \mathbf{F}_q$, for $i = 0, 1, 2, \ldots, q+1$, not all zero, such that

$$\begin{pmatrix} 1 & 1 & 1 & \cdots & 1 \\ y_0 & y_1 & y_2 & \cdots & y_{q+1} \\ y_0^2 & y_1^2 & y_2^2 & \cdots & y_{q+1}^2 \\ \vdots & \vdots & \vdots & \vdots & \vdots \\ y_0^q & y_1^q & y_2^q & \cdots & y_{q+1}^q \end{pmatrix} \begin{pmatrix} w_0 \\ w_1 \\ w_2 \\ \vdots \\ w_{q+1} \end{pmatrix} = \begin{pmatrix} 0 \\ 0 \\ 0 \\ \vdots \\ 0 \end{pmatrix};$$

in fact, the w_i's must all be non-zero, since deleting a column from the above matrix yields a non-singular Vandermonde matrix. Take any integer k satisfying $1 \le k \le q$, and define a vector $w \in \mathbf{F}_{q^2}^{\mathcal{P}}$ by $w((x, y_i, 1)) = w_i x^k$ for $x \in \mathbf{F}_{q^2}$, $i = 0, 1, 2, \ldots, q+1$, and $w(R) = 0$ for all other points R. Again it is easy to see that $w \in C_{\mathbf{F}_{q^2}}(\Pi)^\perp$. Thus, if the $q+1$ points of U that are on $(0, 1, -y_0)^t$ are given by $R_i = (x_i, y_0, 1)$, for $i = 1, 2, \ldots, q+1$, then it follows that

$$\sum_{i=1}^{q+1} w_0 x_i^k = 0.$$

For each integer $k \ge 1$, define

$$\pi_k = \sum_{i=1}^{q+1} x^k,$$

and the functions

$$\pi(z) = \sum_{k=1}^{\infty} \pi_k z^k, \text{ and } \sigma(z) = \prod_{i=1}^{q+1} (1 - x_i z) = \sum_{k=0}^{\infty} \sigma_k z^k.$$

Then $\sigma(z)\pi(z) + z\sigma'(z) = 0$, and from this can be deduced the identities

$$\sum_{m=0}^{n-1} \pi_{n-m}\sigma_m + n\sigma_n = 0,$$

for $n \ge 1$. Thus, since $\pi_k = 0$ for $k = 1, 2, \ldots, q$, it follows that $\sigma_n = 0$ for $n \le q$, and $n \not\equiv 0 \pmod{p}$. Using induction, it follows that $\pi_k = 0$ for $k \ge q+1$, for $k \not\equiv 1 \pmod{p}$. In particular, $\pi_{q^2-2} = 0$ if $p \ne 3$, and $\pi_{q^2-4} = 0$ if $p = 3$. Using $x^{q^2-2} = x^{-1}$ and $x^{q^2-4} = x^{-3}$ for $x \in \mathbf{F}_{q^2}$, we get

$$\sum_{i=1}^{q+1} x_i^{-1} = 0.$$

Let $R_0 = (x_0, y_0, 1)$ be any point on the line PQ, with $R \neq P, Q, R_i$ for $i = 1, 2, \ldots, q + 1$, and compute the cross-ratio $(Q, P; R_i, R_0)$:

$$(Q, P; R_i, R_0) = (0, \infty; x_i, x_0) = \frac{(0 - x_i)(\infty - x_0)}{(\infty - x_i)(0 - x_0)} = \frac{x_i}{x_0}.$$

Thus

$$\sum_{i=1}^{q+1} (Q, P; R_i, R_0) = 0.$$

Interchanging the roles of P and Q, and writing $R = (x, y_0, 1)$ for $Q^\perp \cap PQ$, it follows that

$$0 = \sum_{i=1}^{q+1} (R, Q; R_i, R_0) = \sum_{i=1}^{q+1} (x, 0; x_i, x_0) = \frac{x_0}{x - x_0} \sum_{i=1}^{q+1} \left(\frac{x}{x_i} - 1\right) = \frac{-x_0}{(x - x_0)}.$$

Thus $x = \infty$, and so $P = R \in Q^\perp$. This completes the proof. $\square$

6.8   Translation planes

We will, in fact, be discussing a possibly richer class of planes than translation planes in this section, and a better way to think of the following subject matter is in terms of planes involved with the geometric codes introduced and studied by algebraic coding theorists, i.e. mostly the generalized Reed-Muller codes described in Chapter 5. There is a vast literature on translation planes, and we make no pretence of discussing the topic in depth. We simply wish to indicate how the topic fits naturally into the coding-theoretical approach to planes; for a deeper discussion of translation planes the reader should consult a text on the subject, for example, Lüneburg [197] or Ostrom [224].

Translation planes have prime-power order (see Theorem 6.5.1) as do all the planes we discuss in this section — and they will naturally be given as affine planes. From what we already know about the minimum-weight vectors of the orthogonal of the hull of an affine plane it is clear that we should, when searching for planes of order p^s, look for codes over $\mathbf{F}_p$ of minimum weight p^s and with a sufficiently rich structure of minimum-weight vectors to accommodate $C_p(\pi)$'s for various affine planes π of order p^s. Of course, given an affine plane π, $B = \text{Hull}_p(\pi)^\perp$ is such a code. In fact, there are very standard choices for such codes B and we will see that the generalized Reed-Muller codes of Chapter 5 provide such choices.

We first need some definitions. Let B be an arbitrary code of block length n^2 over $\mathbf{F}_p$ (where p divides n) with minimum weight n.

Definition 6.8.1 *An affine plane π of order n is said to be* **contained** *in B if $C_p(\pi)$ is code-isomorphic to a subcode of B.*

Definition 6.8.2 *An affine plane π of order n is said to be* **linear over** *B if $C_p(\pi)$ is code-isomorphic to a subcode C of B where $B \subseteq C + C^\perp$.*

Example 6.8.1 Taking the Veblen-Wedderburn plane (see Veblen and Maclaglan-Wedderburn [285]) Ψ of order 9, and a "real" line R and setting $\psi = \Psi^R$, then the vectors in Hull$(\psi)^\perp$ of the form v^S, where S is a line or Baer subplane of ψ, generate an $[81, 48, 9]$ ternary code B over which ψ is linear. Moreover, $\omega = \Omega_{dual}^L$ is also linear over B where Ω is the non-desarguesian translation plane of order 9 (see, for example, Lüneburg [197]), Ω_{dual} its dual, and L a line through the translation point. Both Hull $(\psi)^\perp$ and Hull$(\omega)^\perp$ are $[81, 50, 9]$ ternary codes containing B. They are not code-isomorphic since Hull$(\psi)^\perp$ contains 2×306 minimum-weight vectors while Hull$(\omega)^\perp$ contains 2×522.

The code B of the above example is not a standard choice and we turn now to such codes. The notation we are about to introduce will be in force throughout the remainder of this section.

Set $q = p^s$ and let F be an arbitrary subfield of $\mathbf{F}_q$. Let V be a 2-dimensional vector space over $\mathbf{F}_q$; consider V as a vector space over F. If $F = \mathbf{F}_{p^t}$, where $s = th$, then V will have dimension $2h$ over F. In the affine geometry $AG_{2h}(\mathbf{F}_{p^t})$, we consider the design of points and h-flats. By Theorem 5.7.9 we know that the code over $\mathbf{F}_p$ of this design is the generalized Reed-Muller subfield subcode code $\mathcal{A}_{F_{p^t}/F_p}(h(p^t - 1), 2h)$.

For a more convenient notation in this chapter, where the emphasis is on the fields, the dimensions being determined by the fact that V will always be 2-dimensional over the largest field in sight, we will set

$$B(\mathbf{F}_q | \mathbf{F}_{p^t}) = \mathcal{A}_{F_{p^t}/F_p}(h(p^t - 1), 2h). \tag{6.1}$$

For any t dividing s, this is a code of length q^2 over $\mathbf{F}_p$ with minimum weight q, and its minimum-weight vectors are scalar multiples of the vectors v^X, where X is an h-flat of $AG_{2h}(F)$. Its dimension is computable — and given in Theorem 5.7.9. This subspace of $\mathbf{F}_p^V$ provides the "home" for the translation planes whose "kernel" (see Lüneburg [197]) contains F. It is a fair statement that for the majority of all planes discovered so far either the plane or its dual can be seen as contained in one of these codes—at least, an affine part can be so seen.

If F and K are two subfields of $\mathbf{F}_q$ with $F \subseteq K$, then $B(\mathbf{F}_q | K) \subseteq B(\mathbf{F}_q | F)$. At one extreme, $F = \mathbf{F}_q$ and the 2-flats are the lines of the affine plane $AG_2(\mathbf{F}_q)$ with $B(\mathbf{F}_q | \mathbf{F}_q) = C_p(AG_2(\mathbf{F}_q))$. At the other extreme,

$F = \mathbf{F}_p$ and the s-flats are the half-dimensional subspaces over the prime field $\mathbf{F}_p$. In this case $B(\mathbf{F}_q|\mathbf{F}_p) = \mathcal{R}_{F_p}(s(p-1), 2s)$. We wish to organize the class of translation planes of order q via the Galois correspondence we have just set up between the subfields of $\mathbf{F}_q$ and these geometric codes.

Example 6.8.2   (1) For $q = 9$ the only interesting standard choice is $B(\mathbf{F}_9|\mathbf{F}_3) = \mathcal{A}(4,4) = \mathcal{R}_{F_p}(4,4)$ of dimension 50 (see Theorem 5.8.2). It is an $[81, 50, 9]$ ternary code with 2×1170 vectors of weight 9, and is code-isomorphic to $\mathrm{Hull}(\Omega^T)^\perp$ where Ω is the non-desarguesian translation plane of order 9 and T its translation line. $\mathrm{Hull}(\Omega^T)$ has many vectors of weight 18 that are not of the form $v^\ell - v^m$ where ℓ and m are parallel lines. As we shall see this is a feature that plays an important role in the rigidity theorem we will prove; in the terminology that we will introduce, Ω is not "tame".

(2) For $q = 16$ we have $\mathbf{F}_2 \subset \mathbf{F}_4 \subset \mathbf{F}_{16}$ with $C(AG_2(16)) = \mathcal{A}_{F_{16}/F_2}(15, 2) = B(\mathbf{F}_{16}|\mathbf{F}_{16}) \subset B(\mathbf{F}_{16}|\mathbf{F}_4) \subset B(\mathbf{F}_{16}|\mathbf{F}_2) = \mathcal{R}_{F_2}(4, 8)$, and the dimensions of these codes are 81, 129 and 163, respectively. There are precisely two translation planes linear over $B(\mathbf{F}_{16}|\mathbf{F}_4)$: see Kleinfeld [170]. Their codes both have dimension 97. We do not know whether or not they are linearly equivalent.

Our first step is to characterize translation planes through the generalized Reed-Muller codes. We first need to point out an alternative approach to translation planes, first given by André [3] (or see Bruck and Bose [53]) through **spreads**. Thus, with V a $2s$-dimensional vector space over $\mathbf{F}_p$, a spread is a collection of $(p^s + 1)$ s-dimensional subspaces of V with the property that the intersection of any two of the chosen subspaces is $\{0\}$. The translation plane is given by taking the vectors of V to be the point set and the lines to be the subsets corresponding to the spread and all their translates.

Proposition 6.8.1 *Let q be a power of the prime p. Then an affine plane of order q is linear over $B(\mathbf{F}_q|\mathbf{F}_p)$ if and only if it is a translation plane.*

Proof: We first show that $B(\mathbf{F}_q|\mathbf{F}_p)^\perp$ contains every vector of the form $v^X - v^Y$ where X and Y are both translates of the same s-dimensional subspace over $\mathbf{F}_p$. (Actually we know this from Theorem 5.7.9 and, in fact, $\mathcal{A}(s(p-1), 2s)^\perp = \mathcal{A}(s(p-1) - 1, 2s)$ here, but we give an alternative elementary algebraic proof.) Since $B(\mathbf{F}_q|\mathbf{F}_p)$ is generated by vectors of the form v^Z, we need only show that $(v^X - v^Y, v^Z) = 0$ for all Z, where, again, Z is a translate of an s-dimensional subspace. Because of the affine invariance we can assume Z is an s-dimensional subspace; set $X = \mathbf{v} + S$

and $Y = \mathbf{w} + S$ where S is an s-dimensional subspace. If $S \cap Z = \{\mathbf{0}\}$, then $|X \cap Z| = |Y \cap Z| = 1$ and the result is clear. Otherwise, $S \cap Z$ being a subspace of positive dimension, $X \cap Z$ and $Y \cap Z$ are either empty or cosets of $S \cap Z$ and hence of cardinality a positive power of p, whence $(v^X, v^Z) = 0 = (v^Y, v^Z)$; *a fortiori* $(v^X - v^Y, v^Z) = 0$.

Now suppose π is a translation plane of order q. Then $C(\pi)$ is a subcode of $B(\mathbf{F}_q|\mathbf{F}_p)$ and $\mathrm{Hull}(\pi)$ is generated by vectors of the form $v^X - v^Y$, where X and Y are parallel s-flats, since the weight-q vectors of $B(\mathbf{F}_q|\mathbf{F}_p)$ are all of the form v^Z, by Theorem 5.7.9. Hence $\mathrm{Hull}(\pi) \subseteq B(\mathbf{F}_q|\mathbf{F}_p)^\perp$, i.e. $B(\mathbf{F}_q|\mathbf{F}_p) \subseteq \mathrm{Hull}(\pi)^\perp$ and π is linear over $B(\mathbf{F}_q|\mathbf{F}_p)$.

Suppose π is linear over $B(\mathbf{F}_q|\mathbf{F}_p)$ but not a translation plane. Then $B(\mathbf{F}_q|\mathbf{F}_p)^\perp \supsetneq \mathrm{Hull}(\pi)$. Since $\mathrm{Hull}(\pi)$ is generated by vectors of the form $v^\ell - v^m$ where ℓ and m are parallel lines of π, there must be an s-dimensional subspace S with $v^S = v^\ell$ where ℓ is a line of π and an m in π parallel to ℓ but not a translate of S; i.e. $v^m = v^Y$ with $S \cap Y$ empty but Y not a translate of S. If Y is a translate of the subspace T then clearly $S \cap T \neq \{\mathbf{0}\}$ and $S \cap T \neq S$. Let U and U' be subspaces with $S = (S \cap T) \oplus U$ and $V = (S + T) \oplus U'$. Then $U + U' = Z$, an s-dimensional subspace of V, with $Z \cap S \neq \{\mathbf{0}\}$ and $Z \cap T = \{\mathbf{0}\}$. Hence $(v^\ell - v^m, v^Z) = 1$, a contradiction. $\square$

We know from Theorem 5.7.6, following from the work of Delsarte *et al.*, that $\mathcal{A}_{F_{p^t}/F_p}(h(p^t - 1) - 1, 2h)$ is the code generated by vectors of the form $v^X - v^Y$, where X and Y are cosets of the same subspace and that these give the minimum-weight vectors. Now the above proof shows that the hull of any translation plane of order q has minimum weight $2q$. Thus even a non-desarguesian translation plane will always have a hull with the expected minimum weight, but there can be and, it seems, frequently are, minimum-weight vectors that are not of the form $v^\ell - v^m$ where ℓ and m are parallel lines—as, for example, in the hull of the non-desarguesian translation plane of order 9.

This characterization yields an upper bound on $\dim(C_p(\pi))$ when π is any affine translation plane of order q. We will later give a refinement of the following proposition that will take into account the kernel of the translation plane. The **kernel** will, for us, be the largest subfield F of $\mathbf{F}_q$ for which the plane is contained in $B(\mathbf{F}_q|F)$.

Proposition 6.8.2 *If π is an affine translation plane of order $q = p^s$ then*

$$\dim C_p(\pi) \leq q^2 + q - \dim(B(\mathbf{F}_q|\mathbf{F}_p)).$$

Moreover, any two affine translation planes meeting this upper bound are linearly equivalent.

Proof: If π is a translation plane of order q then

$$\mathrm{Hull}_p(\pi) \subseteq B(\mathbf{F}_q|\mathbf{F}_p) \subseteq \mathrm{Hull}_p(\pi)^\perp.$$

Each inequality gives an upper bound. We use the second: $B(\mathbf{F}_q|\mathbf{F}_p) \subset \mathrm{Hull}_p(\pi)^\perp$ implies $\mathrm{Hull}_p(\pi) \subseteq B(\mathbf{F}_q|\mathbf{F}_p)^\perp$ and hence $\dim C_p(\pi) - q \le \dim(B(\mathbf{F}_q|\mathbf{F}_p)^\perp) = q^2 - \dim(B(\mathbf{F}_q|\mathbf{F}_p))$, yielding the inequality.

If $\dim C_p(\pi) = q^2 + q - \dim(B(\mathbf{F}_q|\mathbf{F}_p))$ then $\mathrm{Hull}_p(\pi) = B(\mathbf{F}_q|\mathbf{F}_p)^\perp$; hence any two translation planes with $\dim C_p(\pi) = q^2 + q - \dim(B(\mathbf{F}_q|\mathbf{F}_p))$ have isomorphic hulls or, in other words, they are linearly equivalent. $\square$

Corollary 6.8.1 *If π is an affine translation plane of order $q = p^s$ then*

$$\dim C_p(\pi) \le q^2 + q - \sum_{i=0}^{s-1} (-1)^i \binom{2s}{i} \binom{p(s-i)+s}{2s}.$$

For $p = 2$ this becomes

$$\dim C_2(\pi) \le 2^s(2^{s-1} + 1) - \frac{1}{2}\binom{2s}{s}.$$

Proof: The proof follows from Theorem 5.8.2. $\square$

It can be shown that for $q = p^2$ the bound becomes

$$\dim C_p(\pi) \le \frac{1}{2}p^4 + p^2 - \frac{1}{3}p^3 - \frac{1}{6}p.$$

(Another bound in this case that comes from orthogonality conditions is $\dim C_p(\pi) \le \frac{1}{2}(p^4 + p^2)$ for a plane of order $q = p^2$; this latter bound is, however, for *any* affine plane of order p^2.) Notice that for $p = 3$ the bound is 40. There are precisely six non-desarguesian affine planes of order 9 (see Lam *et al.* [177]) and they all have ternary codes of dimension 40, while $\dim C_3(AG_2(\mathbf{F}_9)) = 36$. For $p = 2$ the bound is 9 and $C_2(AG_2(\mathbf{F}_4))$ has dimension 9. For $p = 5$ the bound is 295. Czerwinski and Oakden [78] have found that there are 21 isomorphism classes of translation planes of order 25. We have computed the dimensions of each of their codes; our computations show that the highest value is 263 and that only 13 distinct values occur: see Assmus and Key [13] but note that because of a typographical error in Czerwinski and Oakden [78] we incorrectly reported 22 isomorphism classes there.

Before going on to more refined upper bounds on the dimensions of the codes of translation planes we pause to answer a rather obvious question:

what about flats of other dimensions with respect to the codes of affine
translation planes?

The incidence vectors of the half-dimensional flats of $AG_{2s}(\mathbf{F}_p)$ generate
$B(\mathbf{F}_q|\mathbf{F}_p)$ and this code is contained in the orthogonal of the hull of any
translation plane of order q. Now we set

$$B(r) = \langle v^X \,|\, X \text{ an } r\text{-flat of } AG_{2s}(\mathbf{F}_p) \rangle.$$

The results of Chapter 5 tell us that

$$B(r) = \mathcal{A}_{F_p/F_p}((2s - r)(p - 1), 2s) = \mathcal{R}_{F_p}((2s - r)(p - 1), 2s)$$

but we will not use this fact, our proof being elementary. Here is the story
for arbitrary dimension r:

Proposition 6.8.3 *If π is a translation plane of order $q = p^s$ and $C(\pi)$
its code, then*

$$B(r) \subseteq \text{Hull}(\pi)^\perp$$

if and only if $r \geq s$. If $r > s$, then $B(r) \subseteq C(\pi)^\perp$.

Proof: Notice first that if $2s \geq r \geq r'$ then $B(r) \subseteq B(r')$, since if X is
an r-flat v^X can be written as a sum of vectors v^Y where the Y are all
translates of some subspace of dimension r'.

Let X be an r-flat and, without loss of generality, suppose that X is a
subspace of V. Let S_i, $i = 0, 1, \ldots, q$, be a spread for π. We show first that
$|X \cap S_i| > 1$ for all i if and only if $r > s$. For suppose $|X \cap S_i| > 1$ for all i.
Then $X - \{\mathbf{0}\} = \cup_{i=0}^{s}(X \cap S_i - \{\mathbf{0}\})$ which is a disjoint union, so $p^r - 1 \geq$
$(p^s + 1)(p - 1)$, i.e. $p^r \geq p^{s+1} - p^s + p$, so that $p^r > p^{s+1} - p^s = p^s(p - 1)$,
and hence $r > s$. Conversely, if $r > s$ and $X \cap S_i = \{\mathbf{0}\}$ for some i, then
$\dim(X + S_i) = r + s > 2s$, which is impossible. Thus $X \cap S_i \neq \{\mathbf{0}\}$ for all i
if and only if $r > s$, which is equivalent to the statement $B(r) \subseteq C(\pi)^\perp$ if
and only if $2s \geq r > s$, since $(v^X, v^{S_i}) \equiv 0 \pmod{p}$ for all S_i and all their
translates.

If $r \leq s$ then there is a value of i for which $X \cap S_i = \{\mathbf{0}\}$. Then
$|X \cap (a + S_i)| = 1$ for all translates $a + S_i$ of S_i if and only if $r = s$. Thus
$v^X \in \text{Hull}(\pi)^\perp$ if and only if $r \geq s$. $\square$

Corollary 6.8.2 *If Π is a finite projective translation plane of order $q =
p^s$, L a translation line and $\pi = \Pi^L$, then, if X is an r-flat for $r > s$, the
vector $v \in \mathbf{F}_p^{\mathcal{P}}$ with $v_Q = 1$ for $Q \in X$, $v_Q = 0$ for $Q \in L$ or $Q \notin X$, is in
$C(\Pi)^\perp$, i.e. $v^{X-L} \in C(\Pi)^\perp$.*

Proof: This follows from the proposition and Theorem 6.3.2 since $C(\pi)^\perp$ is, essentially, the space $\{u | u \in C(\Pi)^\perp$ and $u_Q = 0$ for $Q \in L\}$, into $\mathbf{F}_p^V$. □

We now improve the bound given by Corollary 6.8.1. Recall that for any subfield $F = \mathbf{F}_{p^t}$ of $\mathbf{F}_q$ we view V as a $2h$-dimensional vector space over F, where $s = ht$. We set

$$E(\mathbf{F}_q | F) = \langle v^X - v^Y | X, Y \text{ parallel } h\text{-flats in } V \rangle.$$

From Theorem 5.7.6 we know that

$$E(\mathbf{F}_q | F) = \mathcal{A}_{F_{p^t}/F_p}(h(p^t - 1) - 1, 2h) = \mathcal{A}(h(p^t - 1) - 1, 2h)$$

and that $\mathcal{A}(h(p^t - 1) - 1, 2h)^\perp$ has minimum weight $p^{th} = q$. Also

$$\mathcal{A}(h(p^t - 1) - 1, 2h) \subseteq \mathcal{A}(h(p^t - 1), 2h)^\perp$$

with equality when $t = 1$, $h = s$ — see Theorem 5.7.9.

The refinement for translation planes with kernel $F = \mathbf{F}_{p^t}$ is as follows:

Theorem 6.8.1 *Let π be a translation plane contained in $B(\mathbf{F}_q | F)$, where F is a subfield of $\mathbf{F}_q$. Then*

$$\dim(C(\pi)) \le q + \dim(E(\mathbf{F}_q | F)).$$

Moreover, any two translation planes contained in $B(\mathbf{F}_q | F)$ whose p-ranks meet this bound are linearly equivalent.

Proof: If $C(\pi)$ is isomorphic to $B(\mathbf{F}_q | F)$ then $\text{Hull}(\pi)$ is isomorphic to a subcode H of $E(\mathbf{F}_q | F)$; this follows since $\text{Hull}(\pi)$ is generated by vectors of the form $v^\ell - v^m$, where ℓ and m are parallel lines of π, and, under the isomorphism of $C(\pi)$ onto C, v^ℓ will correspond to a vector v^X that is a generator of $B(\mathbf{F}_q | F)$, these being the only (up to scalar multiples) weight-q vectors of $B(\mathbf{F}_q | F)$. Since $\dim(H) = \dim(\text{Hull}(\pi)) = \dim(C(\pi)) - q$, the required inequality follows. Moreover, equality implies that $H = E(\mathbf{F}_q | F)$, and hence any two affine translation planes contained in $B(\mathbf{F}_q | F)$ and meeting the bound have hulls isomorphic to $E(\mathbf{F}_q | F)$, and hence are linearly equivalent. □

The bound is of course the same as the previous one for planes whose kernel is $\mathbf{F}_p$, since

$$E(\mathbf{F}_q | \mathbf{F}_p) = \mathcal{A}_{F_p/F_p}(s(p-1) - 1, 2s) = \mathcal{A}_{F_p/F_p}(s(p-1), 2s)^\perp.$$

Example 6.8.3 The refined bound for $q = 16$ is 101, rather than 109, when the translation plane is "2-dimensional" i.e. contained in $B(\mathbf{F}_{16}|\mathbf{F}_4)$.

The Galois correspondence between the subfields of $\mathbf{F}_q$ and the codes $B(\mathbf{F}_q|F)$ yields a hierarchy of translation planes corresponding, in the language of the existing literature on translation planes, to the kernel of the translation plane. The theorem shows that this hierarchy reflects itself in the p-rank of the incidence matrix of the translation plane. Roughly speaking, the rank goes down as the kernel gets larger. Of course, the theorem simply gives upper bounds that grow smaller as F gets larger; in practice the p-rank is rather sporadic. As we remarked, we do not have a compact formula for $\dim(\mathcal{A}_{F_{p^t}/F_p}(h(p^t - 1) - 1, 2h))$ for $s = th$ in general, but the formula given in Theorem 5.7.9 is easily computable in any particular case — for example, using the CAYLEY language [66].

A compact general formula, when $s = 2t$, for $\dim(\mathcal{A}_{F_{p^t}/F_p}(2(p^t - 1) - 1, 4))$, would be particularly useful. For example, the dimension of $\mathcal{A}_{F_4/F_2}(5, 4)$ is 85 and yields the bound 101 for 2-dimensional translation planes of order 16 (in this case this simply means that the kernel of the translation plane is the field $\mathbf{F}_4$). There are precisely two such planes (see Kleinfeld [170]) and they both have 2-rank 97. All translation planes of order 16 are known (Dempwolff and Reifart [87]) and all the 2-ranks have been determined (Assmus and Key [13]). The (affine) ranks lie between 81 (the desarguesian plane) and 105 (the plane constructed by Lorimer in [192] and the derived semi-field plane) with, as we have indicated, the 2-dimensional planes having rank 97. Observe that, from dimensional considerations alone, it can be deduced that the kernels of both the plane constructed by Lorimer and the derived semi-field plane are $\mathbf{F}_2$.

We do not, unfortunately, have any instances of planes meeting the refined bounds, but if there were any such planes, they would all be linearly equivalent just as in the case of dimension equal to

$$q + \dim(\mathcal{A}_{F_p/F_p}(s(p - 1) - 1, 2s)).$$

6.9 Tame planes and a rigidity theorem

In this section we will introduce the notion of "tameness" and prove a rigidity theorem for a certain class of projective planes. The theorem in question is not strong enough to establish the conjecture it sets out to prove, but we have been unable to strengthen it.

We have already remarked on the feeling among early workers in the field that the desarguesian projective planes provided the codes with the least redundancy. But the first precise conjecture along these lines was made by Hamada [125] (see Section 4.6) and subsequently he and Ohmori [127] proved a rigidity theorem in characteristic 2 which we will treat in Chapter 7, Corollary 7.5.1. The sweeping conjecture of Hamada [125] proved to be false, as Tonchev [280] observed and, in fact, Delsarte and Goethals [105] had noticed a 2-(31,7,7) counter-example already in 1968 — even before the conjecture was made! Independently, Sachar [251] made a narrower conjecture that is still undecided. We can, however, say something concerning this narrower conjecture which we call the "Hamada-Sachar conjecture". Here is the statement:

Conjecture 6.9.1 *Every projective plane of order p^s, p a prime, has p-rank at least $\binom{p+1}{2}^s + 1$ with equality if and only if it is desarguesian.*

That $\dim C(PG_2(\mathbf{F}_{p^s})) = \binom{p+1}{2}^s + 1$ follows from Theorem 5.7.1 with $m = 2$. Thus the conjecture states that the desarguesian planes have the smallest possible p-rank for planes of order a power of the prime p. We begin with a technical result.

Lemma 6.9.1 *Suppose π and σ are affine planes of order n and p is a prime dividing n. Then if a subcode C of $\mathrm{Hull}_p(\pi)^\perp$ can be chosen with C code-isomorphic to $C_p(\sigma)$ and $\mathrm{Hull}_p(\pi) \subseteq C$, we have that*

$$\dim C_p(\pi) \leq \dim C_p(\sigma)$$

with equality if and only if σ is linearly equivalent to π.

Proof: Without loss of generality we may assume that

$$H = \mathrm{Hull}_p(\pi) \subseteq C_p(\sigma) \subseteq H^\perp.$$

Since the first containment implies $C_p(\sigma)^\perp \subseteq H^\perp$ we have that $\mathrm{Hull}_p(\sigma)^\perp \subseteq H^\perp$ or, in other words,
$$H \subseteq \mathrm{Hull}_p(\sigma).$$

Since

$$\dim H = \dim C_p(\pi) - n \text{ and } \dim \mathrm{Hull}_p(\sigma) = \dim C_p(\sigma) - n$$

we have the required inequality with equality if and only if $H = \mathrm{Hull}_p(\sigma)$; i.e. if and only if π and σ are linearly equivalent. $\square$

The lemma itself is a rigidity theorem, though too general to be of interest. But, in view of the lemma, there are two obstacles in the way to a proof of the conjecture. They are:

(1) If we take $\pi = AG_2(\mathbf{F}_q)$, how restrictive is the assumption that $C(\sigma)$ contains a code isomorphic to $\mathrm{Hull}(AG_2(\mathbf{F}_q))$?

(2) When does linear equivalence imply isomorphism?

The second question leads us to the notion of "tameness", which we now introduce. We wish to know when two projective planes are isomorphic under the assumption that one has an affine part linearly equivalent to an affine part of the other. As the definition makes clear, it is the hull and its minimum-weight vectors that play the important role.

Definition 6.9.1 *A projective plane* Π *of order* n *is said to be* tame *(or, more properly, "tame at* p*") if* $\mathrm{Hull}_p(\Pi)$ *has minimum weight* $2n$ *and the minimum-weight vectors are precisely the scalar multiples of the vectors of the form* $v^L - v^M$, *where* L *and* M *are lines of the plane.*

We then get, immediately, the result we need:

Proposition 6.9.1 *Suppose* Π *and* Σ *are projective planes of order* n *and that* Π *is tame at* p, *where* p *is odd. If for some line* L *of* Π *and some line* M *of* Σ, Π^L *is linearly equivalent at* p *to* Σ^M, *then* Π *and* Σ *are isomorphic.*

Proof: Set $\pi = \Pi^L$ and $\sigma = \Sigma^M$. We may as well assume that $\mathrm{Hull}_p(\pi) = \mathrm{Hull}_p(\sigma)$ in a vector space $\mathbf{F}_p^{n^2}$. Since Π is tame at p and the natural projection, $T \colon \mathrm{Hull}_p(\Pi)^\perp \to \mathrm{Hull}_p(\pi)^\perp$, maps $\{c \in \mathrm{Hull}_p(\Pi) \mid c_Q = 0$ for all Q on $L\}$ isomorphically onto $\mathrm{Hull}_p(\pi)$, the minimum-weight vectors of $\mathrm{Hull}_p(\pi)$ are precisely the scalar multiples of vectors of the form $v^\ell - v^m$ where ℓ and m are parallel lines of π. Since for any affine plane ρ, $\mathrm{Hull}_p(\rho)$ contains (indeed is generated by) all vectors of this form, and since clearly, when p is odd, ℓ and m can be recovered from $v^\ell - v^m$, the lines of π and σ are precisely the same. But Π and Σ are uniquely determined by π and σ, which completes the proof. $\square$

Remark (1) A projective plane need not be tame: the translation plane of order 9 that is non-desarguesian is not tame at $p = 3$.

(2) It is conceivable that a projective plane could be tame at one prime but not at another. For instance, if a projective plane of order $2p$ exists, where p is an odd prime (necessarily congruent to 5 modulo 8) then if that plane had an oval, it would not be tame at 2, but probably would be tame at p.

(3) The weight distributions of the codes of the projective planes of orders
4 and 8 show that they are tame, but, as already mentioned, there are
112 affine planes of order 4 contained in the (16,11) extended binary
Hamming code. They are all, of course, isomorphic but the hull does
not locate a unique plane and it is conceivable that for order 16,
say, there are two non-isomorphic tame planes with affine parts that
are linearly equivalent. It is also possible that the Hamada-Sachar
conjecture is false for q a power of 2, even for translation planes.

The results in Chapter 5 have already allowed us to conclude, through
Corollary 6.4.4, that the classical projective planes are tame.

Theorem 6.9.1 *Desarguesian projective planes are tame.*

We do not have an example of a tame plane that is not desarguesian.
Since the translation plane of order 9 that is not desarguesian is not tame
and the other two planes of order 9 do not appear to be tame, finding a
tame plane that is not desarguesian seems to be a formidable problem and
it may even be true that only the desarguesian planes are tame.

Turning again to the Hamada-Sachar conjecture, we suspect that the
assumption that the hull of $AG_2(\mathbf{F}_q)$ be contained in the code of an affine
plane is not a great restriction; it may be that every translation plane
satisfies this condition. Here is what we can presently prove:

Theorem 6.9.2 *Let σ be an affine plane of odd order $q = p^s$ contained in
$B(\mathbf{F}_q|\mathbf{F}_p)$. Then, if $C(\sigma)$ contains a subcode isomorphic to $\mathrm{Hull}(AG_2(\mathbf{F}_q))$,
we have that*

$$\dim C(\sigma) \geq \binom{p+1}{2}^s$$

with equality if and only if σ is isomorphic to $AG_2(\mathbf{F}_q)$.

Proof: The theorem follows from the lemma, the proposition and the theorems above and the fact that $C(AG_2(\mathbf{F}_q)) \subseteq B(\mathbf{F}_q|\mathbf{F}_p) \subseteq C(AG_2(\mathbf{F}_q))^\perp$. $\square$

6.10 Derivations

In order to give the reader a feel for the construction of one affine plane
from another via a technique called "derivation" we discuss first the so-
called "2-dimensional case", i.e. translation planes of order q^2 contained
in $B(\mathbf{F}_{q^2}|\mathbf{F}_q)$, or, put another way, translation planes of order q^2 whose

kernel contains $\mathbf{F}_q$. We simply indicate how, given two such planes, one can be obtained from the other by replacing some of the lines of one by Baer subplanes of the other.

Notice that we have changed notation: $\mathbf{F}_{q^2}$ is our big field, $F = \mathbf{F}_q$ and we are working in the code $B(\mathbf{F}_{q^2}|\mathbf{F}_q)$. Thus V is a 2-dimensional vector space over $\mathbf{F}_{q^2}$, and a 4-dimensional space over F, and the ambient code is generated by the 2-dimensional flats over F. A simple count shows that the number of such 2-flats is

$$\frac{q^2(q^4 - 1)(q^4 - q)}{(q^2 - 1)(q^2 - q)} = q^2(q^2 + 1)(q^2 + q + 1).$$

It follows from Theorem 5.7.9 that $B(\mathbf{F}_{q^2}|F)$ has precisely

$$(q - 1)q^2(q^2 + 1)(q^2 + q + 1)$$

weight-q^2 vectors. They yield a design with point set V of cardinality q^4, with block size q^2 and $\lambda = q^2 + q + 1$. It is the design given by the code's minimum-weight vectors.

Any affine plane of order q^2 linear over $B(\mathbf{F}_{q^2}|F)$ has $q^4 + q^2$ of these vectors as lines and $q^3(q^2+1)(q+1)$ of them as Baer subplanes. This fact is easy to establish; the more general assertion, concerning Baer subplanes, for arbitrary planes whose order is the square of a prime, is clearly related, but in this more general case it is not clear how many vectors there are in the orthogonal to the hull. The advantage of the coding-theoretical approach is that counting in $B(\mathbf{F}_{q^2}|F)$ is easy and set-theoretical intersections are simple to deal with since they are, essentially, intersections of subspaces. Thus the following assertion about any two such planes should be obvious:

Proposition 6.10.1 *Suppose π and σ are two affine planes of order q^2 that are linear over $B(\mathbf{F}_{q^2}|\mathbf{F}_q)$. Then σ can be obtained from π by using certain of π's lines and certain of its Baer subplanes as the lines of σ.*

We restricted ourselves to affine planes of order q^2 simply to make contact with the existing literature on translation planes. More generally, any two planes of order q that are contained in $B(\mathbf{F}_q|\mathbf{F}_p)$ are part of the design given by the minimum-weight vectors of this code and one can be obtained from the other by a "generalized derivation". There is nothing very deep or mysterious going on here and much of the existing literature on derivations, nets, etc. can be viewed, from the present perspective, as ingenious methods for choosing which lines to ignore and which Baer subplanes, or whatever, to include when going from one affine plane to another. For a detailed account of these geometric matters the reader may wish to consult Ostrom [224].

We record next another simple consequence of the above discussion.

Proposition 6.10.2 *An affine plane that is linear over $B(\mathbf{F}_{q^2}|\mathbf{F}_q)$ has at least*

$$q^3(q^2 + 1)(q + 1)$$

desarguesian Baer subplanes.

Proof: Only the fact that the planes are desarguesian remains to be proved. If v is a weight-q^2 vector of $B(\mathbf{F}_{q^2}|\mathbf{F}_q)$ and H is the hull of the given affine plane of order q^2, then $v \in H^\perp$. Since all such vectors v have supports that are $\mathbf{F}_q$-subspaces or their translates, $|\mathrm{Supp}(v) \cap \ell| = 0, 1, q$ or q^2 for every line ℓ of the plane. If this intersection is of cardinality q^2, $v = av^\ell$; otherwise it is a Baer subplane and moreover its structure is of a 2-dimensional vector space over $\mathbf{F}_q$ with the 1-dimensional $\mathbf{F}_q$-subspaces or their translates as lines; it is, therefore, isomorphic to $PG_2(\mathbf{F}_q)$. $\square$

Caution: The projective completion may very well have more Baer subplanes; for example, $PG_2(\mathbf{F}_{q^2})$ will. This is because some Baer subplanes of a given projective plane Π may meet a given line L in only one point and thus, in $\mathrm{Hull}_p(\Pi^L)$, they will appear as constant vectors of weight $q^2 + q$, rather than as minimum-weight vectors that are Baer subplanes.

Corollary 6.10.1 *Any translation plane of order p^2 has at least*

$$p^3(p^2 + 1)(p + 1)$$

desarguesian subplanes of order p.

For $p = 3$ this is the exact number for the non-desarguesian translation plane of order 9. Curiously, all three non-desarguesian projective planes of order 9 have precisely 1080 Baer subplanes; $PG_2(\mathbf{F}_9)$ has 7560 Baer subplanes. All these Baer subplanes are, of course, isomorphic to $PG_2(\mathbf{F}_3)$. These facts are clearly related to the complete weight enumerator of the codes involved. No one has succeeded in computing the complete weight enumerator of any of the four projective planes of order 9; the complete weight enumerator of $PG_2(\mathbf{F}_3)$ is easy to compute: it was done by Assmus and Salwach [23] and is recorded in Table 2.2.

Ostrom [223] defined a very general notion of derivation for projective planes; with the help of geometric or combinatorial arguments, new planes have been constructed using this notion. The new planes constructed involve the choice of a **Baer segment** of a line as a derivation set, where a Baer segment is simply that part of a line of the ambient plane that is the point set of the line of a Baer subplane. The coding-theoretical analysis of projective planes that we have outlined above leads naturally to a notion of a derivation set that is less general than that proposed by Ostrom, but more general than the derivations involving Baer segments. Our definition involves describing a subset D of points of a line L of a projective plane with code C as a derivation set if there exists a set of constant vectors in $C + C^{\perp}$, all of the same weight, all with D in the support, and any two having at most one common point off L in their supports. In fact, from our definition it follows that for a plane of order p^2, p a prime, the only possible non-trivial derivation sets *are* the Baer segments, and that for planes of prime order there are no non-trivial derivation sets at all.

Let Π be an arbitrary projective plane of order n, and p a prime dividing n. Let L be a line of Π. We want to explain when a subset D of L will be called a "derivation set" for Π.

Set $\pi = \Pi^L$. Then π is also of order n, and we have the natural projection of the orthogonal of Π's hull onto the orthogonal of π's, i.e. of $B(\Pi)$ onto $B(\pi)$. If $\dim(B(\Pi)) = k$, then $B(\pi)$ is an $[n^2, k-1, n]$ code. Besides the incidence vectors of the lines of π, there may be, as we have seen, other minimum-weight vectors in $B(\pi)$, all of which are of the form av^X, where X is a subset of points of π, $|X| = n$, and $a \in \mathbf{F}_p - \{0\}$. We have already described the nature of these minimum-weight vectors of $B(\pi)$ in Theorem 6.3.3 and Lemma 6.4.1. Thus let $v = av^X$ be a weight-n vector of $B(\pi)$, where X is not a line of π, and normalize v so that $a = 1$. Then lines in the same parallel class of π meet X constantly modulo p, and, if r is the number of classes whose lines meet X in a multiple of p points, then there is a unique vector $v' \in B(\Pi)$ with $v' \notin C(\Pi)^{\perp}$ such that $v' = v^{X'}$ for some subset X' of points of Π, v' projects to v, and $\mathrm{wt}(v') = |X'| = n+r$, with $1 < r < n$, $r \equiv 1 \pmod p$. The set $L \cap X'$ is, obviously, of cardinality r.

If $\Pi = PG_2(\mathbf{F}_{q^2})$, X' could be the points of a Baer subplane, with $L \cap X'$ a Baer segment of L. Then X would be an affine Baer subplane of $AG_2(\mathbf{F}_{q^2})$, furnishing a weight-q^2 vector of $B(AG_2(\mathbf{F}_{q^2}))$ that does not come from a line. (Here, of course, q is a power of p, the order of Π being q^2.)

The above discussion motivates our definition of "derivation set".

Definition 6.10.1 *Let D be a subset of points on a line L of a finite projective plane Π of order n. We say D is a **derivation set** for Π if there is a prime p dividing n and a collection $\mathcal{D}$ of vectors of $B(\Pi)$ satisfying the following conditions:*

- *$|\mathcal{D}| = n|D|$;*

- *each vector $v' \in \mathcal{D}$ is of the form $v^{X'}$, where $D \subset X'$ and $\mathrm{wt}(v') = |X'| = n + |D|$;*

- *for distinct v' and w' in $\mathcal{D}$,* $\text{wt}(v' - w') \geq 2n - 2$.

The definition simply indicates that we plan to remove $|D|$ parallel classes of lines, each class having n lines; that each line has n points; and that the "new" lines intersect properly. The point of the definition is that using Π and $\mathcal{D}$ it is easy to construct an affine plane, $\pi(\Pi, \mathcal{D})$, as follows: the points of $\pi(\Pi, \mathcal{D})$ are those of Π^L; the lines of $\pi(\Pi, \mathcal{D})$ are (i) those of π coming from the lines M of Π with $M \cap L \notin D$ and (ii) the subsets X of the points of π coming from $\{v^X | v^X = \text{im}(v'), \ v' \in \mathcal{D}\}$, where $\text{im}(v')$ denotes the image of v' under the natural projection of $B(\Pi)$ to $B(\Pi^L)$.

The proof that this defines a plane follows easily.

It is an immediate consequence of the definition and of what we have already proved concerning the minimum-weight vectors of the orthogonal to the hull of an affine plane, that a derivation set has cardinality congruent to 1 modulo p. Those with cardinality equal to 1 have for $\mathcal{D}$ a parallel class of lines. Those with cardinality greater than 1 may admit various $\mathcal{D}$'s. Notice that the above construction could be carried out using two or more disjoint derivation sets on the line L: for example, the Hall plane of order 16 can be obtained from $PG_2(\mathbf{F}_{16})$ by using two disjoint Baer segments. Perhaps a more exact term for the above notion would be "primitive derivation". However, this notion does explain why Baer segments appeared as derivation sets. The next proposition, which is a direct consequence of our determination of the minimum-weight vectors for planes of prime order and those whose orders are a square of a prime, makes this point.

Proposition 6.10.3 *For a plane of prime order there do not exist non-trivial derivation sets. For a plane whose order is a square of a prime, the only possible non-trivial derivation sets are Baer segments.*

The "classical" derivation [143] uses a Baer segment, D, of L and a collection of Baer subplanes having L as a line and D as the intersection of L with the Baer subplanes; this is, of course, one of our cases, but there are other cases as well. Our definition suggests looking for derivation sets under mild, algebraic-coding-theoretical constraints, whereas the broader definition of Ostrom calls for a set-theoretical search.

Consider, therefore, $B(K|F)$, where $K = \mathbf{F}_q$ and $F = \mathbf{F}_p$. $B(K|F)$ is a code of length q^2 and minimum weight q, and we know all the minimum-weight vectors. If π is an affine translation plane of order q, we have $C(\pi) \subseteq B(K|F) \subseteq B(\pi)$. If we restrict the search to $B(K|F)$, then we would be very close to the methods of Ostrom; searching in $B(\pi)$ would, of course, be superior, but more difficult, since a survey of the weight-q vectors of $B(K|F)$ is more tractable than a survey of those of $B(\pi)$.

We now show that the "classical" Baer subplane derivation, in the case of translation planes, takes place in the subspace $B(K|F)$ of the usually larger $B(\pi)$.

Proposition 6.10.4 *If q is a square and π is an affine translation plane of order q, then every affine Baer subplane of π has point set X satisfying $v^X \in B(\mathbf{F}_q|\mathbf{F}_p)$.*

Proof: Let Π be an arbitrary projective plane, Σ a Baer subplane of Π, and (P, L) a flag of Π that belongs to Σ. Let α be an elation of Π with centre P and axis L. Then, if Q and Q^α are points of Σ where $Q \notin L$, it follows that α is an elation of Σ; i.e. $\Sigma^\alpha = \Sigma$. For, if A is any point of Σ not on L and not on the line QQ^α, we have that AQ and PA are lines of Σ. The point $B = QA \cap L$ is on Σ, and hence $BQ^\alpha \cap PA$ is a point of Σ. But $A^\alpha = BQ^\alpha \cap PA$, and thus A^α is on Σ. We need only show now that every point C of Σ on QQ^α has C^α on Σ; but this follows if we reverse the roles of A and Q, since there must be an A in Σ not on L or QQ^α.

Now let $\pi = \Pi^L$ be an affine translation plane. All elations with axis L and centres on L exist. If X' is the point set of a Baer subplane of Π and L is a line of this Baer subplane, $X = X' \cap \pi$ is an affine Baer subplane of π with $v^X \in B(\pi)$. We want to show that $v^X \in B(\mathbf{F}_q|\mathbf{F}_p)$. Without loss of generality, we can assume $O \in X$, O being the zero vector of V. Suppose $Q \in X$, $Q \neq O$. The line QO meets L at a point P (at infinity) and there is an elation of Π with centre P that moves O to Q. By the above, this elation fixes X setwise. It is defined by the vector (i.e. point) Q' going to $Q' + Q$ and hence X is closed under addition. Thus X is a subspace of V, and $v^X \in B(\mathbf{F}_q|\mathbf{F}_p)$. $\square$

6.11 Ovals and derivation sets

We next turn our attention to the case $p = 2$. Suppose we are given a translation plane; i.e. suppose we are given a spread, $S_0, \ldots, S_q$. It may happen that a subspace T satisfies $\dim(S_i \cap T) = 0$ or 1 for all i. Such a T will have the property that v^T is a weight-q vector of $B(K|F)$ with $|T \cap \ell| = 0, 1$, or 2 for ever line ℓ of the translation plane π. Now if $\pi = \Pi^L$ and v' is the preimage of v^T given by Lemma 6.4.1, then $v' + v^L = v^{X'}$, where X' is an oval ($q + 2$ points with no three collinear) of Π. We show how a set of ovals in a translation plane coordinatized by a nearfield of even order can define the vectors in $\mathcal{D}$, yielding a derivation set D of cardinality $q - 1$. The construction depends on the existence of a hyperbolic oval in the affine translation plane.

Proposition 6.11.1 *Let Π be a projective plane of even order q. Suppose that*

(1) Π has an oval $\mathcal{O}'_0$, and

(2) there are two points, P_0 and Q_0 on $\mathcal{O}'_0$ for which Π is (P_0, Q_0)-transitive.

Then, if L is the line through P_0 and Q_0, $D = L - \{P_0, Q_0\}$ is a derivation set for Π.

Proof: The condition that Π is (P_0, Q_0)-transitive — i.e. that every central collineation with centre P_0 and axis through Q_0 exists — is equivalent to the condition that Π is a translation plane coordinatized by a nearfield. Thus $q = 2^s$, for some s.

We need to produce the set $\mathcal{D}$ of vectors $v^{X'}$ in the binary code $B(\Pi)$ with $|X'| = 2^{s+1} - 1$, and with $X' \supset D$. Since any oval $\mathcal{O}'$ of Π provides a vector $v^{\mathcal{O}'}$ in $C(\Pi)^{\perp}$ of weight $2^s + 2$, the oval $\mathcal{O}'_0$ will yield $v' = v^L + v^{\mathcal{O}'_0}$ of weight $2^{s+1} - 1$ with support containing D. Our goal is to extract a set of $2^s(2^s - 1)$ of these ovals, with $v^{\mathcal{O}'} + v^{\mathcal{Q}'}$ of weight at least $2^{s+1} - 1$ for distinct $\mathcal{O}'$ and $\mathcal{Q}'$. This we accomplish in a purely combinatorial manner.

Let H denote the group of all collineations of Π with centre P_0 and axis passing through Q_0. Then H has order $2^s(2^s - 1)$ and no non-identity element of H can fix $\mathcal{O}'_0$. Thus the orbit Ω of $\mathcal{O}'_0$ under H has size $2^s(2^s - 1)$. If we form the incidence structure $\mathcal{S}$ consisting of the points of Π not on L, and having for blocks the orbit Ω of ovals, together with the lines through P_0 and Q_0 other than L, then we can show that this structure is an affine plane of order 2^s.

Now $\mathcal{S}$ has 2^{2s} points and $2^{2s} + 2^s$ blocks, each of size 2^s. We show that any two points of $\mathcal{S}$ are on exactly one block of $\mathcal{S}$: let P and Q be distinct points of $\mathcal{S}$, not together on a line through P_0 or Q_0. The set of images of $\mathcal{O}'_0$ under the subgroup of elations of H forms a parallel class of blocks, so we can assume that P is on some $\mathcal{Q}'$. If Q is also on $\mathcal{Q}'$, then we have a block through P and Q; if Q is not on $\mathcal{Q}'$ then form the lines PQ_0 and QP_0 and let them intersect at R. Let S be the point of intersection of QP_0 and $\mathcal{Q}'$. Then there is a homology in H with centre P_0 and axis PQ_0 that maps S to Q, and hence maps the oval $\mathcal{Q}'$ to one through P and Q. Thus any two points of $\mathcal{S}$ are on at least one block, and now a count of (pairs of points, block)-intersections, in two ways, yields that any two points are together on exactly one block. The set Ω thus gives the required set of ovals. $\square$

Taking $\Pi = PG_2(\mathbf{F}_{2^s})$ and for the oval $\mathcal{O}'_0$ a conic plus nucleus, with the nucleus either P_0 or Q_0, the derivation occurs inside $B(\mathbf{F}_{2^s}|\mathbf{F}_2)$: the conic given by $x^2 = yz$, with $\langle (0,0,1)' \rangle$ the line L at infinity, contains the point $\langle (0,1,0) \rangle = P_0$ of L, and its nucleus is $Q_0 = \langle (1,0,0) \rangle$, which is again on L. In the affine plane with point set $\{(a,b,1)|a,b \in \mathbf{F}_{2^s}\}$, the conic consists of the points $\{(t,t^2,1)|t \in \mathbf{F}_{2^s}\}$. Since we are in characteristic $p = 2$, this set is an s-dimensional subspace over $\mathbf{F}_2$.

It follows also (and the calculation is similar) that a conic plus nucleus will be furnished by a minimum-weight vector of $B(\mathbf{F}_{2^s}|\mathbf{F}_2)$ if and only if its nucleus is at infinity. Thus, even if we restrict ourselves to ovals coming from conics, the derivation might not be taking place in $B(\mathbf{F}_{2^s}|\mathbf{F}_2)$, but in $B(AG_2(\mathbf{F}_{2^s}))$. Moreover, not all ovals come from conics when $s > 3$, as we have already remarked.

Another similarity with the derivation coming from Baer subplanes is that, in the new projective plane, any line of Π^L that has been discarded, together with the points P_0 and Q_0 on the new line at infinity, becomes an oval of the new plane. We do not know whether or not there is a $\mathcal{D}$ which, for our derivation set of cardinality $2^s - 1$, will yield a new non-desarguesian projective plane. Any $\mathcal{D}$ produced as in the proof of Proposition 6.11.1 will yield a translation plane coordinatized by a nearfield with multiplicative group isomorphic to that of the nearfield of the original plane, since the new projective plane Σ with the new line M at infinity has, in its automorphism group G, all central collineations with centre P_0 and axis through Q_0, and thus has all central collineations with centre Q_0 and axis through P_0. It follows easily that Σ is a translation plane, with M a translation line, Σ^M can be coordinatized by a nearfield and, moreover, this nearfield is unique. Thus when the nearfield is a field, the new plane is desarguesian, however the oval is chosen.

6.12 Other derivation sets

We turn, finally, to other possible derivation sets that might arise from minimum-weight vectors of $B(K|F)$, where $K = \mathbf{F}_q$ and $F = \mathbf{F}_p$. Suppose Π is a translation plane with translation line L and $\pi = \Pi^L$. We have, as before,

$$C(\pi) \subset B(K|F) \subseteq B(\pi).$$

If $v \in B(K|F)$ has weight q but is not a line of π, let $v' = v^{X'}$ be a preimage of v with weight $q + r$, where $r \equiv 1 \pmod{p}$. We have pointed out that X' is a blocking set for Π, and hence $r \geq 1 + \sqrt{q}$. Without loss of generality, we may assume that $v = v^T$, where T is an s-dimensional subspace of V, and $q = p^s$. If $S_0, S_1, \ldots, S_q$ is the spread defining π, then $r = |\{S_i|S_i \cap T \neq \{0\}\}|$. In general it is difficult to count the number of T with given intersection properties, and all sorts of intersections can occur:

a "random" look at the possibilities for a spread of one of the 2-dimensional
translation planes of order 16, for example the one defined by the coordinate
set M_2 of Dempwolff and Reifart [87], gave various intersection patterns for
subspaces in $B(\mathbf{F}_{16}|\mathbf{F}_2)$. Letting the 4-tuple $[x_0, x_1, x_2, x_3]$ correspond to a
subspace T meeting x_i members of the spread in subspaces of dimension i,
for $i = 0, 1, 2, 3$, so that $r = x_1 + x_2 + x_3$, we found instances of: $[12, 0, 5,$
$0]$, $r = 5$ (Baer subplanes); $[10, 3, 4, 0]$, $r = 7$; $[8, 8, 0, 1]$, $r = 9$; $[8, 6, 3, 0]$,
$r = 9$; $[6, 9, 2, 0]$, $r = 11$; $[4, 12, 1, 0]$, $r = 13$; $[12, 15, 0, 0]$, $r = 15$ (ovals).

There is one case for which the count is rather easy: if $S_i \cap T$ is of
dimension $s - 1$ for one (and hence precisely one) S_i. Here we are assuming
$s > 2$, since in the case $s = 2$, all T's, except $S_0, S_1, \ldots, S_q$, yield Baer
subplanes. First notice that if $\dim(S_i \cap T) = s - 1$ and $\dim(S_j \cap T) \geq 2$ for
some $j \neq i$, then, since $S_i \cap S_j = \{0\}$, it follows that $\dim(T) \geq s + 1 > s$,
which is an impossibility. It follows that $\dim(T \cap S_i) = s - 1$ for exactly
one S_i, $\dim(T \cap S_i) = 1$ for exactly p^{s-1} of the S_i, and $T \cap S_i = \{0\}$ for
$p^{s-1}(p-1)$ of the S_i. Here $r = 1 + p^{s-1}$ and the v' has weight $p^s + p^{s-1} + 1$.
Choosing such a T is easy: simply take an $(s - 1)$-dimensional subspace of
one S_i and a 1-dimensional subspace of another and let T be the (necessarily
direct) sum. An easy count gives the number of such subspaces to be
$p(p^s + 1)(p^s - 1)^2/(p - 1)^2$ and hence, counting translates, there are

$$\frac{p^{s+1}(p^s + 1)(p^s - 1)^2}{(p - 1)^2}$$

normalized weight-q vectors of $B(K|F)$ with $r = p^{s-1} + 1$. Each of these
produces a distinct blocking set of cardinality $p^s + p^{s-1} + 1$ in the projective
translation plane.

The following question arises: can there be a derivation set of cardinality
$p^{s-1} + 1$ in $PG_2(\mathbf{F}_{p^s})$? Recall that we must have $n|D|$ vectors in $\mathcal{D}$. Since
$n = q$ for translation planes, the "n" is given by the translates of a T.
Thus we must find $|D| = p^{s-1} + 1$ subspaces, $T_0, T_1, \ldots, T_{p^{s-1}}$, with $T_m \cap$
$S_i = \{0\}$ for any m, implying $T_j \cap S_i = \{0\}$ for all j. Suppose we have
found T_0 with $T_0 \cap S_0$ of dimension $s - 1$ and $T_0 \cap S_i$ of dimension 1
for $1 \leq i \leq p^{s-1}$. Since a k-dimensional subspace of an n-dimensional
space over $\mathbf{F}_q$ has $q^{k(n-k)}$ complementary $(n - k)$-dimensional subspaces,
$T_0 \cap S_0$ has q^{s-1} complementary subspaces in S_0. Next, suppose there is
an s-dimensional subspace T_1 with $T_1 \cap S_0$ complementary to $T_0 \cap S_0$, and
$T_1 \cap S_1$ complementary to $T_0 \cap S_1$. Then $V = S_0 \oplus S_1 = (T_0 \cap S_0 \oplus T_1 \cap S_0) \oplus$
$(T_0 \cap S_1 \oplus T_1 \cap S_1) = (T_0 \cap S_0 \oplus T_0 \cap S_1) \oplus (T_1 \cap S_0 \oplus T_1 \cap S_1) \subseteq T_0 + T_1$.
Thus $T_0 \cap T_1 = \{0\}$ since both T_0 and T_1 are s-dimensional. It is thus
conceivable that there could exist $T_0, T_1, \ldots, T_{p^{s-1}}$ with $T_j \cap S_i = \{0\}$ for
$i > p^{s-1}$ and all j, and with $T_i \cap S_j$ of dimension $s - 1$ for $0 \leq i \leq p^{s-1}$,

and $T_i \cap T_j = \{0\}$ for $i \neq j$. These subspaces and their translates would yield the desired $q(p^{s-1} + 1)$ vectors of $\mathcal{D}$. Put in the language of spreads, $T_0, T_1, \ldots, T_{p^s-1}$, as a partial spread, would replace $S_0, S_1, \ldots, S_{p^s-1}$.

For $q = 2^3$, the first case possible, we have $r = 5$ and, because of the peculiar homogeneity properties of $P\Gamma L_2(\mathbf{F}_8)$ acting on L, any 5-set of L would be suitable if one such could be found. An easy computation shows that there are seven T's available for each 5-subset of L, but, unfortunately, five of these seven cannot be chosen properly. It follows, since $r \geq 1 + \sqrt{8}$, that the only non-trivial derivation sets have $r = 7$ and are given by ovals.

Chapter 7

Hadamard designs

7.1 Introduction

We have just seen how rich a source of designs projective planes provide and how algebraic coding theory allows the organization and discussion of both the planes themselves and the designs they provide. Another venerable source of designs is provided by Hadamard matrices. The symmetric designs the matrices produce are at the other end of the inequality

$$n^2 + n + 1 \geq v \geq 4n - 1$$

for (v, k, λ) designs of order $n = k - \lambda$, and here, i.e. for $v = 4n - 1$, the Bruck-Ryser-Chowla theorem is inoperable; conjecturally, there are designs of all orders.

In this chapter we shall try to organize the designs provided by these matrices using the coding-theoretical approach. Some of the material of the chapter is taken from Assmus and Key [14] and the interested reader may wish to consult that paper. Our aim is to suggest a possible organization via a new notion of equivalence, and much of the chapter will be devoted to making the appropriate definitions and deriving their elementary consequences. We include a classical discussion of Hadamard matrices and their designs, including those arising from difference sets in elementary abelian 2-groups. We state, and give the simple proof of, the Hamada-Ohmori rigidity theorem, construct, and determine the automorphism group of, the unique 12×12 Hadamard matrix, and discuss, in our terms, the five inequivalent 16×16 matrices and the 60 inequivalent 24×24 Hadamard matrices.

A square matrix of ± 1's whose rows are orthogonal is called a **Hadamard** matrix. Thus $H = (h_{ij})$ is a Hadamard matrix provided $h_{ij} = \pm 1$

for all i and j and
$$HH^t = mI_m,$$
where I_m is the $m \times m$ identity matrix, H being also $m \times m$. Since H^t is essentially the inverse of H, it follows that $H^t H = mI_m$; thus the columns of H are orthogonal and H^t is also a Hadamard matrix. We call m the **size** of the Hadamard matrix.

Sylvester (see [269][1]) is the first mathematician known to have studied such matrices, but it was Hadamard in 1893 who discovered that if $X = (x_{ij})$ is an $m \times m$ complex matrix, then

$$|\det(X)|^2 \le \prod_{i=1}^{m} \sum_{j=1}^{m} |x_{ij}|^2.$$

Moreover, normalizing so that $|x_{ij}| \le 1$, the right-hand side of Hadamard's determinantal inequality is bounded by m^m and Hadamard showed that equality occurs if and only if $|x_{ij}| = 1$ and $X\bar{X}^t = mI_m$. He also observed that, for X real, $m = 1, 2$ or $m \equiv 0 \pmod 4$. Thus Hadamard matrices are solutions to an extremal problem in real analysis and they have found numerous applications through both real analysis and combinatorics.

Hadamard (see [113]) was aware of Sylvester's 1867 work and hence knew that such matrices existed whenever m was a power of 2, but he also constructed the 12×12 matrix and one of size 20, and he asked for a determination of those m for which such a matrix of size m existed. That determination has yet to be made.

For $m = 1$ there are two Hadamard matrices, (1) and (-1), and for $m = 2$ it is a simple matter to construct the eight possible matrices: they are all derived from the 2×2 *Sylvester* matrix

$$S = \begin{pmatrix} 1 & 1 \\ 1 & -1 \end{pmatrix}, \tag{7.1}$$

by placing the -1 entry in each of the four positions and by multiplying each of these four matrices by -1.

For $m > 2$, as Hadamard observed (see Theorem 7.2.2 below), m is divisible by 4, and writing $m = 4n$, we call n the **order** of the matrix — for reasons that will become apparent in the following section. It is again quite easy to construct a Hadamard matrix of size 4 (order 1) and to count the number of possibilities,

$$2^4 \times \binom{4}{2} \times 2^2 \times 2 = 768.$$

---

[1]Sylvester, at the end of this paper, promises another, but we have not been able to locate the sequel and it probably never appeared.

That these are all essentially the same Hadamard matrix will be accounted for by the usual concept of Hadamard-matrix equivalence: see Definition 7.3.1 below.

There are various constructions giving infinite classes of Hadamard matrices (for example, the Paley construction), some of which we discuss in this chapter. It is widely believed that Hadamard matrices of all orders exist. This still-unsettled existence question is complicated by the fact that if roots of unity other than ± 1 are allowed, then the $m \times m$ Vandermonde matrix satisfies $|x_{ij}| = 1$ and $X \bar{X}^t = m I_m$; thus existence is not in question for *complex Hadamard* matrices: they exist for all sizes, i.e. for *all* values of m, as both Sylvester and Hadamard indicated. The Hadamard matrices have been enumerated for orders 1 to 7, but little is known about the classification of such matrices in either the real or complex case.

7.2 Hadamard designs

Any 3-design with $v = 4n$, $k = 2n$, and $\lambda = n - 1$ is called a **Hadamard 3-design** because of the association with Hadamard matrices — as we will explain presently. Further, if $\mathcal{T}$ is a Hadamard 3-design, then each of its *derived* designs, $\mathcal{T}_P$, obtained by omitting a point P and all the blocks that are not incident with P, is symmetric. Both $\mathcal{T}_P$ and its complement $\overline{\mathcal{T}}_P$ are called **Hadamard 2-designs**. Their parameters are, respectively, $(4n-1, 2n-1, n-1)$ and $(4n-1, 2n, n)$. The order of $\mathcal{T}$ is $2n$, and the order of each of the derived designs is n. The theorem to follow shows that the 3-design is the *unique* extension of the 2-design; that the 2-design always extends comes from a more general result which we prove first.

Lemma 7.2.1 *Let $\mathcal{D}$ be a 2-$(2k - 1, k - 1, \lambda)$ design with $k > 2$. Then $\mathcal{D}$ extends to a 3-$(2k, k, \lambda)$ design.*

Proof: Let $\mathcal{D} = (\mathcal{P}, \mathcal{B})$ and let ∞ be a new symbol. We construct a new structure $\mathcal{D}^* = (\mathcal{P}^*, \mathcal{B}^*)$, where $\mathcal{P}^* = \mathcal{P} \cup \{\infty\}$ and $\mathcal{B}^*$ consists of new blocks that we now define: for the blocks containing ∞, we must take all the blocks of $\mathcal{B}$ with ∞ adjoined; for the blocks not containing ∞, take the complements (in $\mathcal{P}$) of the blocks $B \in \mathcal{B}$.

Clearly all the new blocks are incident with k points. We need only show that $\mathcal{D}^*$ is a 3-design. Thus let P, Q, R be points of $\mathcal{D}^*$. If one of P, Q, R is ∞, then the three points are on precisely λ blocks. So suppose none of P, Q, R are ∞, i.e. suppose that $\{P, Q, R\} \subset \mathcal{P}$. If there are μ blocks of $\mathcal{B}$ containing all three points, then these blocks define blocks of $\mathcal{B}^*$ containing the three points. By extending the Pascal triangle of intersection numbers (see Figure 1.3) or by simple counting arguments, we see that there are

$b - 3r + 3\lambda - \mu$ blocks of $\mathcal{B}$ that contain none of the three points and hence this number of complements of blocks containing all three points. Thus there are $b - 3r + 3\lambda$ blocks in $\mathcal{B}^*$ containing the three points. But $v = 2k - 1$ for the design $\mathcal{D}$ and so $b = v(v-1)\lambda/(k-1)(k-2) = 2\lambda(2k-1)/(k-2)$ and $r = (v-1)\lambda/(k-2) = 2\lambda(k-1)/(k-2)$. Hence $b - 3r + 3\lambda = \lambda$, as required. $\square$

Theorem 7.2.1 *Let $\mathcal{D}$ be a Hadamard $(4n-1, 2n-1, n-1)$ design. Then $\mathcal{D}$ extends, uniquely up to isomorphism, to a Hadamard 3-design.*

Proof: That $\mathcal{D}$ extends follows from Lemma 7.2.1. Suppose that $\mathcal{D}^* = (\mathcal{P}^*, \mathcal{B}^*)$ is any extension of $\mathcal{D}$; what we must show to prove uniqueness is that, for any block $B^* \in \mathcal{B}^*$, the complement of B^* is also a block of $\mathcal{D}^*$.

For any point P, the design $\mathcal{D}^*_P$ is a symmetric $(4n-1, 2n-1, n-1)$ design, and so any two blocks of this contraction meet in $n-1$ points, i.e. any two blocks of $\mathcal{D}^*$ that contain a point P contain a further $n-1$ points in common. Thus any two distinct blocks of $\mathcal{D}^*$ are either disjoint or meet in n points. Now the number of blocks of $\mathcal{D}^*$ through any point is $4n-1$; it follows that the number of pairs of the form (P, C^*), where P is a point of our block B^* and C^* another block containing P, is $2n(4n-2)$, and hence the number of blocks C^* that meet B^* is $2(4n-2) = 8n-4$. But the total number of blocks of $\mathcal{D}^*$ is $8n-2$, which means there is exactly one block that does not meet B^* at all, and this block is, therefore, the complement of B^*. $\square$

Exercise 7.2.1   (1) In Lemma 7.2.1 we have assumed that $k > 2$. What should this lemma be for $k = 2$? (*Hint:* consider the parameters 1-$(3, 1, 1)$ and show that the affine plane of order 2 is obtained.)

(2) Show that every t-$(2k-1, k-1, \lambda)$ design with t even extends to a $(t+1)$-$(2k, k, \lambda)$ design.

The second exercise above is a generalization of Lemma 7.2.1 and is due to Alltop [2].

We shall next see how Hadamard designs are obtained from Hadamard matrices. In preparation for this we first prove that the size of a Hadamard matrix is either 1 or 2 or else is divisible by 4 and, in proving this, show how the entries in distinct rows (and hence columns) compare.

Theorem 7.2.2 *If H is an $m \times m$ Hadamard matrix with $m > 1$, then m is even and for any two distinct rows of H there are precisely $m/2$ columns in which the entries in the two rows agree. Further, if $m > 2$ then m is*

divisible by 4, *and for any three distinct rows of* H *there are precisely* $m/4$
*columns in which the entries in all three rows agree. The same statements
hold for columns of* H.

Proof: The first statement is an immediate consequence of orthogonality.
More precisely, if r and s are two distinct rows of a Hadamard matrix H,
let $m_{\epsilon,\eta}$ denote the number of columns in which the entry ϵ occurs in r and
η in s. Then

$$m_{1,1} + m_{-1,-1} + m_{1,-1} + m_{-1,1} = m,$$

since there are m entries in each row, and

$$m_{1,1} + m_{-1,-1} - m_{1,-1} - m_{-1,1} = 0,$$

since r and s are orthogonal. Thus $m = 2(m_{1,1} + m_{-1,-1})$ is even, and
r and s agree in $m_{1,1} + m_{-1,-1} = m/2$ columns — and disagree in $m/2$
columns.

Now suppose $m > 2$; then $m = 2\ell$ and any two distinct rows of H
agree in ℓ columns. Let r, s and t be three distinct rows of H. Suppose
s agrees with r in the columns A and disagrees with r in the columns D.
Now t agrees with r in those columns, A' say, of A in which it agrees with
s and in those columns, D' say, of D in which it disagrees with s. Thus
$|A'| + |D'| = \ell$ and $|A'| + |D - D'| = |A'| + |D| - |D'| = \ell$. Since $|D| = \ell$,
$|A'| = |D'|$, ℓ is even and m is divisible by 4. Moreover, all three rows agree
only in the columns of A' and hence in $\ell/2 = m/4$ columns.

Since the transpose of a Hadamard matrix is Hadamard the statements
hold also for the columns of H. $\square$

Definition 7.2.1 *The* **order** *of a* $4n \times 4n$ *Hadamard matrix is* n.

The reason for our[2] definition of the order is that the order of H is then
the order of any symmetric Hadamard 2-design obtained from H, as the
following construction illustrates:

Theorem 7.2.3 *Let* $r = (r_1, r_2, \ldots, r_{4n})$ *be a row of a Hadamard matrix
H of size* $4n$. *With* $\mathcal{P} = \{1, 2, \ldots, 4n\}$, *we set, for any other row s of H,
where* $s = (s_1, s_2, \ldots, s_{4n})$,

$$B_s = \{j | j \in \mathcal{P}, s_j = r_j\}$$

and

$$\bar{B}_s = \mathcal{P} - B_s = \{j | j \in \mathcal{P}, s_j \neq r_j\}.$$

---

[2]Other authors refer to the *size* of a Hadamard matrix as its *order*.

Let
$$\mathcal{B} = \{B_s | s \neq r\} \cup \{\bar{B}_s | s \neq r\}.$$

Then $\mathcal{T}(r, H) = (\mathcal{P}, \mathcal{B})$ is a 3-$(4n, 2n, n - 1)$ design. Conversely, any design with parameters 3-$(4n, 2n, n - 1)$ can be obtained in this way from a Hadamard matrix.

Proof: That the blocks have the right size follows immediately from Theorem 7.2.2, since $m = 4n$ and the number of points in the block B_s is the number of columns in which r and s agree, *viz.* $m/2 = 2n$. Of course, $\bar{B}_s$ also has $2n$ points. Theorem 7.2.2 also shows that two blocks of $\mathcal{B}$ are either disjoint or meet in $n = m/4$ points. By construction, $|\mathcal{B}| = 8n - 2$ and the blocks come in complementary pairs. It now follows that, for any point $P \in \mathcal{P}$, the $4n - 1$ blocks containing P form a symmetric design on $\mathcal{P} - \{P\}$ with $\lambda = n - 1$. Hence we have a 3-design with the announced parameters.

Conversely, suppose we are given a 3-$(4n, 2n, n - 1)$ design $\mathcal{T} = (\mathcal{P}, \mathcal{B})$. Clearly, since any contraction is a symmetric design, any two distinct blocks are either disjoint or meet in n points and the blocks come in complementary pairs. This allows a reversal of the above procedure to produce a Hadamard matrix of size $4n$: a pair of complementary blocks of $\mathcal{T}$, B and $\bar{B}$, forms a row s by setting $s_j = 1$ if $j \in B$ and $s_j = -1$ if $j \in \bar{B}$, where reversing the roles of B and $\bar{B}$ simply changes s to $-s$. The $4n - 1$ complementary pairs, together with a constant row r (i.e. all entries equal to 1 or all entries equal to -1), yield a Hadamard matrix H of size $4n$. $\square$

Note: It can also be checked directly from Theorem 7.2.2 that we have a 3-design, by applying the theorem to the columns of H: for any three points, corresponding to three columns, labelled i, j, k say, whatever the configuration of ± 1's in row r corresponding to these three columns, since the number of rows in which these three columns agree (either all $+1$ or all -1) is n, the number of rows in which this configuration, or its negative, occurs again is $m/4 - 1 = n - 1$, and these rows will define the blocks B_s or $\bar{B}_s$ containing all the three points. (We have implicity used here the fact that multiplying any column of a Hadamard matrix by -1 yields another Hadamard matrix, so that the configuration in row r might as well have been all 1's.)

The theorem now justifies our calling a design with these parameters a Hadamard 3-design, as we have already done above.

Remark Notice that there is a certain arbitrariness involved in whether s or $-s$ is chosen and in the ordering of the points and blocks; this will be eliminated by the usual definition of equivalence of Hadamard matrices, which we discuss in the next section.

Example 7.2.1 The Fano plane defined in Example 1.2.2 (1) is a Hadamard 2-design. Add a new point, "0", to each line and add the complements of each line in the original set, to get the 3-(8,4,1) design. Using the labelling we had in the example, and ordering the points with the new symbol 0 first, the Hadamard matrix obtained in the above way is

$$
\begin{pmatrix}
1 & 1 & 1 & 1 & 1 & 1 & 1 & 1 \\
1 & 1 & 1 & 1 & -1 & -1 & -1 & -1 \\
1 & 1 & -1 & -1 & 1 & 1 & -1 & -1 \\
1 & 1 & -1 & -1 & -1 & -1 & 1 & 1 \\
1 & -1 & 1 & -1 & 1 & -1 & -1 & 1 \\
1 & -1 & 1 & -1 & -1 & 1 & 1 & -1 \\
1 & -1 & -1 & 1 & -1 & 1 & -1 & 1 \\
1 & -1 & -1 & 1 & 1 & -1 & 1 & -1
\end{pmatrix}.
$$

Up to the equivalence we are about to define, the above matrix is the unique Hadamard matrix of order 2.

Exercise 7.2.2 Set $V = \mathbf{F}_2^m$ and define a $2^m \times 2^m$ matrix H, with rows and columns labelled using the vectors of V, by $H = (h_{\mathbf{u},\mathbf{v}})$, where $h_{\mathbf{u},\mathbf{v}} = (-1)^{(\mathbf{u},\mathbf{v})}$. (Here $(\mathbf{u}, \mathbf{v})$ denotes the inner product of the vectors $\mathbf{u}$ and $\mathbf{v}$ in the standard inner product but, of course, $h_{\mathbf{u},\mathbf{v}}$ is viewed as a real number.) Show that H is a symmetric Hadamard matrix.

7.3 Equivalent matrices

The customary definition of equivalence of Hadamard matrices is the following:

Definition 7.3.1 *Two $m \times m$ Hadamard matrices H and K are **equivalent** if and only if there are $m \times m$ monomial matrices M and N whose non-zero entries are ± 1 with $K = MHN$.*

This definition of equivalence is a natural one and corresponds to the observation that the elementary row and column operations that permute rows or columns, or multiply rows or columns by -1, are precisely those elementary operations that preserve the property of being a Hadamard matrix. The monomial matrices of the definition simply effect any sequence of these row and column operations; these matrices are, of course, elements of the $m \times m$ real orthogonal group and they form a subgroup which is the semi-direct product of the symmetric group on m letters and the "diagonal" group $\{\pm 1\}^m$. Denote this group by $\mathcal{M}_m$. More generally, we could let $\mathcal{G}$

be an arbitrary subgroup of the real orthogonal group and define H and K to be equivalent if there were elements P and Q in $\mathcal{G}$ with $PHQ = K$. This is clearly also an equivalence relation. If we were, at one extreme, to take for $\mathcal{G}$ the entire orthogonal group then any two Hadamard matrices would be equivalent since, if H and K have size m, then taking $P = \sqrt{m}H^{-1}$ and $Q = (1/\sqrt{m})K$ effects the equivalence. (This raises the question of whether or not there are other subgroups giving interesting equivalence relations; in this connection the reader may wish to consult Hall [121].)

For a given Hadamard matrix H, the set of all $Q \in \mathcal{M}_m$ for which there is a $P \in \mathcal{M}_m$ with $H = PHQ$ is clearly a subgroup of $\mathcal{M}_m$; it is called the **automorphism group** of H. Clearly $\pm I_m$ is always a normal subgroup of the automorphism group of a Hadamard matrix. The automorphism group of the 2×2 Sylvester matrix is $\mathcal{M}_2$, a group of order eight; the eight 2×2 Hadamard matrices are, of course, simply PH, $P \in \mathcal{M}_2$ where H is any one of them. In the case of the 4×4 Sylvester matrix (see the definition on page 265), the unique Hadamard matrix of order 1, $\mathcal{M}_4$ has order 384 and the automorphism group has order 192. Here both row and column operations are needed to go from any one of the 768 matrices to any other. In fact, the column operation consisting merely of post-multiplication by a diagonal matrix to ensure a row of all 1's, together with all row operations, will suffice to go from any 4×4 matrix to the Sylvester matrix.

We have seen that every Hadamard 3-design yields a Hadamard matrix; moreover, isomorphic designs clearly yield equivalent Hadamard matrices. However, when forming a Hadamard 3-design from a Hadamard matrix, there are $4n$ choices for the row r, so that $4n$ Hadamard 3-designs are produced; the designs need not be isomorphic. These matters were investigated thoroughly by Norman [221] who proved the following:

Theorem 7.3.1 *Two Hadamard matrices H and K are equivalent if and only if there is a row r of H and a row s of K with $T(r, H)$ isomorphic to $T(s, K)$.*

Proof: The proof proceeds by considering the various types of elementary row or column operations that preserve an equivalence class of Hadamard matrices. $\square$

Thus, given a Hadamard matrix H, the set of non-isomorphic 3-designs defined by H depends only on the equivalence class of H. This prompted Norman to define a natural equivalence relation on Hadamard 3-designs:

Definition 7.3.2 *Two Hadamard 3-designs are* **equivalent** *if they produce equivalent Hadamard matrices.*

Norman's theorem then shows that two designs are equivalent if they come from *the same* Hadamard matrix. At the time of writing, the situation as regards the classification of the equivalence classes of Hadamard matrices and designs of a given size m is as follows: for each possible $m \leq 12$ there is only one class of matrices and, for $m = 4, 8$ and 12, a unique 3-design and a unique 2-design; for $m = 16$ there are five equivalence classes of matrices, five 3-designs, and five 2-designs (see Todd [277]); for $m = 20$ there are three equivalence classes of matrices, three 3-designs, and six 2-designs; for $m = 24$ there are 60 equivalence classes of matrices and 130 non-isomorphic 3-designs (see Ito, Leon and Longyear [147], and Kimura [167]). Hence, for order 6, Norman's equivalence relation partitions the 130 Hadamard 3-designs into 60 equivalence classes. The only other order for which something is known is 7: Kimura has recently announced [165] that there are 487 equivalence classes of 28×28 Hadamard matrices.

We next discuss the usual process of *normalization* of a Hadamard matrix. Given a Hadamard matrix H, pick a row r (respectively, column c); we can find matrices in the equivalence class of H with row r (respectively, column c) having all entries equal to 1, or all entries equal to -1: for example, to achieve a row of -1's in row r, simply post-multiply H by a diagonal matrix with the negative of row r along the diagonal. Usually the matrix is normalized in such a way that the first row and first column consist entirely of $+1$'s, but we will see that it is often more convenient to have a row of -1's. Because of a close connection between Hadamard matrices and binary codes and the obvious passage from the multiplicative group of order 2 to the additive group, we usually choose to normalize to a row of -1's and use this row to define the Hadamard 3-design. We make the following definitions:

Definition 7.3.3 *Let $H = (h_{ij})$ be a Hadamard matrix. Then*

$$\log_{-1}(H) = (\log_{-1}(h_{ij})).$$

Let $A = (a_{ij})$ be a matrix of 0's and 1's. Then, with $\exp_{-1}(x) = (-1)^x$,

$$\exp_{-1}(A) = (\exp_{-1}(a_{ij})).$$

Thus if A is an incidence matrix of a symmetric $(4n - 1, 2n - 1, n - 1)$ design, forming $\exp_{-1}(A)$ and bordering with a row and column of -1's yields a Hadamard matrix. For example, for $n = 2$ the negative of the matrix given in Example 7.2.1 is obtained.

7.4   The codes

We are now in a position to examine codes[3] associated with Hadamard designs — and thus with the matrices themselves; we will see that the codes defined (in the usual way) by the Hadamard designs from an equivalence class of Hadamard matrices are equivalent for any prime p and that this equivalence relation is coarser, in general, than Norman's. In fact, the codes obtained are equivalent in a stronger sense than equivalence as defined in Chapter 2. Recall that there we permitted ourselves to take an arbitrary subgroup of the multiplicative group of the field to define code equivalence and used the trivial subgroup for *code isomorphism* and the full multiplicative group for *code equivalence*. It will be convenient here to take the subgroup $\{\pm 1\}$ of the multiplicative group of the field and allow *code equivalence* to be with respect to this subgroup. When dealing with self-dual or even self-orthogonal codes, this is the correct subgroup to take, since otherwise a code equivalent to a self-dual code may fail to be self-dual. Of course, for characteristic 2 we have merely code isomorphism, but for odd characteristic there is a difference whenever the characteristic is not 3. The ambiguity should not cause confusion.

We first collect together as a theorem some simple properties of the codes of Hadamard designs. Recall that for any design $\mathcal{D}$ we denote by $E_p(\mathcal{D})$ the subcode of $C_p(\mathcal{D})$ generated by differences of the incidence vectors of the blocks of the design and that $\widehat{C_p(\mathcal{D})}$ denotes the code of the design extended by an overall parity check.

Theorem 7.4.1 *If $\mathcal{D}$ is a Hadamard $(4n-1, 2n-1, n-1)$ design, $\mathcal{T}$ its extension to a Hadamard 3-design, and p a prime dividing n, then*

(1) $C_p(\mathcal{D}) = C_p(\overline{\mathcal{D}}) + \mathbf{F}_p\mathbf{j}$;

(2) $\mathrm{Hull}_p(\mathcal{D}) = E_p(\mathcal{D}) = E_p(\overline{\mathcal{D}}) = C_p(\overline{\mathcal{D}})$;

(3) $\widehat{C_p(\mathcal{D})} = C_p(\mathcal{T}) \subseteq C_p(\mathcal{T})^{\perp}$, *with equality if p^2 does not divide n;*

(4) $\mathrm{rank}_p(\mathcal{T}) = \mathrm{rank}_p(\mathcal{D}) \leq 2n$;

(5) $\mathrm{rank}_2(\mathcal{T}) > 2 + \log_2 n$;

(6) $C_p(\mathcal{T})^{\perp} = (\widehat{\mathrm{Hull}_p(\mathcal{D})^{\perp}})$.

---

[3]Bose and Shrikhande [46] appear to have been the first authors to examine the connection between codes and Hadamard matrices; they were, however, interested in constant-weight codes and did not consider the codes we are examining here.

Proof: The first two statements follow from Theorem 4.6.3. For (3) note first that the construction of T from $\mathcal{D}$ implies that $\widehat{C_p(\mathcal{D})} = C_p(T)$. Recall that blocks of T are either disjoint or meet in n points and so $C_p(T)$ is self-orthogonal. Finally, when p divides n but p^2 does not, then Theorem 4.6.2 implies that the code $C_p(T)$ is *self-dual*. Clearly, (4) follows from (3).

To prove (5), note first that $\jmath \in C = C_2(T)$. Thus $C^\perp$ has only even-weight vectors. Since for any two points of the design T there is certainly a block containing one of the points but not the other, $C^\perp$ cannot have weight-2 vectors, and hence has minimum weight at least 4. Now the sphere-packing bound (Theorem 2.1.3) applied to $C^\perp$, which has dimension $4n - k$ where $k = \mathrm{rank}_2(T)$, gives $2^{4n-k}(1 + 4n) \leq 2^{4n}$, so that $k > \log_2 4n$, i.e. $k > 2 + \log_2 n$.

To prove (6), notice that the dimensions of the two spaces are the same, so we need only show that $(\widehat{\mathrm{Hull}_p(\mathcal{D})}^\perp) \subseteq C_p(T)^\perp$. But, if $w \in \mathrm{Hull}_p(\mathcal{D})^\perp$, then, for all blocks D of $\mathcal{D}$, $(w, v^D) = a$ with a independent of D. Since also $\sum_{D \in \mathcal{B}} v^D = -\jmath$, we have

$$\left(w, \sum_{D \in \mathcal{B}} v^D\right) = (w, -\jmath) = -\sum_{P \in \mathcal{P}} w(P) = \sum_{D \in \mathcal{B}}(w, v^D) = -a.$$

Thus $\sum_{P \in \mathcal{P}} w(P) = a$. Now, writing $\widehat{w}$ for the extended vector, we want to show that $\widehat{w}$ is orthogonal to every v^T, where now T is any block of T. From what has been said above, this is easily verified by looking at $(\widehat{w}, v^D)$ and $(\widehat{w}, \widehat{\jmath - v^D})$, for D any block of $\mathcal{D}$. $\square$

Now suppose we are given a Hadamard matrix: what can we say about the codes of its related 3-designs? We have the following:

Theorem 7.4.2 *Let T be a Hadamard 3-design defined by the Hadamard matrix H. Set $A = \log_{-1}(H)$. Then*

$$C_2(T) = \mathbf{F}_2\jmath + \langle r + s | r, s \text{ rows of } A\rangle$$

and, for p an odd prime, $C_p(T)$ is equivalent to the row span of H.

Proof: The verification is quite straightforward. $\square$

Had H been normalized to have a row of -1's, then we would have had

$$C_2(T) = \langle r | r \text{ a row of } A\rangle, \tag{7.2}$$

and, for p odd,

$$C_p(T) = \langle r | r \text{ a row of } H\rangle. \tag{7.3}$$

In view of the code equivalences given by the theorem, we will make the following definition:

Definition 7.4.1 *If H is a Hadamard matrix, then the **p-ary code** of H, denoted by $C_p(H)$ or simply $C(H)$, is the p-ary code associated with any 3-design defined by H. Similarly, $\mathrm{rank}_p(H)$ or $\mathrm{rank}(H)$ will denote the rank of the 3-designs defined by H.*

Corollary 7.4.1 *For any prime p and any equivalence class of Hadamard matrices, there is, up to code equivalence, only one p-ary code associated with the 3-designs from the equivalence class.*

Proof: By Theorem 7.3.1, any Hadamard matrix H in an equivalence class will define the full set of Hadamard 3-designs associated with the equivalence class. Since these all give equivalent codes we have the result. $\Box$

Recall that designs with the same paramenters are said to be linearly equivalent if their hulls are isomorphic codes. For the present purposes we have coarsened that equivalence by saying that two Hadamard designs T and T' with parameters 3-$(4n, 2n, n-1)$ are **p-equivalent** whenever $C_p(T)$ and $C_p(T')$ are equivalent codes. Obviously if T and T' are isomorphic designs they are p-equivalent for every prime p; in fact, $C_p(T)$ and $C_p(T')$ are code-isomorphic. Further, T and T' are p-equivalent provided that the Hadamard matrices determined by T and T' are equivalent matrices, as Corollary 7.4.1 above shows. We note that this implies that p-equivalence is a coarser equivalence relation than the equivalence introduced by Norman. We put these observations together as another corollary to Theorem 7.4.2:

Corollary 7.4.2 *Let T and T' be Hadamard 3-$(4n, 2n, n - 1)$ designs. If the Hadamard matrices given by T and T' are equivalent, then T and T' are p-equivalent for every prime p.*

Probably the most celebrated code defined by a Hadamard matrix is the ternary extended Golay code. It is the code over $\mathbf{F}_3$ given by the equivalence class of Hadamard matrices of order 3, there being only one. We shall now give R. J. Turyn's[4] uniqueness proof; the proof has, as a by-product, the construction of its automorphism group, a group whose permutation part is a strictly five-fold transitive group on 12 letters. In fact, Hadamard's description of a 12 × 12 matrix contains the germ of the idea, but Hadamard seems to have missed the conclusion.

---

[4] The proof was first committed to paper by Turyn but he attributes the idea to Marshall Hall or Andrew Gleason; in this connection see also Hall [119] where it is shown that a similar attack would fail for the Mathieu group M_{24}. The reason for the failure is now clearer (see Section 7.11) and, in fact, the Paley-Hadamard matrix of order 6 defines very interesting codes over both $\mathbf{F}_2$ and $\mathbf{F}_3$, but the code over $\mathbf{F}_2$ has many more weight-12 vectors than the matrix provides.

Theorem 7.4.3 *There is exactly one equivalence class of Hadamard matrices of order* 3. *In fact, any five rows of a* 12×12 *Hadamard matrix can be made the first five rows in any desired order and, once this has been done, the other rows are determined (uniquely up to sign and order).*

Proof: Let H be any Hadamard matrix of order 3 (for example, the Paley-Hadamard matrix obtained from the Jacobsthal matrix with $q = 11$ as in Corollary 7.8.1) and choose any five rows in any desired order. We shall show how to bring the matrix into a canonical form in such a way that these five rows are the first five in the given order.

Clearly, after a suitable post-multiplication of H by a monomial matrix, H's first row can be taken to consist entirely of $+1$'s and its second can be taken to be six $+1$'s followed by six -1's. Theorem 7.2.2 shows that the third row of H can be taken, after a suitable permutation of its first six columns and its last six also, as shown below (where each sign stands for three entries):

$$
\begin{array}{cccc}
+ & + & + & + \\
+ & + & - & - \\
+ & - & + & -
\end{array} .
$$

Suppose r is any other row. Let S_1 be the sum of the first three entries, S_2 the sum of the next three, S_3 the sum of the next three and S_4 the sum of the last three. Then, orthogonality implies that $S_1 = -S_2 = -S_3 = S_4$. If S_1 were equal to ± 3, then this row would be (up to sign) of the shape $+ \; - \; - \; +$ and now a further row would necessarily have the sum of its first three entries 0, an impossibility since 3 is not even. It follows that, after suitable change of sign and permutation of columns within each block of three, r must look like

$$
-++ \; +-- \; +-- \; -++ .
$$

Thus within each block of three columns any two further rows must have a dot product of $+3$ or -1 and hence any further pair of rows coincides in exactly one block of three. We now reduce the five given rows so that the fourth and fifth coincide in the first block of three; this may involve changing the signs of the second and third rows. Now, by permuting the columns in the second, third and fourth blocks of three, the first five rows can be made to take the form

$$
\begin{array}{cccccccccccc}
+ & + & + & \; & + & + & + & \; & + & + & + & \; & + & + & + \\
+ & + & + & \; & + & + & + & \; & - & - & - & \; & - & - & - \\
+ & + & + & \; & - & - & - & \; & + & + & + & \; & - & - & - \\
- & + & + & \; & + & - & - & \; & + & - & - & \; & - & + & + \\
- & + & + & \; & - & + & - & \; & - & + & - & \; & + & - & +
\end{array} ,
$$

and now the only remaining column permutation that may be performed is the interchange of the second and third columns.

At this stage each of the rows past the first three can be identified by a quadruple (i, j, k, l), where $1 \leq i, j, k, l \leq 3$ and the entries in the quadruple denote the position within the block of three of the "minority" element: -1 if the sum of the three is $+1$ and $+1$ if the sum of the three is -1. For example, the fourth row corresponds to the quadruple $(1, 1, 1, 1)$ and the fifth to $(1, 2, 2, 2)$. As we have just noted, the orthogonality condition implies that any two quadruples agree in exactly one entry. As Hadamard noted there are nine such quadruples; moreover, it can immediately be verified that from the first two quadruples, the remaining seven can be constructed in precisely two ways: $(1, 3, 3, 3)$ must occur and hence so must either $(2, 2, 3, 1)$ or $(3, 2, 3, 1)$; interchanging the second and third columns normalizes so that, say, $(2, 2, 3, 1)$ occurs. Now no further sign changes or column permutations are possible and the further nine rows are given by the 9×4 array of quadruples. $\square$

Remark   (1) The 9×4 array is the standard one constructed from a projective plane of order 3 via two orthogonal Latin squares: recall that if (i, j, x_{ij}, y_{ij}) where $1 \leq i, j \leq 3$ is the array, then the two squares are simply (x_{ij}) and (y_{ij}).

   (2) We have worked with rows rather than columns for visual reasons, but clearly the automorphism group of the (unique) Hadamard matrix of order 3 has order $2 \times 12 \times 11 \times 10 \times 9 \times 8$ and the permutation part is a strictly five-fold transitive group on 12 letters; it is, of course, the Mathieu group M_{12}.

Corollary 7.4.3 *There is a ternary $[12, 6, 6]$ self-dual code whose full group of automorphisms is $2M_{12}$. Moreover, the supports of the minimum-weight vectors form a 5-$(12, 6, 1)$ design with M_{12} as automorphism group.*

Proof: The code is, of course, simply $C_3(H)$, where H is the matrix of the theorem. The self-duality follows from Theorem 7.4.1(3) and Theorem 7.4.2. The automorphism group of the matrix is clearly a subgroup of the automorphism group of the code. That the minimum weight is 6 follows from the self-orthogonality and the fact that it cannot be 3 because of the five-fold transitivity. The MacWilliams relations now imply that there are precisely 24 vectors of weight 12, *viz.* the rows of the Hadamard matrix and their negatives, and 2×132 minimum-weight vectors. Hence there are 132 supports with no two meeting in five points, since otherwise the minimum weight would be 3. Thus we have a design with the required parameters.

That the automorphism group of the code is precisely $2M_{12}$ follows from the fact that the weight-12 vectors are, essentially, the Hadamard matrix. Clearly M_{12} is a subgroup of the automorphism group of the design and, by maximality, it cannot be larger. $\square$

Remark (1) Signs are important here. The ternary code generated by the incidence matrix of the 5-(12, 6, 1) design is $(\mathbf{F}_3 \jmath)^\perp$, a [12, 11, 2] ternary code. In fact, the incidence matrix of this design yields no interesting codes, as an application of Theorem 2.4.1 easily shows; see also Chapter 8.

(2) One of the earliest papers on the relationship of codes to designs was written by Paige [229] who showed that the binary Golay code was related to the large Mathieu design. Assmus and Mattson [16] investigated the small Mathieu design (the one above) and its relationship to the ternary Golay code; they also looked at more general situations. The reader interested in these early papers may wish to look at Assmus and Mattson [16, 17, 18] and the references given there.

7.5 Kronecker product constructions

Although Hadamard matrices have not been constructed for every possible order,[5] there are several methods of construction that do give infinite classes of Hadamard matrices of various orders: Hall [122] gives a list of these. We will discuss here and in Section 7.8 two of the most important constructions and see how coding theory can assist in a characterization.

The most immediate way to build new Hadamard matrices is by using a Kronecker (or direct) product construction: if $H = (h_{ij})$ is any $k \times k$ Hadamard matrix, and $B_1, B_2, \ldots, B_k$ are any $m \times m$ Hadamard matrices, then the matrix obtained from the Kronecker product, *viz.*

$$H \otimes [B_1, B_2, \ldots, B_k] = \begin{pmatrix} h_{11}B_1 & h_{12}B_1 & \ldots & h_{1k}B_1 \\ h_{21}B_2 & h_{22}B_2 & \ldots & h_{2k}B_2 \\ \vdots & \vdots & & \vdots \\ h_{k1}B_k & h_{k2}B_k & \ldots & h_{kk}B_k \end{pmatrix},$$

is a $km \times km$ Hadamard matrix, as is quite easy to verify: see, for example, Hughes and Piper [144, p. 104]. This is a standard method for creating new Hadamard matrices, especially in the case $B_1 = B_2 = \ldots = B_n = B$,

---

[5] As far as we are aware, at the moment (August 1991) the first open order is 107; i.e. a 428×428 Hadamard matrix has yet to be constructed.

when $H \otimes [B_1, B_2, \ldots, B_n] = H \otimes B$, which is the more familiar form of the Kronecker product. Of course, since the transpose of any Hadamard matrix is also Hadamard, the transpose of $H \otimes [B_1, B_2, \ldots, B_k]$ is Hadamard.

Starting with the 2×2 Sylvester matrix (see Equation(7.1)), form $S \otimes S$ and continue to build in this way. Now it turns out that different representatives of an equivalence class of matrices will, in general, give inequivalent Hadamard matrices through a Kronecker product. The first place that this can happen is when $k = 2$ and $m = 8$: indeed, by choosing different members of the (unique) equivalence class of Hadamard matrices of size 8 and taking Kronecker products with S, together with one transpose, all five inequivalent Hadamard matrices of size 16 (order 4) are obtained. We will look at this case, along with all the associated codes, as an example below.

To begin with, we can say something about the possible relationships that hold between the codes of the associated designs in the binary case:

Theorem 7.5.1 *Suppose that H, $B_1, \ldots, B_k$, and $C_1, \ldots, C_k$ are normalized Hadamard matrices, that H has size k, and that the B_i and C_i have size m, where $k \geq 2$, $m \geq 4$. Then*

$$C_2(H \otimes [B_1, \ldots, B_k]) \subseteq C_2(H \otimes [C_1, \ldots, C_k])^{\perp}.$$

Proof: This is easy to see by noting that, since we have normalized all the matrices, $C_2(H \otimes [B_1, B_2, \ldots, B_k])$ will be the space spanned by $\log_{-1}(H \otimes [B_1, B_2, \ldots, B_k])$, possibly together with the all-one vector $\jmath$. This is also a Kronecker product, with entries $\log_{-1}(B_i)$ or $J + \log_{-1}(B_i)$. Then the conditions $k \geq 2$ and $m \geq 4$ will ensure the result. $\square$

If $B_i = C_i$ in the theorem, then all we get is the self-orthogonality of the code of the Hadamard 3-designs. But taking different members of an equivalence class can give a dimensional classification of the designs and matrices from Kronecker constructions. We can start with $H = S$, the 2×2 Sylvester matrix (Equation (7.1)). Recall that the rank is simply the dimension of the code involved (Definition 7.4.1).

Theorem 7.5.2 *Let S be the 2×2 Sylvester matrix and let K and L be normalized Hadamard matrices of size $m \geq 4$. Then, over the field $F = \mathbf{F}_2$:*

(1) $\mathrm{rank}_2(S \otimes [K, L]) + \dim(\langle \log_{-1}(K), \jmath \rangle \cap \langle \log_{-1}(L), \jmath \rangle) = \mathrm{rank}_2(K) + \mathrm{rank}_2(L) + 1;$

(2) $C_2(S \otimes [K, K]) \subseteq C_2(S \otimes [K, L]).$

Proof: This is quite straightforward. $\square$

Now we look at some specific cases, starting with the most important, i.e. the Sylvester matrices, which will give the designs of points and hyperplanes of affine geometries over $\mathbf{F}_2$ and, consequently, the first-order Reed-Muller codes.

Starting with $S_1 = S$, recursively define S_m for $m \geq 2$ via

$$S_m = S \otimes [S_{m-1}, S_{m-1}] = S \otimes S_{m-1}. \tag{7.4}$$

Then S_m is a $2^m \times 2^m$ Hadamard matrix; it is the **Sylvester matrix** and gives the unique Hadamard 3-design of points and hyperplanes $((m-1)$-flats) of the affine geometry of dimension m over $\mathbf{F}_2$. Thus $C_2(S_m) = \mathcal{R}(1,m)$. We state and prove this as a theorem:

Theorem 7.5.3 *The $2^m \times 2^m$ Sylvester matrix S_m yields the Hadamard 3-design consisting of points and $(m-1)$ flats of the affine geometry $AG_m(\mathbf{F}_2)$ and thus $C_2(S_m) = \mathcal{R}(1,m)$ and $C_2(S_m)^\perp = \mathcal{R}(m-2,m)$.*

Proof: Since $\mathrm{rank}_2(S_1) = 2$, by induction and Theorem 7.5.2, we have $\mathrm{rank}_2(S_m) = m+1$. Now S_m defines a 3-$(2^m, 2^{m-1}, 2^{m-2} - 1)$ design, which has exactly $2^{m+1} - 2$ blocks; adding the all-one vector and the zero vector then accounts for all the 2^{m+1} vectors of the code of S_m. This is a characterizing property of the design of points and $(m-1)$-flats of $AG_m(\mathbf{F}_2)$. $\square$

A corollary to this, part of which was proved by Hamada and Ohmori [127], yields a rigidity theorem for designs with these parameters.

Corollary 7.5.1 *The binary code of a 3-$(2^m, 2^{m-1}, 2^{m-2} - 1)$ design has dimension at least $m+1$ and has at least $2^{m+1} - 2$ vectors of weight 2^{m-1}. If either equality holds so does the other and this occurs if and only if the design is the design of points and hyperplanes in the affine geometry of dimension m over $\mathbf{F}_2$.*

Proof: Simply counting the minimum number of vectors in the code, using the characteristic functions of the blocks, $\jmath$ and the zero vector, shows that the dimension cannot possibly be less than $m+1$. If it is precisely $m+1$ then we have the Reed-Muller code, from the theorem. Suppose, finally, that the only weight-2^{m-1} vectors of the code are those coming from the blocks. Then, for any two blocks B and C, since they meet in 0 or $n = 2^{m-2}$ points, $v^B + v^C$ is either $\jmath$ or the zero vector or v^D where D is a block. Thus the space spanned by the v^B's can only consist of characteristic functions of blocks, along with $\jmath$ and 0, which means it has dimension $m+1$. $\square$

Now it is clear from the two theorems above that we can employ recursive constructions to obtain 3-$(2^m, 2^{m-1}, 2^{m-2} - 1)$ designs for $m \geq 3$, whose binary codes contain $\mathcal{R}(1, m)$ and are contained in its orthogonal, $\mathcal{R}(m - 2, m)$. We illustrate this for $m = 4$ (since for $m = 3$ the design is unique) in the next section.

Exercise 7.5.1 Show that the Hadamard matrix constructed in Exercise 7.2.2 is equivalent to the Sylvester matrix of the same size.

7.6   Size 16 and special n-tuples

The Hadamard matrix S_4 has rank 5 over $\mathbf{F}_2$ and is constructed from S_3 as described above. It has been shown (see Todd [277]) that there are precisely five inequivalent 16×16 Hadamard matrices and five non-isomorphic 3-$(16, 8, 3)$ designs: see also Bhat and Shrikhande [38] and Grey [110]. On the other hand, there is only one equivalence class of 8×8 Hadamard matrices, so if we use the Kronecker construction, we must use equivalent matrices. This is done in the following way: we start with S_3 as given, a symmetric matrix with first row and first column entries all equal to 1. Then we can find three other members of S_3's equivalence class, K_1, K_2, K_3, having the property that each has first row $\jmath$ and that K_1 has three other rows in common with S_3, K_2 has one other row in common with S_3 and K_3 has no other row in common with S_3. (This can, possibly, be more easily seen in the underlying Fano plane, by simply constructing three new Fano planes with three, one or zero lines in common with the original.) Arrange matters to make the first column of each K_i all 1's.

Now let $B_i = S_1 \otimes [S_3, K_i]$ for $i = 1, 2, 3$; let B_0 denote S_4 and let B_4 denote the transpose of B_3. Let the Hadamard 3-design defined by the first row of B_i be denoted by $\mathcal{D}_i$, for $i = 0, 1, 2, 3, 4$. Then, since rank$(S_3) = 4$, Theorem 7.5.2 shows that rank$(B_0) = 5$, rank$(B_1) = 6$, rank$(B_2) = 7$, and rank$(B_3) = 8$, and that $C(B_0) \subseteq C(B_i) \subseteq C(B_0)^\perp$, for $0 \leq i \leq 3$. This clearly accounts for four of the five inequivalent matrices. For the last, we look at B_4. Now S_3 was defined to be symmetric and the transpose, $(K_3)^t$, of K_3 is just a matrix obtained from S_3 by permuting its rows. Since $C(S_3)$ is a self-dual code, we get $C(B_4) \subset C(S_4)^\perp$; since S_4 is symmetric and $C(S_4) \subset C(B_3)$, we get also $C(S_4) \subset C(B_4)$, i.e. again

$$C(B_0) \subseteq C(B_4) \subseteq C(B_0)^\perp.$$

Of course, rank(B_4) is also 8, and it is not clear that either the designs or the codes are not isomorphic. To see that they are not, we look at the weight-4 vectors that occur in the codes of the new designs. All the codes are self-orthogonal, generated by weight-8 vectors, and hence are doubly-even. They all contain $\jmath$, and all, except $C(B_0)$, must have vectors of weight other than 8, i.e. weight-4 vectors must occur.

In fact, an analysis of the vectors arising from the defining Hadamard matrix shows that $C(B_1)$ can be obtained by adding a single weight-4 vector of $C(B_0)^{\perp}$ to $C(B_0)$. All the complements in the three 3-flats that contain this 2-flat must occur too, so $C(B_1)$ has weight enumerator

$$X^0 + 4(X^4 + X^{12}) + 54X^8 + X^{16}.$$

Similarly, for $C(B_2)$, two weight-4 vectors, corresponding to two 2-flats through a line or 1-flat, are added, from which a further 2-flat through that line will also appear. With the complements as well, the weight enumerator of $C(B_2)$ becomes

$$X^0 + 12(X^4 + X^{12}) + 102X^8 + X^{16}.$$

For $C(B_3)$ seven weight-4 vectors (corresponding to seven planes, or 2-flats, through a line) are added, giving in all 28 weight-4 vectors, and weight enumerator

$$X^0 + 28(X^4 + X^{12}) + 198X^8 + X^{16}.$$

(All of these weight enumerators can quite easily be obtained by examining the codewords that must appear when the defining Hadamard matrix is used to generate the code.)

For $C(B_4)$, the weight enumerator is the same as that for $C(B_3)$. However, the supports of the weight-4 vectors do not have the same configuration as in the case of B_3, but form a 3-$(8, 4, 1)$ design. Thus the codes are distinct, the designs are non-isomorphic and the matrices are inequivalent. The two binary, doubly-even, self-dual codes arising from B_3 and B_4 are the only such codes. They are both extremal (see page 86) and hence it is no accident that their weight enumerators are identical; the weight enumerator is that of the direct sum of two $[8, 4]$ extended Hamming codes. There is another instance of both a matrix and its transpose yielding, somewhat mysteriously, extremal codes: see the remarks at the end of Section 7.11. The reader will also find there an instance of a Hadamard matrix whose transpose yields a binary code with a distinct weight enumerator.

The occurrence of weight-4 vectors in the codes of the design can be interpreted more generally in terms of **special n-tuples** in the 3-$(4n, 2n, n-1)$ design $\mathcal{T}$, using the terminology of Bhat and Shrikhande [38]: a *special n-tuple* of $\mathcal{T}$ is a set of n points that is the intersection of three blocks of $\mathcal{T}$. The *characteristic number* of $\mathcal{T}$ is the number of special n-tuples of $\mathcal{T}$. Bhat and Shrikhande classified the 3-$(16, 8, 3)$ designs through their characteristic numbers, using intersection properties of the special n-tuples in case the characteristic numbers were the same. We will now show how these special n-tuples manifest themselves in the codes related to the designs.

Proposition 7.6.1 *Let $T = (\mathcal{P}, \mathcal{B})$ be a 3-$(4n, 2n, n-1)$ design and let X be a special n-tuple. If $X = B_1 \cap B_2 \cap B_3$, where $B_i \in \mathcal{B}$ for $i = 1, 2, 3$, then*

(1) *$B_i - X$ is a special n-tuple for $i = 1, 2, 3$ and hence $\mathcal{P}$ is the disjoint union of these four special n-tuples;*

(2) *a block of T that neither contains X nor is disjoint from it meets X in $n/2$ points; thus special n-tuples can exist, for $n > 1$, only when n is even;*

(3) *if p is a prime dividing $n/2$ then $v^X \in C_p(T)^\perp$;*

(4) *if $n = 2p$ where p is a prime, then, conversely, the special n-tuples are precisely the supports of the constant weight-n vectors of $C_p(T)^\perp$ and, moreover, when p is odd, $C_p(T) = C_p(T)^\perp$.*

Proof: (1) $\mathcal{P} - B_i$ is a block for each i, and $B_i - X = B_i \cap (\mathcal{P} - B_j) \cap (\mathcal{P} - B_k)$, where B_j and B_k are the other two blocks containing X.

(2) Blocks in T meet in 0 or n points. Let $B \in \mathcal{B}$ and suppose $B \neq B_i$ for $i = 1, 2, 3$, and let $x = |B \cap X|$. If $x \neq 0$ then $|B| = 2n = x + 3(n - x)$, i.e. $2x = n$. Since if $n > 1$ there will exist blocks other than the B_i meeting X, we must have n even for special n-tuples to exist.

(3) If p divides $n/2$ and $B \in T$, then $|B \cap X| \equiv 0 \pmod{p}$ by (2) above. Hence $v^X \in C_p(T)^\perp$.

(4) Suppose $X \subset \mathcal{P}$ with $|X| = n$ and $v^X \in C_p(T)^\perp$. Then $|B \cap X| \equiv 0 \pmod{p}$ for every $B \in T$. Thus, $|B \cap X|$ must meet X in 0, p or $2p = n$ points. Let x_i be the number of blocks meeting X in i points. Then counting incidences gives:

$$x_0 + x_p + x_{2p} = 8n - 2 = 16p - 2,$$

$$px_p + 2px_{2p} = n(4n - 1) = 2p(8p - 1),$$

$$p(p - 1)x_p + 2p(2p - 1)x_{2p} = n(n - 1)(2n - 1) = 2p(2p - 1)(4p - 1).$$

These equations imply that $x_{2p} = 3$ and hence that X is a special n-tuple. That $C_p(T)^\perp = C_p(T)$ follows from the fact that p divides the order n to the first power (see Theorem 7.4.1). $\square$

Consider again the five non-isomorphic 3-$(16, 8, 3)$ designs, $\mathcal{D}_i$, for $i = 0$ to 4. Here $n = 4 = 2p$ and the proposition on n-tuples applies. Thus the weight-4 vectors in $C(\mathcal{D}_i)^\perp$ simply give the supports of the special 4-tuples in $\mathcal{D}_i$: 140 for $\mathcal{D}_0$, 78 for $\mathcal{D}_1$, 44 for $\mathcal{D}_2$, and 28 for $\mathcal{D}_3$ or $\mathcal{D}_4$, the latter being in the codes of $\mathcal{D}_3$ and $\mathcal{D}_4$, since they are self-dual.

7.7 Geometric constructions

The geometric device used in Section 7.6 to produce new designs, which amounted to using weight-8 vectors in the Reed-Muller code $\mathcal{R}(2,4)$, can easily be generalized to weight-$2^{(m-1)}$ vectors in $\mathcal{R}(m-2,m)$. It is most likely that, for any dimension between $m+1$ and 2^{m-1}, a design of that rank whose blocks are the supports of weight-2^{m-1} vectors of this Reed-Muller code can be found. Certainly non-isomorphic designs of the same dimension can be produced in many ways: see Dillon [92]. We will describe here just one of these methods.

Recall that $\mathcal{R}(m-2,m)$, viewed as the code of the design of points and 2-flats of the m-dimensional affine geometry over $\mathbf{F}_2$, contains amongst its weight-2^t vectors, for any t such that $2 \le t \le m$, vectors corresponding to the characteristic functions of all t-flats. These will produce weight-2^{m-1} vectors in $\mathcal{R}(m-2,m)$, whose supports are not hyperplanes, in the following way: if T is a t-dimensional subspace and X is an $(m-1)$-dimensional subspace not containing T, then the vector $v^T + v^X = v^{X \oplus T}$ is a weight-2^{m-1} vector with support $X \oplus T$, the symmetric difference of sets X and T. If $t = m - 1$ then $X \oplus T$ is a hyperplane, but if $t < m - 1$ we get new vectors of weight 2^{m-1} that are not the characteristic functions of hyperplanes. Adding v^T to $\mathcal{R}(1,m)$ will then give a code of dimension $m + 2$, and it is only a matter of deciding which weight-2^{m-1} vectors to take for the blocks to produce a 3-$(2^m, 2^{m-1}, 2^{m-2} - 1)$ design with this as its code. The following theorem describes the obvious choice:

Theorem 7.7.1 *Let T be a subspace of dimension t of the affine geometry of dimension m over $\mathbf{F}_2$, where $2 \le t < m - 1$. Let $\mathcal{B}$ denote the set of subspaces of dimension $m - 1$. Then the collection of subsets*

$$\mathcal{X} = \{X \oplus T | T \not\subset X \in \mathcal{B}\} \cup \{X | T \subset X \in \mathcal{B}\},$$

together with the complements of these subsets, forms the blocks of a 3-design whose code is $\langle v^T, \mathcal{R}(1,m) \rangle$.

Proof: The proof is most easily carried out projectively since then complements can be avoided and only a 2-design need be produced. We leave the details to the reader. □

Exercise 7.7.1 Consider the design of points and hyperplanes of the projective space $PG_{m-1}(\mathbf{F}_2)$ and let T be a subspace of projective dimension $(t-1)$ with $1 < t < m - 1$. Let $\mathcal{B}$ consist of those hyperplanes containing T together with the complements of $T \oplus S$ when S is a hyperplane not containing T. Letting $\mathcal{P}$ be the point set of the projective space, show that

$(\mathcal{P}, \mathcal{B})$, with incidence that of the projective space, is a symmetric 2-design and that its binary code contains the binary code of the design of points and hyperplanes. (*Hint:* for a hyperplane S with T not in S choose points P and Q with $P \in S \cap T$ and $Q \in T, Q \notin S$; with R the third point of the line determined by P and Q and with S' any subspace of codimension 1 in S but not containing P, consider the hyperplane generated by S' and the point Q and the hyperplane generated by S' and the point R, and use these to show that the incidence vector of T is in the binary code of the constructed design.)

Observe that the code of this new design is $\mathbf{F}_2 v^T \oplus \mathcal{R}(1, m)$ and that its weight distribution is easily calculated. It is clear that different values of t will give different codes and hence different designs; there are also many other possibilities for constructions in this way, some of them involving a variety of subspaces and producing designs of larger 2-rank.

Exercise 7.7.2 Find the weight enumerator of the binary code of the extension of the design constructed in Exercise 7.7.1, i.e. find the weight distribution of the code $\langle v^T, \mathcal{R}(1, m) \rangle$ where T is a subspace of the affine space.

We have yet to find an example of a 3-$(2^m, 2^{m-1}, 2^{m-2} - 1)$ design whose binary code C cannot be placed in the range

$$\mathcal{R}(1, m) \subseteq C \subseteq \mathcal{R}(m - 2, m). \tag{7.5}$$

For such a code C, the dimension of C *does* lie within the range. In fact, not only is the dimension at least $m + 1$ but $C = \mathcal{R}(1, m)$ if the rank is $m + 1$, by Corollary 7.5.1. The dimension is upper-bounded by 2^{m-1} since C is self-orthogonal; it is achieved, of course, if and only if C is self-dual. Theorem 7.5.1 makes it clear that an enormous number of these Hadamard 3-designs do have binary codes that can be placed in the given range; moreover, the extended quadratic-residue codes, when $2^m - 1$ is a prime, can be placed in this range. The situation is somewhat analogous to the case of affine planes where the majority of planes appear to lie in a similar range and are translation planes: in this case the codes are not necessarily binary — see Section 6.8.

Confining ourselves to those Hadamard 3-designs whose codes contain $\mathcal{R}(1, m)$, we will still be treating a large subclass of the designs with parameters 3-$(2^m, 2^{m-1}, 2^{m-2} - 1)$ and possibly all such designs. Moreover, as Theorem 7.7.1 makes clear, for each d with $1 \le d \le m - 2$ there will be a design whose code is contained in $\mathcal{R}(d, m)$ but is not contained in

$\mathcal{R}(k, m)$ for any $k < d$. Natural questions immediately arise. (1) For the self-orthogonal codes coming from such designs, does every possible minimum weight occur? That is to say, is there, for every multiple of 4 less than or equal to 2^{m-1}, a design whose code has this minimum weight? (2) Does every possible dimension occur? (3) Is there anything special about the Hadamard matrices that give rise to extremal codes? (At one extreme, maximal minimum weight, we have the Sylvester matrices, but what about the other extreme? Is there anything special about the matrices giving rise to self-dual codes with the maximal minimum weight?) Of course, what would be really desirable is a helpful characterization of those doubly-even, self-orthogonal codes that are binary codes of Hadamard matrices; this would amount to a characterization up to linear equivalence of the designs. There are, for example, things that can be said about the vectors of a given weight in such a code; an instance of this is how the vectors of weight 4 must be deployed in the case $m = 4$.

All dimensions and minimum weights occur in the case $m = 4$ as we saw in Section 7.6. Moreover, the two $[16, 8]$ self-dual codes obtained are the only two that exist. This case is too small to be decisive, but the next case, $m = 5$, should be a better testing-ground since here there is no shortage of designs and, further, there are five extremal $[32, 16]$ codes (i.e. $[32, 16, 8]$ codes).

The Reed-Muller codes already give a very rough classification of the matrices and it might be interesting to investigate those whose codes are contained in $\mathcal{R}(2, m)$, for example.

7.8 The Paley construction

The construction of Paley [230] gives Hadamard matrices of size $q+1$, where $q = p^r \equiv 3 \pmod 4$ with p prime, i.e. of order $n = (q + 1)/4$ where q is a prime power.[6] There are generalizations of this construction, and these can be found in Hall [122, Chapter 14].

To discuss this construction we need some facts about the *squares*, usually called *quadratic residues*, of $\mathbf{F}_q$. Recall the following result:

Lemma 7.8.1 *If $q = p^r$, where p is an odd prime, then exactly half the non-zero elements of $\mathbf{F}_q$ are squares. Moreover, -1 is a square if and only if $q \equiv 1 \pmod 4$.*

Proof: An element $y \in \mathbf{F}_q$ is a square if $y = x^2$ has a solution in $\mathbf{F}_q$; thus if $p = 2$, all the elements are squares. When p is odd, the map of $\mathbf{F}_q^\times$ into

---

[6] For q prime the construction was given in 1898 by Scarpis [254].

itself given by $x \mapsto x^2$ is a homomorphism with kernel $\{\pm 1\}$ and hence the image, the subgroup of squares, has order $(q-1)/2$. For -1 to be in this subgroup, $(q-1)/2$ must be even; i.e. $q \equiv 1 \pmod 4$. $\square$

Now we define a function, the *Legendre symbol*, which is a character on the multiplicative group $\mathbf{F}_q^\times$ of $\mathbf{F}_q$.

Definition 7.8.1 *If q is a power of an odd prime, then χ, the **Legendre symbol**, is the following mapping:*

$$\chi : F \to \{0, 1, -1\},$$

where $\chi(0) = 0$ and

$$\chi(x) = \begin{cases} 1 & \text{if } x \text{ is a non-zero square} \\ -1 & \text{if } x \text{ is a non-square} \end{cases}.$$

It is trivial to verify that $\chi(xy) = \chi(x)\chi(y)$ for all $x, y \in \mathbf{F}_q^\times$, i.e. that χ is a character of the group $\mathbf{F}_q^\times$.

Next we define a matrix whose entries are from the set $\{0, 1, -1\}$; it will lead to the Hadamard matrices defined by Paley, the so-called *Paley-Hadamard* matrices.

Definition 7.8.2 *Using the elements of $\mathbf{F}_q$ as row and column labels, define a $q \times q$ matrix, $Q = (q_{xy})$, called a **Jacobsthal matrix**, by*

$$q_{xy} = \chi(y - x).$$

Remark (1) Clearly $q_{xx} = 0$ for all x and $q_{yx} = \chi(-1)q_{xy}$. Thus Q is a *symmetric* matrix if -1 is a square in $\mathbf{F}_q$ and *skew-symmetric* if not. Hence it is symmetric if and only if $q \equiv 1 \pmod 4$.

(2) If $q = p$ is an odd prime, then it is customary to use the natural order of $\mathbf{F}_p$ when writing the matrix. Then $q_{i+1,j+1} = q_{i,j}$ (the suffixes taken modulo p) and so the Jacobsthal matrix will be a circulant. The notation has been chosen so that, in this case, one cycles to the right.

Example 7.8.1 For $q = p = 7$, the squares are 1, 2, and 4 and the matrix is

$$Q = \begin{pmatrix} 0 & 1 & 1 & -1 & 1 & -1 & -1 \\ -1 & 0 & 1 & 1 & -1 & 1 & -1 \\ -1 & -1 & 0 & 1 & 1 & -1 & 1 \\ 1 & -1 & -1 & 0 & 1 & 1 & -1 \\ -1 & 1 & -1 & -1 & 0 & 1 & 1 \\ 1 & -1 & 1 & -1 & -1 & 0 & 1 \\ 1 & 1 & -1 & 1 & -1 & -1 & 0 \end{pmatrix}.$$

We need another lemma:

Lemma 7.8.2 *If χ denotes the Legendre symbol on $\mathbf{F}_q$ and c is a non-zero element of $\mathbf{F}_q$, then*

$$\sum_{b \in \mathbf{F}_q} \chi(b)\chi(b+c) = -1.$$

Proof: Since $\chi(0) = 0$, the term in the sum for $b = 0$ can be ignored. For $b \neq 0$, we can solve $b + c = bz_b$ for z_b where $z_b \neq 1$. Thus

$$\sum_{b \in \mathbf{F}_q} \chi(b)\chi(b+c) = \sum_{b \neq 0} \chi(b)\chi(bz_b) = \sum_{b \neq 0}(\chi(b))^2\chi(z_b)$$

$$= \sum_{z \neq 1}\chi(z) = 0 - \chi(1) = -1,$$

which gives the result. $\square$

With the usual meaning for J, I and $\jmath$, we have the following two results:

Theorem 7.8.1 *Let Q be a Jacobsthal matrix defined by $\mathbf{F}_q$, where q is an odd prime power. Then $QJ = JQ = 0$ and $QQ^t = qI - J$.*

Proof: The first assertion follows simply from the fact that half the non-zero elements of $\mathbf{F}_q$ are squares. Now set $P = (p_{xy}) = QQ^t$ and recall that, by the remarks above, $Q^t = (-1)^{(q-1)/2}Q$. Then

$$p_{xx} = \sum_{y \in \mathbf{F}_q} q_{x,y}^2 = \sum_{y \in \mathbf{F}_q} \chi(y-x)^2 = q - 1;$$

$$p_{xy} = \sum_{z \in \mathbf{F}_q} q_{x,z}q_{y,z} = \sum_{z \in \mathbf{F}_q} \chi(z-x)\chi(z-y),$$

$$= \sum_{b \in \mathbf{F}_q} \chi(b)\chi(b+c), \text{ where } b = z - x \text{ and } c = x - y,$$

$$= -1 \text{ for } x \neq y$$

by Lemma 7.8.2. This completes the proof. $\square$

Corollary 7.8.1 *If $q \equiv 3 \pmod 4$ and Q is a Jacobsthal matrix for $\mathbf{F}_q$, then*

$$H = \begin{pmatrix} 1 & \jmath \\ \jmath^t & Q - I \end{pmatrix}$$

is a Hadamard matrix of size $q + 1$.

Proof: First notice that H is a matrix with all entries equal to 1 or -1. Also

$$HH^t = \begin{pmatrix} 1 & J \\ J^t & Q - I \end{pmatrix} \begin{pmatrix} 1 & J \\ J^t & Q^t - I \end{pmatrix}$$
$$= \begin{pmatrix} q+1 & 0 \\ 0^t & J + (Q-I)(Q^t - I) \end{pmatrix}.$$

The theorem shows that $HH^t = (q+1)I$ and so H is a Hadamard matrix, of *Paley type*. □

Notice that different orderings of the field elements will simply give equivalent Hadamard matrices. The Hadamard 2-designs defined by the Paley matrices are customarily known as the *Paley designs*: these are actually difference-set designs, where the group G is the additive group of the field $\mathbf{F}_q$, and the difference set D is the set of non-zero squares in the multiplicative group of the field. Lander [181] has a fuller discussion of these designs and their automorphism groups.

When q is a prime, the codes of the Paley matrices over a field $\mathbf{F}_p$, where p is a quadratic residue modulo q, are the extended quadratic-residue codes of Section 2.10. A proof of this can be found in Pless [238, Chapter 6]; it follows easily from a consideration of the *idempotent* generator for a cyclic code. Thus they are extended cyclic codes and, in case $q + 1 = 2^m$ and $p = 2$, we can say something about these designs and matrices in relation to our observation (7.5).

Theorem 7.8.2 *If P is a Paley-Hadamard matrix of size $q + 1 = 2^m$, where q is a prime power and $m \geq 3$, then q is a prime and*

$$\mathcal{R}(1, m) \subseteq C_2(P) \subseteq \mathcal{R}(m - 2, m).$$

Proof: We first show that if $q = p^n = 2^m - 1$, then q is a prime, i.e. $n = 1$. If n is even then $p^n \equiv 1 \pmod 4$, contradicting $p^n = 2^m - 1$. If $n > 1$ and odd, then $2^m = p^n + 1 = (p + 1)(p^{n-1} - p^{n-2} + \cdots - p + 1)$, where the second factor is odd, giving a contradiction. Thus we may take q to be a prime and write $q = p$.

Recall that $C_2(P) = C_2(\mathcal{T}) = \widehat{C_2(\mathcal{D})}$ where $\mathcal{T}$ is a 3-design and $\mathcal{D}$ a 2-design given by the Paley-Hadamard matrix. Also, $C_2(\mathcal{D}) = \mathcal{Q}(p, 2)$, the binary quadratic-residue code. (See Definition 2.10.1. Notice here that 2 is a quadratic residue modulo p since $p \equiv -1 \pmod 8$: see Lemma 2.10.1.) Moreover, $\mathcal{R}(m - 2, m) = \widehat{\mathcal{H}_m}$: see Example 5.2.3.

Let □ be the set of squares in $\mathbf{F}_p^\times$. Then $\mathcal{Q}(p, 2) = (g_0(X))$ where

$$g_0(X) = \prod_{r \in \square} (X - \omega^r),$$

where ω is a primitive p^{th} root of unity in $\mathbf{F}_{2^m}$. Also, from Theorem 2.8.1 (or see Corollary 5.5.1), $\mathcal{H}_m = (m(X))$, where

$$m(X) = (X - \omega)(X - \omega^2)(X - \omega^{2^2}) \ldots (X - \omega^{2^{m-1}}).$$

Since $\square \supseteq \{1, 2, 2^2, \ldots, 2^{m-1}\}$, $m(X)$ divides $g_0(X)$ and hence $(g_0(X)) \subseteq (m(X))$. The same inclusion holds for the extended codes, and since by Theorem 2.10.1 $\widehat{\mathcal{Q}(p, 2)}$ is self-dual, we have the stated result. $\square$

7.9 Bent functions

We introduce here certain functions in the Reed-Muller codes $\mathcal{R}(m, 2m)$ that have a close connection with difference-set designs. These designs are, in turn, closely related to designs determined by so-called *regular* Hadamard matrices, to be introduced in Section 7.10. These functions were discovered by Rothaus [248], who called them **bent** functions.

For a full account of these functions and their relationship to Hadamard matrices and designs the reader should consult Dillon [94, 95, 93]. Our account will be brief and based on that given by MacWilliams and Sloane [203, Chapter 14]. We will use the notation of Chapter 5. We first recall the definition of the Hadamard transform, given in Section 2.7, and make this more general:

Definition 7.9.1 *Let H be a Hadamard matrix of size n and let X be a real vector of length n. The **Hadamard transform** of X is the vector $\hat{X} = XH$.*

Each Hadamard matrix defines a Hadamard transform, but it is the Sylvester matrix whose transform we are interested in; it is the transform already defined in Equation (2.11).

We introduce an alternative to the "exp" notation of Definition 7.3.3 that will be more convenient in the current context: for $f \in \mathbf{F}_2^V$ and $\mathbf{u} \in V$, write

$$F(\mathbf{u}) = (-1)^{f(\mathbf{u})}$$

and view F as a real-valued function on V.

Let $V = \mathbf{F}_2^m$. For $f \in \mathbf{F}_2^V$, H the Hadamard matrix defined by Exercise 7.2.2 — with $\mathbf{uv}^{\text{th}}$ entry $(-1)^{(\mathbf{u}, \mathbf{v})}$ — and F as above, the Hadamard transform of F is given by

$$\hat{F}(\mathbf{u}) = \sum_{\mathbf{v} \in V} (-1)^{(\mathbf{u}, \mathbf{v})} F(\mathbf{v}) = \sum_{\mathbf{v} \in V} (-1)^{(\mathbf{u}, \mathbf{v}) + f(\mathbf{v})}. \tag{7.6}$$

Since $\hat{F} = FH$ and $HH^t = H^2 = 2^m I$, we can solve for F to obtain $F = \frac{1}{2^m}\hat{F}H$, or

$$F(\mathbf{u}) = \frac{1}{2^m} \sum_{\mathbf{v} \in V} (-1)^{(\mathbf{u},\mathbf{v})} \hat{F}(\mathbf{v}). \tag{7.7}$$

It follows from Equation (7.6) that

$$\hat{F}(\mathbf{u}) = 2^m - 2d(f,(\mathbf{u},\ )),$$

where d denotes the Hamming distance in the space $\mathbf{F}_2^V$ and $(\mathbf{u},\ )$ is the linear function on V given by the inner product. Thus

$$d(f,(\mathbf{u},\ )) = \frac{1}{2}\{2^m - \hat{F}(\mathbf{u})\}$$

and

$$d(f, 1 + (\mathbf{u},\ )) = \frac{1}{2}\{2^m + \hat{F}(\mathbf{u})\},$$

where 1, as usual, denotes the function that is identically 1. Since the Reed-Muller code $\mathcal{R}(1,m)$ consists of all polynomial functions of degree at most 1, i.e. all functions of the form $(\mathbf{u},\ )$ and $1 + (\mathbf{u},\ )$, these equations show that the codeword closest, in the Hamming metric, to f in $\mathcal{R}(1,m)$ is that for which $|\hat{F}(\mathbf{u})|$ is largest.

Lemma 7.9.1 *For any $f \in \mathbf{F}_2^V$ and $\mathbf{v} \in V$*

$$\sum_{\mathbf{u} \in V} \hat{F}(\mathbf{u})\hat{F}(\mathbf{u}+\mathbf{v}) = \begin{cases} 2^{2m} & \text{if } \mathbf{v} = 0 \\ 0 & \text{if } \mathbf{v} \neq 0 \end{cases}.$$

Further, Parseval's equation holds:

$$\sum_{\mathbf{u} \in V} \hat{F}(\mathbf{u})^2 = 2^{2m}.$$

Proof:

$$
\begin{aligned}
\sum_{\mathbf{u} \in V} \hat{F}(\mathbf{u})\hat{F}(\mathbf{u}+\mathbf{v}) &= \sum_{\mathbf{u} \in V}\sum_{\mathbf{w} \in V} (-1)^{(\mathbf{u},\mathbf{w})} F(\mathbf{w}) \sum_{\mathbf{z} \in V}(-1)^{(\mathbf{u}+\mathbf{v},\mathbf{z})} F(\mathbf{z}) \\
&= \sum_{\mathbf{w},\mathbf{z} \in V} (-1)^{(\mathbf{v},\mathbf{z})} F(\mathbf{z})F(\mathbf{w}) \sum_{\mathbf{u} \in V}(-1)^{(\mathbf{u},\mathbf{w}+\mathbf{z})} \\
&= \sum_{\mathbf{w},\mathbf{z} \in V} (-1)^{(\mathbf{v},\mathbf{z})} F(\mathbf{z})F(\mathbf{w}) 2^m \delta_{\mathbf{w},\mathbf{z}} \\
&= 2^m \sum_{\mathbf{w} \in V} (-1)^{(\mathbf{v},\mathbf{w})} F(\mathbf{w})^2 \\
&= 2^m \sum_{\mathbf{w} \in V} (-1)^{(\mathbf{v},\mathbf{w})} \\
&= 2^{2m} \delta_{\mathbf{v},0}.
\end{aligned}
$$

Parseval's equation follows. $\square$

It follows from Parseval's equation that if $|\hat{F}(\mathbf{u})|$ is a constant for all $\mathbf{u}$, then this constant is $2^{m/2}$. This is only possible if m is even, since $\hat{F}(\mathbf{u})$ is certainly an integer.

Definition 7.9.2 *A function* $f \in \mathbf{F}_2^V$ *is* **bent** *if* $\hat{F}(\mathbf{u}) = \pm 2^{m/2}$ *for all* $\mathbf{u} \in V$.

From the discussion above, it should be clear that bent functions are those that are furthest, in the Hamming metric, from the codewords of $\mathcal{R}(1, m)$. That is, the bent functions are those that are "most non-linear", and, presumably, that is why Rothaus called them "bent".

Example 7.9.1 (1) For $m = 2$ the function $f(x_1, x_2) = x_1 x_2$ is bent.

(2) For $m = 4$ the function $f(x_1, x_2, x_3, x_4) = x_1 x_2 + x_3 x_4$ is bent.

Lemma 7.9.2 *A function* $f \in \mathbf{F}_2^V$ *is bent if and only if the* $2^m \times 2^m$ *matrix with rows and columns labelled by V and with entry at the* $(\mathbf{u}, \mathbf{v})^{\text{th}}$ *position* $(-1)^{f(\mathbf{u}+\mathbf{v})}$ *is Hadamard.*

Proof: Let K denote the matrix with entries $(-1)^{f(\mathbf{u}+\mathbf{v})}$. The entries are all clearly ± 1, so K is Hadamard if and only if any two rows are orthogonal. Now using Equation (7.7) we get, for any $\mathbf{v} \in V$,

$$\sum_{\mathbf{u} \in V} (-1)^{f(\mathbf{u})}(-1)^{f(\mathbf{u}+\mathbf{v})} = \sum_{\mathbf{u} \in V} F(\mathbf{u}) F(\mathbf{u} + \mathbf{v})$$

$$= \frac{1}{2^m} \sum_{\mathbf{w} \in V} (-1)^{(\mathbf{w}, \mathbf{v})} \hat{F}(\mathbf{w})^2.$$

Now if f is bent, then $\hat{F}(\mathbf{w})^2 = 2^m$, so that the sum is 2^m if $\mathbf{v} = \mathbf{0}$ and 0 otherwise. Thus K is a Hadamard matrix.

Conversely, suppose that K is Hadamard. We need to show that $\hat{F}(\mathbf{u}) = \pm 2^{m/2}$ for all $\mathbf{u} \in V$. Since K is Hadamard, the above equations imply that

$$\sum_{\mathbf{w} \in V} (-1)^{(\mathbf{w}, \mathbf{v})} \hat{F}(\mathbf{w})^2 = \begin{cases} 2^{2m} & \text{if } \mathbf{v} = \mathbf{0} \\ 0 & \text{if } \mathbf{v} \neq \mathbf{0} \end{cases}.$$

With H as in Exercise 7.2.2, as before, this equation shows that

$$\hat{F}^2 H = (2^{2m}, 0, \ldots, 0).$$

Since H can be taken to be symmetric (by labelling rows and columns similarly), $H^2 = 2^m I_{2^m}$, and it is now a simple matter to deduce that $\hat{F}(\mathbf{u}) = \pm 2^{m/2}$ for all $\mathbf{u}$ and thus that f is bent. $\square$

In order to be able to deduce a bound in the next section for the 2-rank of certain difference-set designs, we need finally to show that the bent functions have degree at most $m/2$, i.e. that they are in the Reed-Muller code $\mathcal{R}(m/2, m)$.

Theorem 7.9.1 *If $m > 2$ and f is bent, then $f \in \mathcal{R}(m/2, m)$.*

Proof: Let C be any $[m, k]$ code in V. Then Lemma 2.7.1 applied to $F(\mathbf{u})$ gives

$$\sum_{\mathbf{u} \in C^\perp} F(\mathbf{u}) = \frac{1}{2^k} \sum_{\mathbf{u} \in C} \hat{F}(\mathbf{u}). \tag{7.8}$$

Viewing f as real-valued, $F(\mathbf{u}) = 1 - 2f(\mathbf{u})$ and $\hat{F}(\mathbf{u}) = 2^{m/2}(-1)^{\hat{f}(\mathbf{u})} = 2^{m/2}(1 - 2\hat{f}(\mathbf{u}))$. Equation (7.8) gives

$$\sum_{\mathbf{u} \in C^\perp} f(\mathbf{u}) = 2^{m-k-1} - 2^{\frac{1}{2}m-1} + 2^{\frac{1}{2}m-k} \sum_{\mathbf{u} \in C} \hat{f}(\mathbf{u}), \tag{7.9}$$

where these are, of course, real sums with all entries integral.

Let $\mathbf{a} \in V$ and define the code C, which depends on $\mathbf{a}$, as follows, using the notation for $I_\mathbf{a}$ and K introduced in Lemma 5.2.1:

$$C = \{\mathbf{c} | \mathbf{c} \in V \text{ and } I_\mathbf{c} \subseteq I_{\mathbf{a}+\jmath}\} = \langle e_i | i \in K - I_\mathbf{a} \rangle,$$

so that

$$C^\perp = \{\mathbf{c} | \mathbf{c} \in V \text{ and } I_\mathbf{c} \subseteq I_\mathbf{a}\} = \langle e_i | i \in I_\mathbf{a} \rangle.$$

Then $|C^\perp| = 2^{\text{wt}(\mathbf{a})}$ and Equation (7.9) becomes

$$\sum_{I_\mathbf{c} \subseteq I_\mathbf{a}} f(\mathbf{c}) = 2^{\text{wt}(\mathbf{a})-1} - 2^{\frac{1}{2}m-1} + 2^{\text{wt}(\mathbf{a})-\frac{1}{2}m} \sum_{I_\mathbf{c} \subseteq I_{\mathbf{a}+\jmath}} \hat{f}(\mathbf{c}). \tag{7.10}$$

Comparing this equation with those in Exercise 5.2.1, we see that the coefficient of $x_1^{a_1} \ldots x_m^{a_m}$ is the left-hand side of Equation (7.10), taken modulo 2. Now if $\text{wt}(\mathbf{a}) > m/2$ and $m > 2$, then the right-hand side of this equation is even, and the coefficient is zero. Thus the degree of f is at most $m/2$. $\square$

The bent functions will emerge in the following section in the context of difference-set designs from elementary abelian 2-groups.

7.10 Regular Hadamard matrices

The symmetric designs we have obtained from Hadamard matrices of order n have had $v = 4n - 1$. However, there is another way to obtain symmetric designs, but now on $4n$ points, from Hadamard matrices of a particular type, *viz.* those that have constant row and column sum, sometimes known as **regular** Hadamard matrices.

Theorem 7.10.1 *Suppose H is a Hadamard matrix of order n with the property that the sum of the entries in each row is constant. Then n is a square and the constant is either $2\sqrt{n}$ or $-2\sqrt{n}$ with the sum of the entries in each column summing to the same constant. Setting $n = N^2$, then $\log_{-1}(H)$ is an incidence matrix of either a $(4N^2, 2N^2 - N, N^2 - N)$ design or a $(4N^2, 2N^2 + N, N^2 + N)$ design, depending on whether the constant is positive or negative.*

Proof: Suppose the sum of the entries in each row is a. Then $HJ = aJ$, so that $(HJ)^t = JH^t = aJ$ and $JH^tH = 4nJ = aJH$. Thus $JH = (4n/a)J$ and the column sum is constant and equal to $4n/a$. Further, $JHJ = aJ^2 = 4naJ = ((4n)^2/a)J$. Thus $a^2 = 4n$, and $a = \pm 2\sqrt{n}$. Set $n = N^2$.

Suppose a row r has x entries equal to 1 and y entries equal to -1. Then $x + y = 4n$ and $x - y = a$, so that $x = (4n + a)/2$ and $y = (4n - a)/2$ are independent of the row. Consider two distinct rows, r and s, of H, and compare the column entries in the two rows as we did in Theorem 7.2.2, using the same notation. Then

$$m_{-1,-1} + m_{-1,1} = m_{-1,-1} + m_{1,-1} = y$$

so that

$$m_{-1,1} = m_{1,-1}.$$

Since

$$m_{1,1} + m_{-1,-1} = m_{-1,1} + m_{1,-1} = 2n,$$

we get $m_{1,-1} = n$ and $m_{-1,-1} = y - n$. Thus any two rows have $y - n$ entries of -1 in common. Exactly the same holds for columns.

Construct the incidence structure $\mathcal{D} = (\mathcal{P}, \mathcal{B})$ with point set $\mathcal{P}$ the column labels as usual and with $4n$ blocks defined by the entries -1 in each row. This is equivalent to taking $\log_{-1}(H)$ as incidence matrix. Thus the block size is $y = (4n - a)/2 = 2N^2 \mp N$, where $a = \pm 2N$. By the above, any two blocks meet in $\lambda = y - n = N^2 \mp N$ points. Since all this holds for columns as well, we certainly have a symmetric 2-design, with parameters

$$(4N^2, 2N^2 \pm N, N^2 \pm N),$$

and order $N^2 = n$. $\square$

As before, the procedure is reversible. In fact something a little more general holds for designs with parameters $(4n, k, n)$.

Theorem 7.10.2 *Let $\mathcal{D}$ be a symmetric design of order n with $v = 4n$. Then $n = N^2$ and $\mathcal{D}$ has parameters $(4N^2, 2N^2 \pm N, N^2 \pm N)$. Further, if A is an incidence matrix for $\mathcal{D}$, then $\exp_{-1}(A)$ is a Hadamard matrix of order N^2 with constant sum $\mp 2N$, and the associated Hadamard 3-designs are obtained by picking a block B_0 of $\mathcal{D}$ and taking the set of all symmetric differences, $B_0 \oplus B$ for $B \neq B_0$, and their complements.*

If p divides N^2, then $C_p(\mathcal{D})$ is self-orthogonal. If N is even and T is a Hadamard 3-design obtained from $\mathcal{D}$, then $C_2(T) \supseteq E_2(\mathcal{D})$ with equality if and only if $\jmath \in E_2(\mathcal{D})$.

Proof: Since $\mathcal{D}$ is a symmetric design, if k is the number of points per block, then k satisfies

$$(4n - 1)(k - n) = k(k - 1).$$

Solving this quadratic in k gives $k = 2n \pm \sqrt{n}$, as required. The rest is clear. $\square$

Non-isomorphic designs $\mathcal{D}$ may yield equivalent Hadamard matrices; put another way, an equivalence class of Hadamard matrices may yield many of these designs. This is true for the Sylvester matrices where the number of inequivalent designs grows very fast as the size of the Sylvester matrix grows. In contrast to the Hadamard designs, the codes of these designs will not necessarily (indeed not usually) be equivalent, even though they come from the same equivalence class of matrices. This accounts for part, at least, of the subtlety of this aspect of the theory.

Symmetric designs with parameters $(4N^2, 2N^2 \pm N, N^2 \pm N)$ may arise as *difference-set designs*, in which case the difference set is called a *Hadamard difference set*, because of the connection with Hadamard matrices. Moreover, Mann [207, p. 72] has shown that any non-trivial symmetric design on 2^m points and of order n must have $2^m = 4n$, and hence must have parameters of the above form. Thus if G is a (finite) 2-group, the difference set *must* be Hadamard. Our particular interest here will be in the case when G is an elementary abelian 2-group.

Thus let G be the elementary abelian 2-group of order 2^{2m}. Then G is effectively the vector space $V = \mathbf{F}_2^{2m}$ and, with our earlier notation, writing $F = \mathbf{F}_2$, $F^G = F^V$ is the vector space of all polynomial functions in $2m$ boolean variables. This is precisely the setting for the Reed-Muller codes. In this case the difference sets in G have characteristic functions that are the bent functions, as was first shown by Dillon [94]. For completeness we give a proof of this result:

Theorem 7.10.3 *Any difference set in an elementary abelian 2-group of order 2^{2m} defines a set of bent functions that are equivalent under the group* *of translations in the affine group* $AGL_{2m}(\mathbf{F}_2)$*; conversely, any bent function uniquely defines a difference-set design.*

Proof: We will make use of Lemma 7.9.2. Suppose first that $\mathcal{D}$ is a difference-set design given by the set D in the elementary abelian 2-group G, written additively as the vector space V. Then, by Mann's result mentioned above, $\mathcal{D}$ must have the parameters of Theorem 7.10.2, and if A is an incidence matrix for $\mathcal{D}$, then $\exp_{-1} A$ is a Hadamard matrix. Suppose we have labelled the columns of A by the vectors $\mathbf{u} \in V$ and the rows in the same order by the translates $\mathbf{u} + D$: thus the $(\mathbf{u}, \mathbf{v})^{\text{th}}$ entry of A, *viz.* $a_{\mathbf{u},\mathbf{v}}$, is 1 if $\mathbf{u} + \mathbf{v} \in D$ and 0 otherwise. Define a function $f : V \to F$ (where $F = \mathbf{F}_2$) by

$$f(\mathbf{u}) = \begin{cases} 1 & \text{if } \mathbf{u} \in D \\ 0 & \text{if } \mathbf{u} \notin D. \end{cases}$$

Then, to show that f is bent, by Lemma 7.9.2 we need only show that, if $H = (h_{\mathbf{u},\mathbf{v}}) = \exp_{-1}(A)$, then $h_{\mathbf{u},\mathbf{v}} = (-1)^{f(\mathbf{u}+\mathbf{v})}$. But $h_{\mathbf{u},\mathbf{v}} = -1$ if and only if the corresponding entry $a_{\mathbf{u},\mathbf{v}} = 1$, i.e. if and only if $(\mathbf{u} + \mathbf{v}) \in D$. Thus $h_{\mathbf{u},\mathbf{v}} = (-1)^{f(\mathbf{u}+\mathbf{v})}$, as required, and f is bent. (Notice that if a translate $\mathbf{a} + D$ were used to define the function f, we would get a function obtained by a translation on the underlying affine geometry of dimension $2m$, and the function would still be bent.)

Conversely, suppose f is bent. Form the matrix H with $(\mathbf{u}, \mathbf{v})^{\text{th}}$ entry $(-1)^{f(\mathbf{u}+\mathbf{v})}$ as usual. Since this is a Hadamard matrix by Lemma 7.9.2, and since the matrix has constant row (and column) sum, we know from Theorem 7.10.1 that $A = \log_{-1}(H)$ is an incidence matrix of a symmetric design with the correct parameters. One of the blocks is given by the set $D = \{\mathbf{u} | f(\mathbf{u}) = 1\}$, which is the row corresponding to the vector $\mathbf{0}$. Now the entries of -1 in the row of H corresponding to the vector $\mathbf{a}$ occur when $f(\mathbf{a} + \mathbf{u}) = 1$, i.e. when $\mathbf{u} \in \mathbf{a} + D$. Thus the matrix A has for its rows the incidence vectors of the translate design of D, making this a difference-set design. $\square$

Recall that bent functions correspond to cosets of the first-order Reed-Muller code of highest possible weight. The minimum-weight vectors in such a coset also yield a symmetric design, but not usually the design given by the corresponding difference set.

Some of the results we discuss in this section were developed with an eye to investigating the curiously difficult problem of determining when the all-one vector is in the binary code of a symmetric design — in the case where all the parameters are even. We do not even have an example for which the all-one vector is not in E. By Theorem 2.4.2, that could only happen for such a design $\mathcal{D}$ if $C_2(\mathcal{D})$ and $C_2(\overline{\mathcal{D}})$ had different dimensions with the smaller of the two codes equal to E. This fact allows us to restate an interesting result due to Dillon and Schatz [96].

Theorem 7.10.4 *Suppose $\mathcal{D}$ is a $(2^{2m}, 2^{2m-1} - 2^{m-1}, 2^{2m-2} - 2^{m-1})$ design with both $C_2(\mathcal{D})$ and $C_2(\overline{\mathcal{D}})$ of dimension $2m + 2$. Then there is a difference set D in the elementary abelian 2-group corresponding to the first-order Reed-Muller code $\mathcal{R}(1, 2m)$ with*

$$C_2(\mathcal{D}) = C_2(\overline{\mathcal{D}}) = \langle v^D \rangle \oplus \mathcal{R}(1, 2m)$$

and the design $\mathcal{D}$ is the design of minimum-weight vectors of this binary code.

Notice that the theorem shows that an equivalence class of Hadamard matrices may yield non-isomorphic designs; in fact this is true of the Sylvester matrix. In terms of the matrices this means that an equivalence class may contain many matrices with constant row sums, each such matrix yielding a design and all these designs distinct — that is, not isomorphic.

The assumption in the theorem that both $\mathcal{D}$ and $\overline{\mathcal{D}}$ have the same rank could be eliminated if there were a positive answer to the following question:

Question 7.10.1 *Does the code of a design with parameters $(2^{2m}, 2^{2m-1} - 2^{m-1}, 2^{2m-2} - 2^{m-1})$ contain the all-one vector?*

The rank given in the theorem is the smallest possible rank for a design with parameters $(2^{2m}, 2^{2m-1} - 2^{m-1}, 2^{2m-2} - 2^{m-1})$. We know of no design with parameters $(4k^2, 2k^2 - k, k^2 - k)$ whose codes over any field do not contain the all-one vector. The reader may want to consult Assmus [7] for another discussion of the all-one problem and Theorem 7.10.4. Notice that the design given by the theorem is not necessarily isomorphic to the design given by the difference set, i.e. the development or translate design. (The paper [96] is concerned mostly with symmetric designs that have the so-called *symmetric-difference property*, first introduced by Kantor [150]. See also Jungnickel and Tonchev [148].)

Suppose next that we are given a design that *is* the design given by a difference set in an elementary abelian 2-group. Thus $\mathcal{D}$ has parameters $(2^{2m}, 2^{2m-1} - 2^{m-1}, 2^{2m-2} - 2^{m-1})$ and is given by the translates of the difference set. In this case we know, from Corollary 4.6.2, that the all-one vector is present in its binary code. Thus the rank of $\mathcal{D}$ is $1 + \dim(E)$. If C is the binary code of $\mathcal{D}$, then since the difference set is given by a bent function, we have, by Theorem 7.9.1, that $C \subseteq \mathcal{R}(m, 2m)$ and E is in the code orthogonal to $\mathcal{R}(m, 2m)$, *viz.* $\mathcal{R}(m - 1, 2m)$. This provides an upper bound for the dimension of C. The upper bound is of the same nature as those given in Chapter 6 in that any two designs whose ranks meet the bound are linearly equivalent. In this case it means that the binary codes of the designs are isomorphic. Hence we have, using Theorem 5.3.3:

Theorem 7.10.5 *Let $\mathcal{D}$ be the design given by a difference set in an elementary abelian 2-group of order 2^{2m}. Then the 2-rank of $\mathcal{D}$ is at most $2^{2m-1} + 1 - \frac{1}{2}\binom{2m}{m}$. Moreover, any two such designs with 2-ranks meeting this bound have isomorphic binary codes.*

For $m = 2$ the upper bound is 6. This explains the fact that although there are three designs with parameters $(16, 6, 2)$ and all of them are difference sets in groups of order 16, only the one of rank 6 is a difference set in an elementary abelian 2-group: see Assmus and Salwach [23]. The proposition also yields a proof of the uniqueness of a $(16, 6, 2)$ design of 2-rank 6 since the minimum-weight vectors of the code are precisely the vectors given by the blocks of the design. The two 16×16 Hadamard matrices of rank 8 cannot be produced by such a design since all three designs contain the all-one vector and hence can only produce matrices of 2-rank at most 7. It follows from rank considerations alone, therefore, that three of the five matrices are produced by $(16, 6, 2)$ designs and two are not. More precisely, neither of the two equivalence classes of 16×16 Hadamard matrices of rank 8 possess matrices with constant row sums while the other three equivalence classes do; moreover, in each of these three cases the design can be chosen to be a difference-set design.

It must be said, however, that the bound does not appear to be useful for $m > 2$ although it is better than the bound coming from self-orthogonality, that bound being 2^{2m-1}. But this fact does, at least, imply that no Hadamard matrix of size 2^{2m} and rank 2^{2m-1} can be produced by an elementary Hadamard difference set.

That the design of points and hyperplanes of the affine geometry has the bounding dimension is well-known, and it is easy to show that in the case of equality it is the classical design: see Corollary 7.5.1. This result was first observed by Hamada and Ohmori: we called it a "rigidity" theorem because of the topological analogy; it says that among the structures $\mathcal{D}$ of a certain class the invariant $C_2(\mathcal{D})$ determines the structure uniquely. It is a rather easy example of such a theorem in design theory since it is really the dimension of the invariant that determines the structure — provided the dimension is as small as possible. The analogous result for translation planes (see Chapter 6) is not so easy and, at present at least, more than the dimension of the invariant is necessary for the characterization.

As we remarked before (see Equation (7.5)), we have not been able to find a Hadamard 3-$(2^{m+1}, 2^m, 2^{m-1} - 1)$ design whose binary code does not contain a copy of a first-order Reed-Muller code of the appropriate length, and for $m = 3$ all the designs do, as we have seen. We thus ask the following:

Question 7.10.2 *Does the binary code of a 3-$(2^{m+1}, 2^m, 2^{m-1} - 1)$ design always contain a copy of $\mathcal{R}(1, m + 1)$?*

One must be careful about what is being asked here. Even if the Hadamard 3-design is obtained using a design coming from a difference set in an elementary abelian 2-group, the Reed-Muller code need not be the one associated with *that* 2-group. We have checked that the designs coming from the four inequivalent bent functions, viewed as difference sets, of $\mathcal{R}(3, 6)$ each contain a copy of a first-order Reed-Muller code. Only the quadratic bent function contains the "natural" Reed-Muller code.

7.11   Hadamard matrices of size 24

The number of equivalence classes of Hadamard matrices of size $4n \le 28$ is known: there is only one equivalence class for each $n < 4$ and the designs are also unique (up to isomorphism) since the matrices have transitive automorphism groups. As we have seen, the sharply 5-transitive Mathieu group and the celebrated ternary Golay code occur for $n = 3$. For $n = 4$ there are five matrices, each with a transitive automorphism group, and hence five 3-designs and five 2-designs; we discussed this case in detail in Section 7.6. For $n = 5$ there are three matrices, three 3-designs and six 2-designs. In this section we examine the case $n = 6$; the binary and ternary codes arising from the 3-design (and hence from the matrices) will be self-dual and of maximum rank 12.

There are exactly 60 equivalence classes of Hadamard matrices of size 24 (see Ito *et al.* [147] and Kimura [167]) and, up to isomorphism, 130 Hadamard 3-$(24, 12, 5)$ designs. Each such design produces a binary doubly-even $[24, 12]$ code and also a self-dual ternary $[24, 12]$ code. We look at the binary case, since these codes have been classified. In fact, following an unpublished result of Conway, it has been shown by Pless and Sloane in [239] that there are precisely nine binary doubly-even $[24, 12]$ codes. We first prove that three of the nine cannot occur as the binary code of a Hadamard 3-$(24, 12, 5)$ design.

Proposition 7.11.1 *If C is a binary doubly-even $[24, 12]$ code with the property that there exist seven coordinate places such that the projection onto these seven coordinates is the $[7, 4]$ Hamming code, then C is not the binary code of a Hadamard 3-$(24, 12, 5)$ design.*

Proof: Suppose that C is a binary doubly-even $[24, 12]$ code that is generated by the characteristic functions of the blocks of a Hadamard 3-$(24, 12, 5)$ design and let S be a set of seven coordinates with the property that C', the projection of C to S, is a $[7, 4]$ Hamming code. We must now reach a contradiction. Every 3-subset of S is covered by five blocks of the design, each of which, viewed in C, is a weight-12 vector. These weight-12 vectors project to non-zero vectors of C'. Since two blocks of the design meet either in six points or not at all, the all-one vector of C' is the projection of at most one block. Therefore each of the other non-zero vectors of C' must be the projection of at least four blocks, since no 3-subset of S is covered by two of the 14 weight-3 and weight-4 vectors of C'. But then there would be at least $4 \times 14 = 56$ blocks of the design. Since there are only 46, we have the contradiction. $\square$

Corollary 7.11.1 *At most six of the nine binary doubly-even $[24, 12]$ codes can be the binary codes of Hadamard 3-$(24, 12, 5)$ designs.*

Proof: Each of the two decomposable codes is built from a Hamming code and is thus eliminated by the proposition above. Moreover, B_{24} (in the notation of Pless and Sloane) is also so built (but with *glue*) and the proposition eliminates it also. $\square$

We have used CAYLEY [66] on the Birmingham University VAX to classify the 130 Hadamard 3-$(24, 12, 5)$ designs through their binary codes by finding the codes associated with each of the 60 equivalence classes of matrices. All of the six possible $[24, 12]$ codes appeared. The results are given in Table 7.1, where the equivalence classes of matrices are as listed in Ito *et al.* [147], with $1, 2, \ldots, 59$ representing $H1, H2, \ldots, H59$ and K representing the 60th class (found by Kimura).[7] Also we use the notation of Pless and Sloane [239] for the six codes that occur. Each entry in the table also gives information about the equivalence class of the transpose of that entry: the superscript denotes the matrix equivalence class of the transpose, and the subscript denotes the 2-equivalence class of the transpose. Thus 5_E^9 in the row corresponding to A_{24} indicates that $H5$ has code A_{24} and that its transpose is in the class of $H9$ which has binary code E_{24}. The 12 classes that contain both a matrix and its transpose are entered in boldface and the superscripts and subscripts are omitted.

Remark (1) G_{24} is the $[24, 12]$ extended Golay code. The two Hadamard matrices, $H8$ and $H59$ (the latter representing the class of the

---

[7]Notice that our Hi denotes the transpose of Hi in [147], since we construct our Hadamard matrices from the rows of an incidence matrix of the design.

Code	Hadamard Matrices
A_{24}	5_E^9, **10**, 18_C^{11}, 19_C^{12}, 36_D^{13}, 37_D^{14}, 51_F^{15}, 52_F^{16}
C_{24}	3_E^{17}, 11_A^{18}, 12_A^{19}, **20**, **22**, **25**, **27**, **33**, 38_D^{21}, 39_D^{24}, 40_D^{28},
	41_D^{30}, 53_F^{23}, 54_F^{26}, 55_F^{29}, 56_F^{31}
D_{24}	2_E^{34}, 7_E^{35}, 13_A^{36}, 14_A^{37}, 21_C^{38}, 24_C^{39}, 28_C^{40}, 30_C^{41}, 32_D^{42}, 42_D^{32},
	43_D^{44}, 44_D^{43}, **45**, **46**, **48**, 57_F^{47}
E_{24}	**1**, 9_A^5, 17_C^3, 34_D^2, 35_D^7, 49_F^4, 50_F^6, 58_G^8
F_{24}	4_E^{49}, 6_E^{50}, 15_A^{51}, 16_A^{52}, 23_C^{53}, 26_C^{54}, 29_C^{55}, 31_C^{56}, 47_D^{57}, **K**
G_{24}	8_E^{58}, **59**

Table 7.1: Binary codes of the 24×24 Hadamard matrices

Paley-Hadamard matrix) that generate this binary code have transitive automorphism groups and hence only one 3-$(24, 12, 5)$ design is involved with each case. The Paley-Hadamard matrix also generates the extremal $[24, 12, 9]$ ternary extended quadratic-residue code. There are two extremal $[24, 12, 9]$ ternary codes (see Leon, Pless and Sloane [184] and note that we have checked that K's ternary code has weight-6 vectors and hence cannot be extremal) and curiously it is not $H8$ that generates the other, but $H8$'s transpose, $H58$. We do not have an explanation for this fact.

(2) Observe that 2-equivalence distinguishes among $H8$, $H58$ and $H59$ and between the two extremal $[24, 12, 9]$ ternary codes, thus yielding a non-group-theoretical proof that these ternary codes are inequivalent: if, for an arbitrary 24×24 Hadamard matrix H, both H and its transpose generate G_{24}, then H is the Paley-Hadamard matrix; on the other hand if H generates G_{24} and its transpose does not, then H is $H8$ and its transpose is $H58$. In all, 12 of the 24×24 Hadamard matrices are characterized by their two 2-equivalence classes (of the matrix and its transpose) but note that 2-equivalence will not distinguish between $H18$ and $H19$, for example. It might have been possible that 2-equivalence and 3-equivalence, together with transpose information, could completely characterize each of the 24×24

Hadamard matrices, but recent work of Lam, Thiel and Pautasso [178] on the ternary codes indicates that this is not the case. Since there are at most ten ternary self-dual $[24, 12]$ codes (see Leon *et al.* [184], but note that the authors of that paper were not aware of the error in Ito *et al.* [147], nor of Kimura's matrix), there must be matrices (generating the codes C_{24} and D_{24}) that are both 2-equivalent and 3-equivalent.

(3) Since $H58$ and $H59$ both generate extremal ternary codes, these codes do not have any weight-6 vectors at all. It follows from Proposition 7.6.1 (4) that the two associated 3-designs do not have special 6-tuples. Put another way, any three distinct blocks of either design intersect in at most five points.

(4) It is an easy computational matter to decide the 2-equivalence class of a 3-$(24, 12, 5)$ design or of a 24×24 Hadamard matrix, since the six $[24, 12]$ binary codes involved have different weight distributions — indeed, different numbers of weight-4 vectors.

(5) The next binary case to consider is order 8, i.e. 3-$(32, 16, 7)$ designs. This case has been considered intractable by combinatorialists, since there are millions of these designs (see Norman [221]). But the $[32, 16]$ binary doubly-even codes have been classified (see Conway and Pless [74]), and since the binary code of a 32×32 Hadamard matrix is doubly-even (but not necessarily self-dual), it follows (see MacWilliams, Sloane and Thompson [204]) that it is contained in one of the 85 binary doubly-even $[32, 16]$ codes. Even an examination of the five extremal binary doubly-even $[32, 16, 8]$ codes might prove interesting (see also Koch [173]). In this case we have only 2-equivalence.

(6) The 20×20 (or 3-$(20, 10, 4)$) case must be treated over $\mathbf{F}_5$. Leon, Pless and Sloane [183] have investigated the self-dual codes over $\mathbf{F}_5$, but the $[20, 10]$'s were not classified. It is possible that the designs and matrices reduce to two 5-equivalence classes. The 28×28 case will have to be treated over $\mathbf{F}_7$; we do not have any coding-theoretical information on this case, but Tonchev [279] has classified those matrices having transitive automorphism groups with the help of coding theory: there are seven. Kimura has recently announced that there are precisely 487 Hadamard matrices of order 7: see Kimura [165, 166] and Kimura and Ohmori [168].

7.12 Hadamard matrices from Steiner systems

The construction of Hadamard designs that we describe here was first discovered by Shrikhande and Singh [263] and later rediscovered by Goethals and Seidel [106, Theorem 4.5].

Theorem 7.12.1 *Let A be an incidence matrix of a Steiner system with parameters 2-$(v, k, 1)$. Then the symmetric matrix $B = AA^t - kI$ is an incidence matrix of a 2-design $\mathcal{E}$ if and only if $v = 2k^2 - k$, in which case this design is a symmetric Hadamard $(4k^2 - 1, 2k^2, k^2)$ design. Further, $\mathcal{E}$ is self-dual and the equivalence class of Hadamard matrices defined by $\mathcal{E}$ contains a symmetric Hadamard matrix with constant entries on the diagonal.*

Proof: Let $\mathcal{D}$ be the Steiner system. The number of 1's occurring in the ith row of B is the number of blocks of $\mathcal{D}$ meeting the ith block, i.e. $r(k - 1)$, where $r = (v - 1)/(k - 1)$ is the replication number for $\mathcal{D}$. To see if B is the incidence matrix of a 2-design, we need to check that for any two columns of B the number of rows that have an entry 1 in both columns is independent of the two columns. This is equivalent to checking that the number of blocks of $\mathcal{D}$ that meet both of two given blocks is independent of the choice of the given blocks. There are two cases: either the blocks have a point in common, in which case this number is $(k - 1)^2 + (r - 2)$, or they are disjoint and the number is k^2. These numbers are equal precisely when $r = 2k + 1$. It now follows that $\mathcal{D}$ is a 2-$(2k^2 - k, k, 1)$ design and that $\mathcal{E}$ is a $(4k^2 - 1, 2k^2, k^2)$ design. The order of $\mathcal{D}$ is $2k$ and the order of $\mathcal{E}$ is k^2.

Clearly, both Hadamard 2-designs, $\mathcal{E}$ and $\overline{\mathcal{E}}$, are self-dual, B being symmetric. A $4k^2 \times 4k^2$ Hadamard matrix of the required form is obtained by forming $\exp_{-1}(B)$ and then bordering it with a row and column of 1's. $\square$

Consider now the codes of the Hadamard designs that can be constructed in this way, especially in relationship to those of the Steiner systems. Since the order of the Steiner system is $2k$ and the order of both $\mathcal{E}$ and its complement is k^2, we take primes p dividing k.

Theorem 7.12.2 *Let $\mathcal{D}$ be a 2-$(2k^2 - k, k, 1)$ design and let $\mathcal{E}$ be the symmetric $(4k^2 - 1, 2k^2, k^2)$ design that it produces. Then for any prime p dividing k, there is an exact sequence*

$$0 \to \text{Hull}_p(\mathcal{D}) \to C_p(\mathcal{D}) \to C_p(\mathcal{E}) \to 0.$$

In particular,

$$\text{rank}_p(\mathcal{E}) = \text{rank}_p(\mathcal{D}) - \dim(\text{Hull}_p(\mathcal{D})).$$

Proof: If A is an incidence matrix for $\mathcal{D}$, then A^t maps the space of row vectors, $\mathbf{F}_p^{2k^2-k}$, into the space of row vectors, $\mathbf{F}_p^{4k^2-1}$. Then $C_p(\mathcal{D})$ is mapped onto $C_p(\mathcal{E})$ and the kernel of the mapping is $C_p(\mathcal{D})^\perp$. Thus the kernel of the restriction of the mapping to $C_p(\mathcal{D})$ is $\mathrm{Hull}_p(\mathcal{D})$. $\square$

Corollary 7.12.1 *If $\mathcal{T}$ is the Hadamard 3-design defined by $\mathcal{E}$, then*

$$\mathrm{rank}_p(\mathcal{T}) = \mathrm{rank}_p(\mathcal{D}) - \dim(\mathrm{Hull}_p(\mathcal{D})) + 1.$$

Proof: Since p divides k,

$$\sum_E v^E = -\jmath,$$

where the sum is over all the blocks of $\overline{\mathcal{E}}$. Thus $\jmath \in C_p(\overline{\mathcal{E}})$, and $C_p(\mathcal{E}) \subseteq C_p(\overline{\mathcal{E}})$. Equality cannot hold, since $C_p(\mathcal{E}) \subseteq \langle \jmath \rangle^\perp$ but $C_p(\overline{\mathcal{E}}) \not\subseteq \langle \jmath \rangle^\perp$. Since $\widehat{C_p(\overline{\mathcal{E}})} = C_p(\mathcal{T})$, the corollary follows. $\square$

The following result of Goethals and Seidel [106] leads to Hadamard matrices with constant row sum and hence to symmetric designs on $4k^2$ points.

Theorem 7.12.3 *Let $\mathcal{D}$ be a 2-$(2k^2 - k, k, 1)$ design with the property that the blocks of $\mathcal{D}$ can be partitioned into two sets, $\mathcal{B}_1$ and $\mathcal{B}_2$, of size $k(2k-1)$ and $(k+1)(2k-1)$, respectively, such that each point appears k times in $\mathcal{B}_1$ and $(k+1)$ times in $\mathcal{B}_2$. Let $\mathcal{E}$ be a Hadamard $(4k^2 - 1, 2k^2, k^2)$ design defined by $\mathcal{D}$ as in Theorem 7.12.1, let $\overline{\mathcal{E}}$ be its complement, and let $\mathcal{T}$ be the extension of $\overline{\mathcal{E}}$ to a Hadamard 3-design. If H is any Hadamard matrix defined by $\mathcal{T}$, then in the equivalence class of H there is a matrix K with constant row sum, so that K defines a $(4k^2, 2k^2 \pm k, k^2 \pm k)$ design, for which $\mathcal{T}$ is the design defined by the symmetric differences.*

Proof: Let $\mathcal{P}$ be the point set of $\mathcal{D}$. Then $(\mathcal{P}, \mathcal{B}_1)$ is a 1-design $\mathcal{C}_1$ with $r = b_1 k/v = k(2k-1)k/k(2k-1) = k$ and $(\mathcal{P}, \mathcal{B}_2)$ is a 1-design $\mathcal{C}_2$ with $r = b_2 k/v = (k+1)(2k-1)k/k(2k-1) = (k+1)$. Let

$$A = \begin{pmatrix} A_1 \\ A_2 \end{pmatrix},$$

where A_i is an incidence matrix for $\mathcal{C}_i$, for $i = 1, 2$. Then

$$AA^t = \begin{pmatrix} A_1 \\ A_2 \end{pmatrix} \begin{pmatrix} A_1^t & A_2^t \end{pmatrix} = \begin{pmatrix} A_1 A_1^t & A_1 A_2^t \\ A_2 A_1^t & A_2 A_2^t \end{pmatrix} = \begin{pmatrix} B & C \\ C^t & D \end{pmatrix},$$

where B is a $k(2k-1) \times k(2k-1)$ matrix, C is a $k(2k-1) \times (k+1)(2k-1)$ matrix and D is a $(k+1)(2k-1) \times (k+1)(2k-1)$ matrix. Then

$$AA^t - kI = \begin{pmatrix} B - kI & C \\ C^t & D - kI \end{pmatrix} = \begin{pmatrix} B_1 & C \\ C^t & D_1 \end{pmatrix} = M,$$

where, from the definition and since $r = k$ for $\mathcal{C}_1$ and $r = k+1$ for $\mathcal{C}_2$, we have, in each row of

B_1: $k(k-1)$ entries equal to 1, k^2 equal to 0;

C: $k(k+1)$ entries equal to 1, $k^2 - 1$ equal to 0;

D: k^2 entries equal to 1, $k^2 + k - 1$ equal to 0;

C^t: k^2 entries equal to 1, $k^2 - k$ equal to 0.

As before, M is the incidence matrix for the 2-design and gives a Hadamard matrix by forming $E = \exp_{-1}(M)$ and bordering with a first row and column of 1's. This is clearly equivalent to

$$H = \begin{pmatrix} -1 & 1 \dots 1 \\ 1 & \\ \vdots & -E \\ 1 & \end{pmatrix}.$$

Then

$$K = \begin{pmatrix} I_{k(2k-1)+1} & 0 \\ 0 & -I_{(k+1)(2k-1)} \end{pmatrix} H \begin{pmatrix} -I_{k(2k-1)+1} & 0 \\ 0 & I_{(k+1)(2k-1)} \end{pmatrix},$$

is equivalent to H and has constant row sum $2k$. The last statement of the theorem follows directly from Theorem 7.10.1. $\square$

We next examine examples of the above construction; this amounts to finding Steiner systems with the appropriate parameters. We then use the theorem to help determine the designs $\mathcal{E}$ that can occur and hence the classes of Hadamard matrices that arise. We start with a well-known class of designs which, following Wertheimer [294], we will call *oval designs*. Bose and Shrikhande [47] first described these designs in the geometric setting that we will give here. (Refer to Section 8.4 for further properties of these designs and their codes.)

Definition 7.12.1 *Let* Π *be a projective plane of order* $n = 2k$ *and let* $\mathcal{O}$ *be an oval of* Π. *The* **oval design** $W(\Pi, \mathcal{O})$ *has as points the set of all lines of* Π *exterior to* $\mathcal{O}$ *and as blocks the set of all points of* Π *not on the oval* $\mathcal{O}$; *incidence is given by the incidence in the ambient plane.*

Proposition 7.12.1 *The incidence structure* $W(\Pi, \mathcal{O})$ *is a* 2-$(2k^2-k, k, 1)$ *design of order* $n = 2k$.

Proof: Simple counting arguments yield the proposition. $\square$

Notice that the order of the oval design is the same as the order of the plane from which it arises. There are many such designs since all known planes of even order have ovals. These designs have been fairly extensively studied, particularly when Π is desarguesian and $\mathcal{O}$ is a regular oval, i.e. a conic plus nucleus, and their p-ary codes, for p dividing k (but actually for $p = 2$ since all known planes of even order have order a power of 2), have received some attention: see [161, 200, 294, 293]. It is not difficult to see, simply by considering the projection of the code of the plane onto the coordinate positions of the oval, that

$$\operatorname{rank}_p(W(\Pi, \mathcal{O})) \leq \operatorname{rank}_p(\Pi) - (n+1).$$

If Π is desarguesian with $n = 2^m$, then $\operatorname{rank}_2(\Pi) = 3^m + 1$ and hence

$$\operatorname{rank}_2(W(\Pi, \mathcal{O})) \leq 3^m - 2^m.$$

We give the easy proofs of these results in Section 8.4. If Π is non-desarguesian there is no known general formula for the dimension, although some bounds are known, in particular for translation planes: see Section 6.8.

What can we say about the Hadamard designs and matrices that arise in this way? Before looking at special cases, we remark that Maschietti [208] rediscovered this construction in a slightly different context: for the points take all those points of Π not on the oval, and for the blocks construct one for each point of Π off $\mathcal{O}$, given as the sum (modulo 2) of the exterior lines through the given point. This gives the symmetric $(4k^2 - 1, 2k^2, k^2)$ design directly and it is exactly the same as the design $\mathcal{E}$ we get from $W(\Pi, \mathcal{O})$.

Also notice that each point on $\mathcal{O}$ defines a distinct resolution for the design $W(\Pi, \mathcal{O})$ by taking the secants through the point to define the parallel classes of blocks. Using such a resolution to label the blocks for the incidence matrix A will then give a special form for the incidence matrix B of $\mathcal{E}$, consisting of $(2k + 1)$ matrix blocks of the zero matrix of size $(2k - 1) \times (2k - 1)$ along the diagonal. These resolutions also mean that the oval designs satisfy Theorem 7.12.3 and hence define the symmetric designs associated with constant-sum Hadamard matrices.

In the case where Π is desarguesian with $n = 2^m$ and $\mathcal{O}$ is regular, we will use the notation as in Buekenhout, Delandtsheer and Doyen [58], i.e. $W(2^m)$ for $W(\Pi, \mathcal{O})$. Since all the known even-order planes have $n = 2^m$, all our codes will be binary for the remainder of this section. The first non-trivial case is for $n = 4$, $k = 2$. Here $W(4)$ is the unique 2-(6, 2, 1) design, i.e. all 2-subsets of a 6-set. Clearly $C(W(4))$ has dimension 5 and has $\langle \jmath \rangle$ for its hull, so $\mathrm{rank}_2(\mathcal{T}) = 5$. Since $\mathcal{T}$ is a 3-(16, 8, 3) design of rank 5 it must be (see Section 7.6) the design of points and hyperplanes of a 4-dimensional affine space over $\mathbf{F}_2$. Moreover, $C(\mathcal{T}) = \mathcal{R}(1, 4)$. This was noticed by Maschietti [208], but he showed that $C(\mathcal{T}) \neq \mathcal{R}(1, 2m)$ for $m \geq 3$.

For $n = 8$ and $k = 4$ the plane is still necessarily desarguesian and the oval regular. The design $W(8)$ is well-known to be the familiar smallest Ree unital, a particularly interesting 2-(28, 4, 1) design. The corresponding Hadamard 3-design $\mathcal{T}$ is a 3-(64, 32, 15) design. The dimension of $C(W(8))$ is 19 and its hull has dimension 7: see Section 8.3. So $\mathrm{rank}(\mathcal{T}) = 13$, showing once again that $\mathcal{T}$ is not the design of points and hyperplanes of the 6-dimensional affine space over $\mathbf{F}_2$ and its binary code is not $\mathcal{R}(1, 6)$. In fact, in order to compare this design with those mentioned in Section 7.10 with respect to bent functions, we computed the weight enumerator of the code $C(\mathcal{T})$. It is

$$X^0 + 588(X^{24} + X^{40}) + 1680(X^{28} + X^{36}) + 3654X^{32} + X^{64}.$$

The observation that the weight-24 vectors cannot form a 1-design (since 8 does not divide 588) shows that $\mathcal{T}$ cannot have a transitive automorphism group and hence that it does not arise from a (64, 28, 12) design coming from a difference set — although it does arise, as the discussion above makes clear, from a (64, 28, 12) design and hence yields a design with these parameters that is not the design of a difference set of any group. In contrast, the three (16, 6, 2) designs all arise from difference sets: see Section 4.4.

For higher values of n the 3-designs were too large to examine with CAYLEY. However, computations related to oval designs made in Key and Mackenzie [161] and Mackenzie [200] suggest a general formula for the dimension of the hull in the case of a regular oval. These computations led to the following (see [200]):

Conjecture 7.12.1 *If $\mathcal{T}$ is the Hadamard design constructed by the method of Theorem 7.12.1 from $W(2^m)$ then $\mathrm{rank}_2(\mathcal{T}) = 2^{m-1}m + 1$.*

The situation for non-desarguesian planes, or for non-regular ovals in desarguesian planes, seems to be quite different. For example, the non-regular oval in the desarguesian plane of order 16, i.e. the so-called *Hall* oval (discovered by Lunelli and Sce [198], but see Hall [120] or Hirschfeld [136, p. 177]) gives a 3-(256, 128, 63) design of 2-rank 65 (in contrast to that from the regular oval, which has 2-rank 33).

We remark that the Ree unital is by no means the only unital on 28 points; apart from the hermitian unital, Brouwer [50] has constructed many others and examined their codes. All of these give 3-$(64, 32, 15)$ designs in the way described, but Brouwer's computations on the codes of the unitals show that none of these 3-designs has rank less than 13. We have not checked these sporadic designs to try to determine whether or not they might provide a negative answer to Question 7.10.2.

A class of examples of 2-$(2k^2 - k, k, 1)$ designs for all k would yield Hadamard matrices for all sizes of the form $4k^2$; we are not aware of such a class. Apart from $k = 2^m$ we know of no further infinite classes; however, such designs are known to exist for all values of k up to 8: see Hall [122, pp. 408-417]. For $k = 3$ the 2-designs are the Steiner triple systems on 15 points, of which there are 80. For the codes to assist in classification, we need p to divide k, i.e. we must choose $p = 3$. The 3-rank of these Steiner triple systems is always 14 (see Doyen *et al.* [99]) and the hull is $\mathbf{F}_3 \jmath$, the 1-dimensional code generated by the all-one vector. Thus, the 3-$(36, 18, 8)$ designs will all have 3-rank 14 and their 3-ranks will not distinguish them. We do not know if the corresponding Hadamard matrices are equivalent or not: cf. Goethals and Seidel [106], where a finer equivalence relation seems to be intended.

Chapter 8

Steiner systems

8.1 Introduction

Steiner systems are t-(v, k, λ) designs with $\lambda = 1$ and $t \geq 2$. Thus, every t-subset must reside in a *unique* block of the design. Such systems were of great interest in the early, recreational period leading up to Steiner's "Combinatorische Aufgabe" [268]; in general they are not easy to find. For $t = 2$ and 3, however, there are many infinite classes of such designs, some of which we have already encountered in earlier chapters, but for $t = 4$ and 5 only a finite number are known. None have been constructed for $t \geq 6$ and the question of the existence, or not, of Steiner systems for large values of t remains an open problem in design theory. In fact, even for general λ there were only a finite number of 5-designs known until Alltop [2] in 1972 produced an infinite class.

The question of the existence of non-trivial t-designs for general λ and $t \geq 6$ was, at one time, linked with the Schreier conjecture — that the only finite 6-transitive permutation groups are the alternating or symmetric groups. This conjecture has since been proved using the classification theorem for finite simple groups. In fact, it had been suggested in 1965 that a proof of the Schreier conjecture might be obtained by showing that there were no non-trivial t-designs for $t \geq 6$: see Hughes [142]. However, in 1983, Magliveras and Leavitt [205] found a 6-design (with $\lambda = 36$); a few years later Teirlinck [271] showed that non-trivial t-designs exist for all values of t. As far as we know there is not yet any direct proof of the Schreier conjecture and, of course, Teirlinck's result seems to indicate that design theory will be of little help in finding a direct proof.

A curious, and as yet unexplained, phenomenon is that although coding theory did play a role in uncovering new 5-designs (see Assmus and Mattson [18] or Pless [237]) it seems to be of little help in finding or treating t-designs for $t \geq 6$. Roughly speaking, powerful group-theoretical methods did most of the discovering and classification of designs in the early stages of the theory. Coding-theoretical methods — those that we have discussed, which are, essentially, linear-algebraic — have extended and deepened the understanding of designs, but yet the best existence theorems seem to be of a recursive and set-theoretical nature.

In this chapter we will discuss some of the classes of Steiner 2-designs and 3-designs that are amenable to a discussion via coding theory. As regards the Steiner 4-designs and 5-designs, the binary Golay code $\mathcal{G}_{23}$ and its extension $\mathcal{G}_{24}$ arise as the codes of the Witt designs, which are designs with parameters 4-$(23, 7, 1)$ and 5-$(24, 8, 1)$, respectively. These designs and their codes — and the automorphism groups M_{23} and M_{24}, the large Mathieu groups — have been heavily documented and studied, but we will give a description here as well, since the structure is so unique and beautiful. The ternary extended Golay code does not, as we mentioned earlier, arise in a simple way as the code of the associated Steiner system, the Witt 5-$(12, 6, 1)$ design, and the theory that we have developed does not immediately apply; as we have explained, the connection is more subtle: see Corollary 7.4.3 and the following remark. The Steiner 5-designs discovered by Denniston [88] do not seem to be amenable to a coding-theoretical discussion but it is conceivable that, once again, the connection with coding theory might be more subtle.

Notice that the symmetric Steiner 2-designs are precisely the finite projective planes, and their codes have been thoroughly discussed in Chapter 6, as have the Steiner 2-$(k^2, k, 1)$ designs, i.e. the affine planes.

8.2 Steiner triple and quadruple systems

A **Steiner triple system** is a Steiner 2-design with block size 3, i.e. a 2-$(v, 3, 1)$ design. It follows immediately from the parameters and divisibility conditions that $v \equiv 1$ or $3 \mod 6$. These systems exist for all values of $v > 1$ that satisfy one of these congruence relations, a result due to Kirkman [169] — or see Moore [216]. Up to $v = 15$ all the systems have been constructed: for $v = 3, 7, 9$ they are unique (see the examples given below); for $v = 13$ there are two ; for $v = 15$ there are 80: see the table in Beth *et al.* [37], for example. Divisibility conditions do not exclude the possibility of an extension of any Steiner triple system to a 3-$(v + 1, 4, 1)$ design, i.e. a **Steiner quadruple system**. No example of a Steiner triple system that does not extend to a quadruple system is known; whether or not all Steiner triple systems extend was one of Steiner's original questions and is still an open question. For every permissible value of v there is a Steiner triple system that does extend to a quadruple system: see Hanani [131, 130]. Thus Steiner quadruple systems exist for all values of $v > 2$ satisfying $v \equiv 2$ or $4 \mod 6$.

The replication number, $r = \lambda_1$, of a Steiner triple system is $(v-1)/2$ and the order is $(v-3)/2$. The following examples have been treated already — as exercises — in Chapters 2 and 5 but we include them again for the convenience of the reader.

Example 8.2.1 (1) The design of points and lines of the projective geometry $PG_m(\mathbf{F}_2)$ is a 2-$(2^{m+1} - 1, 3, 1)$ design of order $2(2^{m-1} - 1)$. The design extends to the affine-geometry design of points and planes of $AG_{m+1}(\mathbf{F}_2)$, but there might be other extensions: see de Vries [287].

(2) The design of points and lines of the affine geometry $AG_m(\mathbf{F}_3)$ is a 2-$(3^m, 3, 1)$ design of order $\frac{3}{2}(3^{m-1} - 1)$. The design always extends — in different ways in general — to a 3-$(3^m + 1, 4, 1)$ design: see Beth *et al.* [37, p. 170] or Key and Wagner [163].

The designs in these two classes of examples have doubly-transitive automorphism groups, since they are geometric designs. In fact they are the only Steiner triple systems with doubly-transitive automorphism groups, a property that was conjectured by Hall [118] and proved by Key and Shult [162] using the classification theorem for finite simple groups.

The codes associated with Steiner triple or quadruple systems have not been extensively studied, except in particularly well-known cases, including the two classical examples quoted above: in the first example the binary code is the punctured Reed-Muller code $\mathcal{R}(m-1, m+1)^*$, i.e. the Hamming code $\mathcal{H}_{m+1}$, and in the second example the ternary code is the generalized Reed-Muller code $\mathcal{R}_{F_3}(2(m-1), m)$, as defined in Chapter 5. Both these classes of codes have the property that the scalar multiples of the incidence vectors of the blocks of the Steiner system form the minimum-weight vectors of the code.

Some properties and conditions on the possible p-ranks have been obtained by Doyen, Hubaut and Vandensavel [99]. Their result is as follows:

Theorem 8.2.1 *Let $\mathcal{D}$ be a 2-$(v, 3, 1)$ design and let p be a prime. Then*

(1) $\mathrm{rank}_p(\mathcal{D}) = v$ *if* $p \geq 5$;

(2) $\mathrm{rank}_2(\mathcal{D}) \geq v - \log_2(v + 1)$, *with equality if and only if $\mathcal{D}$ is the design of points and lines of the projective geometry $PG_m(\mathbf{F}_2)$, where* $v = 2^{m+1} - 1$:

(3) $\text{rank}_3(\mathcal{D}) \geq v - \log_3(v) - 1$, *with equality if and only if* $\mathcal{D}$ *is the design of points and lines of the affine geometry* $AG_m(\mathbf{F}_3)$, *where* $v = 3^m$.

This theorem is another instance of a *rigidity theorem*, since it characterizes the geometric designs of Example 8.2.1 as being the unique Steiner triple systems with smallest rank. The proof uses a classification of Steiner triple systems through the occurrence of subsystems obtained by Teirlinck [270], where also some general results are obtained about the p-rank of 2-designs with block size 3 and 3-designs with block size 4. Dehon [79] then extends these results to t-designs with block size $(t+1)$, and, in particular, shows that for Steiner quadruple systems, only the binary codes might be of rank less than v.

Exercise 8.2.1 Let $\mathcal{D}$ be a Steiner triple system and p a prime; set $C = C_p(\mathcal{D})$. If the minimum weight of C is 3, show that the vectors of weight 3 are the scalar multiples of the incidence vectors of the blocks. (*Hint:* Argue from first principles or use Theorem 8.2.1 (1) above to show that $p = 2$ or 3.) In the case $p = 2$, show that C must be single-error-correcting and that it is a Hamming code with $\mathcal{D}$ one of the designs in Example 8.2.1 (1). (See Exercise 2.5.3 (2).) Similarly show that if the binary code of a Steiner quadruple system has minimum weight equal to 4, then the design must be the design of points and planes of an affine geometry over $\mathbf{F}_2$.

Hamada [126] went some way towards obtaining another type of rigidity theorem for *Hall triple systems*, i.e. Steiner triple systems in which every triangle of points generates the affine plane $AG_2(\mathbf{F}_3)$ of order 3. (Here we take the system *generated* by a set of points to be the minimal Steiner triple system inside the original system that contains the given set of points. Note also that if every triangle generates a projective plane of order 2 — i.e. a Fano plane — then the system is necessarily the design of points and lines of $PG_m(\mathbf{F}_2)$.) Using a correspondence of Young [301] between Hall triple systems and certain Moufang loops, Hamada obtained some results concerning the 3-rank of Hall triple systems that led to a conjecture of a rigidity theorem using a characterization due to Bénéteau [32].

8.3   Unitals

A **unital** or **unitary design** $\mathcal{U}$ is a Steiner system with parameters 2-$(m^3+1, m+1, 1)$, where $m \geq 2$. The name is derived from the classical case of the hermitian unital, as discussed in Chapter 3, where the set of absolute points and non-absolute lines of a unitary polarity of the desarguesian plane

$PG_2(\mathbf{F}_{q^2})$ forms such a design with $m = q$. It follows that unitals do exist on $q^3 + 1$ points for any prime power q and, until recently, none in fact were known for q not a prime power; Mathon [210] and Bagchi and Bagchi [28] have, however, independently constructed unitals with $m = 6$. The article of Piper [236] raised other questions concerning unitals, some of which were answered by Brouwer [50], who produced a large collection of unitals with $m = 3$ that gave negative answers to many questions.

A better-behaved class of unitals was discovered by Lüneburg [194] who showed how the Ree groups define unitals on which the groups act 2-transitively. We will describe this construction below, assuming properties of the Ree groups. Thus, since Lüneburg's construction, there have been two known infinite classes of unitals with doubly-transitive automorphism groups. Using the classification theorem for finite simple groups, Kantor [151] showed that these are the only unitals with doubly-transitive automorphism groups. There are, however, many other unitals: see, for example, Baker and Ebert [29], Brouwer [50], Buekenhout [57], Ganley [102], and Metz [215]. Many of these "sporadic" unitals embed in projective planes as we have seen in Section 6.6.

We describe now the Ree unital $\mathcal{R}(q)$, where $q = 3^{2m+1}$, as it was originally described by Lüneburg [194]. Let $G = R(q)$, the Ree group of order $(q^3 + 1)q^3(q - 1)$ (see Ree [246], Tits [276], or Ward [289]). Then G has a doubly-transitive representation on $q^3 + 1$ points, *viz.* by conjugation on the set $\mathcal{P}$ of all Sylow 3-subgroups. For any P, Q in $\mathcal{P}$, the pointwise stabilizer $G_{P,Q}$ is cyclic of order $q - 1$ and thus contains a unique involution, t. It follows that t fixes exactly $q + 1$ points of $\mathcal{P}$. For each involution t we identify a block of our structure to be precisely the points that are fixed by the involution t. Clearly then there is a unique block containing any two points. Since G acts doubly-transitively on $\mathcal{P}$, the structure defined by $(\mathcal{P}, \mathcal{B})$, where $\mathcal{B} = \{t | t$ an involution of $G\}$ and incidence is as given — i.e. P is on t if t is in the normalizer, $N_G(P)$, of P — is a 2-design with parameters 2-$(q^3 + 1, q + 1, 1)$. It is called a **Ree unital**. It admits G as a 2-transitive automorphism group; the full automorphism group is larger, for $m \geq 1$, admitting also the field automorphism.

Thus we have two classes of unitals with 2-transitive automorphism groups: the hermitian (or classical) unital $\mathcal{H}(q)$ for any prime power q, with $P\Gamma U_3(\mathbf{F}_{q^2})$ acting inside $P\Gamma L_3(\mathbf{F}_{q^2})$, and the Ree unital, $\mathcal{R}(q)$ where $q = 3^{2m+1}$, with $R(q)$, extended by the field automorphism of order $2m+1$, acting. The smallest hermitian unital, $\mathcal{H}(2)$, is a 2-(9,3,1) design and is hence the unique affine plane of order 3. The smallest Ree unital obtained from the smallest Ree group, $R(3)$, which is also isomorphic to $P\Gamma L_2(\mathbf{F}_8)$ represented on 28 points, is a 2-(28,4,1) design and it is, like its group, a little different from the general case. In fact, we come across it again as an oval design, in the next section.

There are many results that characterize these special unitals by the existence, or non-existence, of certain geometric configurations, by embeddability properties in a projective plane, and so on: O'Nan configurations (see O'Nan [222] and Wilbrink [297]) provide an example of this.

Here we are asking how the associated codes can assist in the classification. First notice that the order of a unital $\mathcal{U}$ with parameters 2-$(q^3 + 1, q + 1, 1)$ is $q^2 - 1 = (q - 1)(q + 1)$. Thus, in principle, primes dividing either $q + 1$ or $q - 1$ might yield interesting codes. Mortimer [219] obtained, however, the following result, using a theorem of Klemm [171]:

Theorem 8.3.1 *Let $\mathcal{U}$ be a hermitian or Ree unital on $q^3 + 1$ points. Then the p-rank of $\mathcal{U}$ is less than q^3 if and only if p divides $(q + 1)$.*

More recently, using modular character theory, Hiss [138] has been able to give explicit formulas for the ranks of the Ree unitals. His theorem is as follows:

Theorem 8.3.2 *Let $q = 3^{2m+1}$ and let p be a prime dividing $q + 1$. Then the dimension of the code over $\mathbf{F}_p$ of the Ree unital, $\mathcal{R}(q)$, is given by*

(1) $\mathrm{rank}_p(\mathcal{R}(q)) = q^3 - \frac{1}{2\sqrt{3}}q^{5/2} - \frac{1}{2}q^2 + \frac{1}{2}q + \frac{1}{2\sqrt{3}}q^{1/2}$ *for $p \neq 2$;*

(2) $\mathrm{rank}_2(\mathcal{R}(q)) = q^3 - \frac{1}{\sqrt{3}}q^{5/2} + \frac{1}{\sqrt{3}}q^{1/2}$.

Hiss makes the comment in his paper that work of Geck on the unitary groups should shortly result in an analogous result for the hermitian unitals. Andriamanalimanana [4] computed the p-rank of $\mathcal{H}(q)$ for $q \leq 5$ and suggested a possible formula for the p-rank when p divides $(q + 1)$, namely $q(q^3 + 1)/(q + 1)$, which is b/q, where b is the number of blocks of the unital.

Not a great deal seems to be known about the codes arising from unitals. The existence of ovals for the Ree or hermitian unitals would, of course, by Theorem 2.4.4, yield that the minimum weight of the binary code, in case 2 divides $(q + 1)$ — which is always satisfied for the Ree unitals — would be the block size $q + 1$. (Ovals in the even-order case would be $(q^2 + 1)$-arcs, from Definition 1.5.2.) This is true for the 2-(28,4,1) design $\mathcal{R}(3)$, which has ovals, and has the added property that the minimum-weight vectors of the binary code of the design are precisely the characteristic functions of the blocks. The binary code of the hermitian unital $\mathcal{H}(3)$ also has minimum weight $q + 1 = 4$, but there are no ovals, and there are more weight-4 vectors in the binary code than the characteristic functions of the blocks. In fact, the classical unitals have ovals only for $q = 2$, when the unital is the affine plane of order 3; we will give a proof of this below. For the

Ree unitals an attempt was made in Assmus and Key [10] to find ovals through a construction using the group-theoretical properties of $R(q)$ that are analogous to the properties of $R(3)$ that give an oval. There it was found that the analogous construction yielded arcs of size $3q + 1$, which is the oval size, $q^2 + 1$, only when $q = 3$. It seems unlikely that ovals exist for $q > 3$. In the next section, oval designs will be discussed and ovals for these designs found for the case in which the design comes from a regular oval in the desarguesian plane. As the small Ree unital is a particular case of an oval design, it emerged that this unital might fit more naturally into that collection of designs, at least with respect to this property.

The picture for the binary codes in the case of the hermitian and Ree unitals of order $q^2 - 1 = 8$ was discussed in Assmus and Key [10] and some interesting properties emerged: writing $\mathcal{H} = \mathcal{H}(3)$ and $\mathcal{R} = \mathcal{R}(3)$, computations (with CAYLEY [66]) yielded that $C_2(\mathcal{R})$ can be embedded in $C_2(\mathcal{H})$, and that $C_2(\mathcal{H})^{\perp} \subset C_2(\mathcal{H})$, and $H_2(\mathcal{H}) = H_2(\mathcal{R})$. The automorphism group of $C_2(\mathcal{H})$ is larger than that of $\mathcal{H}$, being the symplectic group, $Sp_6(2)$. The weight-4 vectors of the code $C_2(\mathcal{H})$ include the characteristic functions of designs forming the Ree and hermitian unital; the subgroups $P\Gamma L_2(\mathbf{F}_8)$ and $P\Gamma U_3(\mathbf{F}_9)$ of $Sp_6(2)$ acting on this code will isolate weight-4 vectors that make up the blocks of copies of each of these unitals. The full set of weight-4 vectors also define a design: a 2-(28,4,5) design which is a particular case of a class of designs with parameters 2-$(q^3 + 1, q + 1, q + 2)$, for q odd, found by Hölz [140] for the hermitian unitals and analogously, for $q = 3^{2n+1}$, by Assmus and Key [10] for the Ree unitals. The blocks of both these classes of designs meet in zero, one or two points. The inclusions for the case $q = 3$ are shown in Figure 8.1. The weight enumerator of $C_2(\mathcal{H})^{\perp}$ is simply

$$X^0 + 63(X^{12} + X^{16}) + X^{28},$$

showing that $\mathcal{H}$ does not have ovals. On the other hand, $C_2(\mathcal{R})^{\perp}$ has weight enumerator

$$X^0 + 84(X^{10} + X^{18}) + 63(X^{12} + X^{16}) + 216X^{14} + X^{28},$$

showing the existence of ovals. Tonchev [282] has recently shown by computer analysis that the only unitals included amongst the weight-4 vectors of $C_2(\mathcal{H})$ are copies of the hermitian and Ree unitals. Thus, in the language of Section 2.4, this linear equivalence class contains only the two doubly-transitive designs. The reader will have perhaps observed that the rigidity theorems we have discussed have always yielded the design with a doubly-transitive automorphism group as the unique design with the smallest rank and here we have a case where there are two such designs with, now, a common linear equivalence class that contains no other design. Brouwer

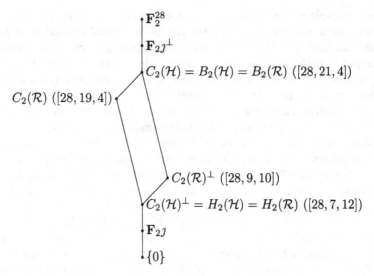

Figure 8.1: The binary codes of $\mathcal{H}(3)$ and $\mathcal{R}(3)$

[50] conjectured that the 2-rank of 19 for $\mathcal{R}$ characterizes the Ree unital amongst those on 28 points. We do not know what happens in the next instance of two doubly-transitive unitals with the same parameters.

Exercise 8.3.1 Show that the designs of Hölz [140] or Assmus and Key [10] have contractions at a point whose parameters satisfy the conditions for a generalized quadrangle.[1]

To show that we cannot in general assume that the binary code of the classical unital $\mathcal{H}(q)$ has minimum weight equal to the block size $q+1$, we prove that $\mathcal{H}(2)$, i.e. $AG_2(\mathbf{F}_3)$, is the only hermitian unital that has ovals. Notice that the order of $\mathcal{H}(q)$ is $q^2 - 1$, so for q odd the oval size is $q^2 + 1$, and for q even, the oval size is q^2: see Definition 1.5.2. For the ambient projective plane of order q^2, the oval size is $q^2 + 1$ and $q^2 + 2$, respectively.

Theorem 8.3.3 *The hermitian unital $\mathcal{H}(q)$ has ovals only for $q = 2$.*

The proofs for q odd (the interesting case) and q even are quite different: for q odd we give the proof of Andriamanalimanana [4] and, for q even, that

---

[1]Thas [272] has shown that these are the generalized quadrangles of Ahrens and Szekeres [1] in the Hölz case; those from the Ree unitals are not extensions of generalized quadrangles except in the case $q = 3$ ($v = 28$), when the designs are the same: see Key [157].

of Assmus and Key [10]. There are other proofs in the literature, notably one for q even in Fisher *et al.* [101].

The following classical result, which can be found in Schmidt [256], is used in the even-order case:

Lemma 8.3.1 *For $n \geq 2$ let $f(x_1, x_2, \ldots, x_n)$ and $g(x_1, x_2, \ldots, x_n)$ be polynomials over $\mathbf{F}_q$ of degree k and m, respectively, and having no common factor. Then the number of common roots of f and g in $\mathbf{F}_{q^2}$ is at most $q^{n-2} km \min\{k, m\}$.*

Lemma 8.3.2 *If q is odd, then $\mathcal{H}(q)$ has no ovals.*

Proof: (Andriamanalimanana) For $q = 3$ the weight enumerator of the orthogonal binary code of the design gives the result, as mentioned above.

Let $q \geq 5$. The oval size is $q^2 + 1$ in this case, so an oval $\mathcal{O}$ for $\mathcal{H}(q)$ would be an oval for the plane $PG_2(q^2)$. Hence, by Segre's theorem (Theorem 3.7.1), $\mathcal{O}$ is a conic, and, taking homogeneous coordinates, the points of $\mathcal{O}$ satisfy an irreducible equation of degree 2. These points are also absolute points of a unitary polarity, and thus satisfy an equation of degree $q + 1$. Using Lemma 8.3.1, the number of common solutions is at most $4q^2(q + 1)$. Thus the number of points of $\mathcal{H}(q)$ on the conic is at most $(4q^2(q + 1) - 1)/(q^2 - 1)$, which for $q \geq 5$ is less than $q^2 + 1$. The contradiction proves the result. $\square$

Segre's theorem does not hold for q even, and we replace it by a geometric argument that rests on the completion of q^2-arcs in $PG_2(\mathbf{F}_{q^2})$.

Lemma 8.3.3 *If q is even, $\mathcal{H}(q)$ has ovals only when $q = 2$.*

Proof: For q even, the order $q^2 - 1$ is odd, so the oval size is q^2. For $q = 2$, $\mathcal{H}(2) = AG_2(\mathbf{F}_3)$, and the quadrangles are ovals for the design. Now take $q \geq 4$, and suppose that $\mathcal{H}(q)$ has an oval $\mathcal{O}$, i.e. a q^2-arc. Then $\mathcal{O}$ is a q^2-arc in the plane $PG_2(\mathbf{F}_{q^2})$, and hence is contained in an oval $\mathcal{O}'$ of the plane: see Hirschfeld [136, p. 197]. Thus $\mathcal{O}' = \mathcal{O} \cup \{P, Q\}$, where P and Q are points of $PG_2(\mathbf{F}_{q^2})$ that are not on the unital, and are hence non-absolute points of the unitary polarity, σ, that defines $\mathcal{H}(q)$.

Since $\mathcal{O}'$ is a $(q^2 + 2)$-arc in $PG_2(\mathbf{F}_{q^2})$, it can have no tangents. For $R \in \mathcal{O}$, R^σ is an absolute line, and contains R and no other absolute point. As it must meet $\mathcal{O}'$ again, it can only meet it at P or Q. Thus either $P \in R^\sigma$ or $Q \in R^\sigma$, i.e. either $R \in P^\sigma$ or $R \in Q^\sigma$. Now P^σ and Q^σ are non-absolute lines, and thus form blocks of the unital. Every point R of $\mathcal{O}$ is on one of P^σ or Q^σ, and thus if there are more than four points on $\mathcal{O}$, at least three of them will be together on a block, which contradicts the assumption that $\mathcal{O}$ is an oval. $\square$

The proof of the Theorem 8.3.3 now follows.

Thus we are not able to use Theorem 2.4.4 in the even-order case for the binary codes, and we can deduce nothing more than that the minimum weight is at most $q + 1$, as is always the case for codes from designs.

8.4   Oval designs

The oval design, $W(\Pi, \mathcal{O})$, which is derived from a projective plane Π of even order n and an oval $\mathcal{O}$ of Π, was described in Definition 7.12.1; recall that its block set is the set of points of the plane that are not on the oval and its points are the lines of the plane that are exterior to the oval, with incidence given by the ambient projective plane. Its parameters are $2\text{-}(\frac{1}{2}n(n-1), \frac{1}{2}n, 1)$ and its order is n. We mentioned in Section 7.12 the following result, which we will now prove:

Theorem 8.4.1 *Let Π be a projective plane of even order n and let $\mathcal{O}$ be an oval of Π. Let $W(\Pi, \mathcal{O})$ be the associated oval design. If p is any prime, then*

$$\sqrt{(n^3 - 2n^2 - 1)/2} \le \mathrm{rank}_p(W(\Pi, \mathcal{O})) \le \mathrm{rank}_p(\Pi) - (n+1),$$

and, if $p \ne 2$, then

$$\sqrt{(n^3 - 2n^2 - 1)/2} \le \mathrm{rank}_p(W(\Pi, \mathcal{O})) \le \mathrm{rank}_p(\Pi) - (n+2).$$

If $\Pi = PG_2(2^m)$ and $p = 2$, then

$$\sqrt{(2^{3m} - 2^{2m+1} - 1)/2} \le \mathrm{rank}_2(W(\Pi, \mathcal{O})) \le 3^m - 2^m.$$

Proof: Notice that the lower bound quoted is simply the bound of Hillebrand of Section 2.4, which applies, for any field, since our design is a Steiner 2-design.

Let $F = \mathbf{F}_p$ and set $\Pi = (\mathcal{P}, \mathcal{L}, \mathcal{I})$, $W = W(\Pi, \mathcal{O})$. Define the map φ where

$$\varphi : \begin{cases} F^{\mathcal{P}} & \to & F^{\mathcal{O}} \\ w & \mapsto & w|_{\mathcal{O}} \end{cases},$$

for any $w \in F^{\mathcal{P}}$, i.e. $\varphi(w)$ is the restriction of the function w to the set $\mathcal{O}$. Then $\varphi(C_F(\Pi)) = F^{\mathcal{O}}$ for p odd and $\varphi(C_F(\Pi)) = \langle \jmath_{\mathcal{O}} \rangle^{\perp}$ for $p = 2$. Further,

$$\ker(\varphi) \supseteq \langle v^x | x \text{ an exterior line} \rangle$$

and, since $\langle v^x | x \text{ an exterior line} \rangle = C_F(W^t)$, the first two inequalities follow.

The desarguesian plane of order 2^m has 2-rank $(3^m + 1)$, from Theorem 6.4.2, giving the last inequality. $\square$

All known planes have prime-power order, so that for the known even-order planes only the binary codes need be considered. For the general plane, and the general oval, the associated binary codes have not been studied, although there were unsuccessful attempts to construct 2-$(45, 5, 1)$ and 2-$(66, 6, 1)$ designs with enough parallelism to allow the retrieval of the projective planes (of orders 10 and 12, respectively) which might have produced them as oval designs. We now know, of course, that in the 2-$(45, 5, 1)$ case this was futile: see Lam [176]. The most sophisticated attack along these lines was made by Thompson [274] where the representation theory of finite groups was brought into play by studying the action of the symmetric group of the oval on the obvious involutions of the oval defined by its exterior points. An earlier, and purely combinatorial, attack was made by Bruck [52]. In both cases the authors examined, as an exercise, the oval that would have been produced had there been a projective plane of order 6.

Exercise 8.4.1 Use elementary counting arguments to show that *any* projective plane of order 4 has a 6-arc, i.e. an oval. Now prove that the oval uniquely determines the plane and hence show that there is a unique projective plane of order 4, namely $PG_2(\mathbf{F}_4)$.

The upper bounds given are not always sharp: see Mackenzie [200], where computations (using CAYLEY on the Clemson University VAX) showed that, for Π the translation plane of order 16 given by the coordinate set m_4 in Dempwolff and Reifart [87] and an oval found in the plane through a short computer search, $\mathrm{rank}_2(\Pi) = 98$, $n + 1 = 17$, but $\mathrm{rank}_2(W(\Pi, \mathcal{O})) = 80$. However, with Π desarguesian of order 2^m for $m \le 5$, computations, using a regular oval, gave equality in the stated relation.

In fact something more can be said about the binary code of $W(\Pi, \mathcal{O})$ when Π is desarguesian and $\mathcal{O}$ is regular, i.e. a conic together with its nucleus. In this case, as we already mentioned in Section 7.12, we write $W(n) = W(\Pi, \mathcal{O})$. Wertheimer [293, 294] constructed the designs and examined the binary codes as part of a general construction from quadrics. We will give a particular construction here that goes back to Lüneburg [193]. It was also described by Kantor [149] and it emphasizes the large group acting on the oval in the plane — and thus on the design and on the code. The same construction was used by Assmus and Prince [21] in a discussion of biplanes associated with projective planes, the biplane's blocks being given by ovals.

Let $G = PSL_2(\mathbf{F}_n)$, where $n = 2^m \geq 4$, and let H be a dihedral subgroup of G of order $2(n+1)$. Let G act in the usual way on the set $\mathcal{P}$ of right cosets of H. Since G is simple, this representation is certainly faithful and $|\mathcal{P}| = \frac{1}{2}n(n-1) = v$. This will be our point set. Then for any point $P \in \mathcal{P}$, the stabilizer G_P is dihedral and if $P = Hg$, then $G_P = g^{-1}Hg$. For the block set $\mathcal{B}$ we take the set of involutions in G and we define P to be incident with a block x if the involution x fixes the point P, i.e. if $x \in G_P$. Since G_P is dihedral, it contains $(n+1)$ involutions, so each point is on $r = (n+1)$ blocks. There is one conjugacy class of involutions in G, so each involution fixes the same number, k, of points and thus each block is incident with the same number, k, of points. We have thus a 1-design, with $bk = vr$, where b is the number of involutions in G, i.e. $(n^2 - 1)$. This gives $k = \frac{1}{2}n$. To show that we have a 2-design, notice that for distinct points P and Q, $|G_P \cap G_Q| \leq 2$, so that there is at most one block through two distinct points. Now, counting points on blocks through P, we have P on $n+1$ blocks, each with $\frac{1}{2}n - 1$ points other than P and no point on more than one block with P. This gives $(n+1)(\frac{1}{2}n - 1) = v - 1$ and hence every point is on a block with P and we have a Steiner 2-design which we will denote by $W(n)$. This design has G as an automorphism group, which is $1\frac{1}{2}$-transitive on points, where the stabilizer of a point P has all orbits other than $\{P\}$ of equal length $(n+1)$, since we have just shown that $|G_{P,Q}| = 2$ for $P \neq Q$. The action of G on points (cosets of H) is through right multiplication and on blocks (involutions) the action is by conjugation. The semilinear group, $P\Gamma L_2(\mathbf{F}_n)$ also preserves the design (see Example 3.7.4) which is, in fact, the oval design $W(\Pi, \mathcal{O})$ obtained from a regular oval $\mathcal{O}$ in $\Pi = PG_2(\mathbf{F}_{2^m})$.

Since $n = 2^m$, $W(2^m)$ is a 2-$(2^{m-1}(2^m - 1), 2^{m-1}, 1)$ design of order $n = 2^m$ and we take $m \geq 2$ for non-triviality. The binary code of the design is preserved by $P\Gamma L_2(\mathbf{F}_{2^m})$ and has minimum weight at most 2^{m-1}. We use Theorem 2.4.4 to show that this is precisely the minimum weight by constructing a 2-design of ovals for $W(2^m)$, for $m \geq 3$. Notice that for $m = 2$, $W(4)$ is a 2-(6,2,1) trivial design, with the full set $\mathcal{P}$ an oval, so we will exclude this case from consideration. For $m = 3$, $P\Gamma L_2(\mathbf{F}_8)$ is 2-transitive on the points of the design. The theorem is as follows:

Theorem 8.4.2 *The design* $\mathcal{W} = W(2^m)$, *for* $m \geq 3$, *has a collection of ovals that form a* 2-$(2^{m-1}(2^m - 1), 2^m + 2, 2^m + 2)$ *design,* $\mathcal{W}'$. *The minimum weight of the binary code of* $\mathcal{W}$ *is* 2^{m-1} *and the minimum weight of the binary code of* $\mathcal{W}'$ *is* $2^m + 2$.

We prove this through a series of lemmas, for which we need some extra notation. Assume that $n = 2^m \geq 8$. For any $P \in \mathcal{P}$, G_P has $\frac{1}{2}n - 1$ orbits of length $(n + 1)$, which we denote by $\mathcal{P}_i$, for $i = 1$ to $\frac{1}{2}n - 1$. We will show that, for $n \geq 8$, $\{P\} \cup \mathcal{P}_i$ is an oval for $\mathcal{W}(n)$ for each P and i. Let $\mathcal{T}$ denote the set of involutions in G_P, i.e. the set of blocks containing P.

Lemma 8.4.1 *For every block x and any point P not on x, there is a unique involution $t \in \mathcal{T}$ such that t fixes x.*

Proof: An element t fixes the block x if $x^t = t^{-1}xt = x$. The number of blocks not containing the point P is $(n^2 - 1) - (n + 1) = 2(\frac{1}{2}n - 1)(n + 1)$. Each $t \in \mathcal{T}$ fixes $\frac{1}{2}n$ points and has $\frac{1}{2}n(\frac{1}{2}n - 1)$ transpositions. If a block x is fixed by t and $P \notin x$, then clearly no point on x can be fixed by t, so each $t \in \mathcal{T}$ fixes $\frac{1}{2}n(\frac{1}{2}n - 1)/\frac{1}{4}n = n - 2$ blocks other than its pointwise fixed block, i.e. each $t \in \mathcal{T}$ fixes $(n - 2)$ blocks that do not contain P.

Now $|G_{P,x}| = 1$ or 2, so at most one involution in $\mathcal{T}$ can fix any given block. Now count the members of the set $S = \{(x, t)|x \in \mathcal{B}, t \in \mathcal{T}, P \notin x, x^t = x\}$ in two ways: counting involutions first gives $|S| = (n + 1)(n - 2)$; counting blocks first gives $|S| = \sum_{x \not\ni P} a_x$, where a_x is the number of involutions in $\mathcal{T}$ that fix x, i.e. $a_x = 0$ or 1. Since there are exactly $(n + 1)(n - 2)$ such blocks x, we must have $a_x = 1$ for all $x \not\ni P$, proving the assertion. $\square$

Lemma 8.4.2 *For any $P \in \mathcal{P}$ and each $i = 1$ to $\frac{1}{2}n - 1$, $\{P\} \cup \mathcal{P}_i$ is an oval for $W(n)$ for $n \geq 8$.*

Proof: For a fixed P and i, write $\mathcal{O} = \{P\} \cup \mathcal{P}_i$. Notice first that $\mathcal{O}$ has the correct size, i.e. (n+2), for an oval of $W(n)$.

We first show that if Q and R are two points in $\mathcal{P}_i$, then P, Q, R are not together on a block of $W(n)$. Thus let $g \in G_P$ satisfy $Qg = R$. Suppose P, Q, R are together on a block t, which thus is also an involution fixing all three points. Then t^g is also an involution fixing all three points and so $t^g = t$. Since G_P is dihedral of order $2(n + 1)$, $C_{G_P}(t) = \langle t \rangle$, and so g cannot centralize t.

Now take $Q \in \mathcal{P}_i$. There are $(n + 1)$ blocks through Q, one of which passes through P. The other n do not contain P, and thus by Lemma 8.4.1 there is an involution $t \in \mathcal{T}$ for each of these n blocks that fixes the block. Since $t \in \mathcal{T}$, and $\mathcal{P}_i$ is fixed by t, $Qt \in \mathcal{P}_i$ and hence each of these n blocks must meet $\mathcal{P}_i$ again. But there are exactly n other points and each is certainly on a block with Q. Thus the blocks from Q to each of these are all distinct and $\mathcal{P}_i$ is an arc. From the above observation, $\mathcal{O}$ is an oval for $W(n)$. $\square$

Lemma 8.4.3 *The design $W(n)$ is resolvable into parallel classes.*

Proof: Any Sylow 2-subgroup of G is elementary abelian, of order $n = 2^m$, and contains $n - 1$ involutions. Since any two Sylow 2-subgroups intersect trivially, the involutions are partitioned in this way. Each involution t fixes $n - 1$ blocks and these form a parallel class, these being the blocks that correspond to the involutions in the Sylow 2-subgroup that contains t. $\square$

Lemma 8.4.4 *The incidence structure with point set $\mathcal{P}$ and block set $\mathcal{B}' = \{\{P\} \cup \mathcal{P}_i | P \in \mathcal{P}, 1 \leq i \leq \frac{1}{2}n - 1\}$, is a 2-$(\frac{1}{2}n(n - 1), n + 2, n + 2)$ design, denoted by $W'(n)$.*

Proof: The structure is clearly a 1-design with $b = |\mathcal{B}'| = \frac{1}{2}n(n-1)(\frac{1}{2}n-1)$, block size $n + 2$, $r = \frac{1}{2}n - 1 + (\frac{1}{2}n(n - 1) - 1) = \frac{1}{2}n^2 - 2$.

Let Q and R be two points. Then certainly there is a block $\{Q\} \cup \mathcal{Q}_i$ with $R \in \mathcal{Q}_i$ and a block $\{R\} \cup \mathcal{R}_j$ with $Q \in \mathcal{R}_j$. These blocks cannot be identical: for if $\{Q\} \cup \mathcal{Q}_i = \{R\} \cup \mathcal{R}_j = \mathcal{O}$, then $G_\mathcal{O}$ is 2-transitive on $\mathcal{O}$, so that $(n + 2)(n + 1)$ divides the order of G, which is impossible.

Now Q and R will be together in an orbit $\mathcal{P}_k$, for some P and k, if and only if there is an involution $t \in \mathcal{T}$ (with notation as before) which includes (P, Q) as a transposition. Let x be the unique block of $W(n)$ that contains P and Q. Then G_x is a Sylow 2-subgroup of G, elementary abelian of order n, and each involution in G_x fixes a unique block pointwise. There are $n-1$ parallel blocks corresponding to these involutions, x being one of them, and x is fixed by $n - 2$ involutions other than the one that fixes it pointwise. Further, G_x is transitive on the points of x, since there are $\frac{1}{2}n$ points on x, and for any Q on x, $|G_{x,Q}| = 2$, so $|Q(G_x)|$, the length of the orbit of Q under G_x, is $\frac{1}{2}|G_x| = \frac{1}{2}n$. Thus every transposition (Q, R) occurs and there are $n - 2$ involutions available, and $\frac{1}{2}n - 1$ transpositions, so each transposition occurs twice, with different blocks fixed pointwise. So there are two blocks of $W(n)$, which are parallel, giving $2(\frac{1}{2}n) = n$ points for which Q and R occur in the same orbit, i.e. giving n blocks of $\mathcal{B}'$. Thus the total number of blocks in $\mathcal{B}'$ through both Q and R is $n + 2$, showing that $\lambda = n + 2$ and that $W'(n)$ is a 2-design. $\square$

Now the proof of the theorem is immediate, by simply using Theorem 2.4.4. $\square$

Remark For $n = 8$, $W(8) = \mathcal{R}(3)$, the smallest Ree unital, which has been known to have ovals for some time: see, for example, Brouwer [50].

The resolutions of $W(n)$ that correspond to the description through the secants, as discussed in Section 7.12, can be described through this construction in the following way: the involutions in G either commute, in which case they are together in a unique Sylow 2-subgroup, or they generate a dihedral group of order dividing $2(n+1)$ or $2(n-1)$. If the former, then they are contained in a point stabilizer (through this representation) and are thus not parallel (as blocks). If the latter, then the set of $n-1$ involutions in this dihedral group forms a parallel class of blocks: for if two of the involutions share a common fixed point, then these two involutions will generate a dihedral subgroup of order dividing both $2(n+1)$ and $2(n-1)$, which is not possible. To find resolutions containing this parallel class of blocks, let D be this dihedral group and consider the set of $n+1$ Sylow 2-subgroups of G. Any two have only the identity in common and any one Sylow 2-subgroup can intersect D trivially or in a subgroup of order 2. Thus precisely two Sylow 2-subgroups intersect D trivially. Let S be one of these and let t be an involution in S. We claim that D and D^t have no involutions in common: for if they do, and $d, d^t \in D$, then $\langle d, d^t \rangle$ is a dihedral subgroup of D that is normalized by $t \notin D$, which is impossible: see Dickson [89]. For the same reason D^t and D^s have no involutions in common, for any distinct t and s in S. Since also S meets D^t trivially, the resolution is defined by the involutions in S and in each of D^t for $t \in S$. The other Sylow 2-subgroup intersecting D trivially will also give a resolution in this way. If X is the number of resolutions obtained in this way, then counting (resolution, dihedral subgroup) "incident" pairs gives $Xn = \frac{1}{2}n(n+1)2$ (since there are $\frac{1}{2}n(n+1)$ dihedral subgroups of order $2(n-1)$, from Dickson [89], since there is a single conjugacy class of them and they are self-normalized), i.e. $X = n+1$. Together with the resolution defined by the full set of Sylow 2-subgroups, we get $n+2$ resolutions, analogous to the $n+2$ resolutions obtained from the points of the oval, with the nucleus playing a special part.

This also indicates the way in which the plane and oval can be retrieved from this construction: take the points to be the set of involutions of G, along with the set of resolutions. For the lines of the plane take all the dihedral subgroups of order $2(n+1)$ or $2(n-1)$ and all the Sylow 2-subgroups. Incidence is defined through an involution t being on a line L if $t \in L$; a resolution $\mathcal{R}$ is on a line L if the line appears in the resolution, through the defined construction. Then it is simple to show that the dual conditions for a projective plane hold, with G acting as automorphism group fixing the oval and the nucleus.

8.5 Some Steiner 3-designs

We consider first **inversive planes**, i.e. extensions of affine planes.

Definition 8.5.1 *A* **finite inversive plane** *is a Steiner 3-design with parameters* 3-$(m^2 + 1, m + 1, 1)$ *where* $m \geq 2$.

Since the contraction of an inversive plane of block size $m + 1$ at any point is a 2-$(m^2, m, 1)$ design, it is an affine plane and thus, since affine planes of prime-power order are the only ones known at present, the only inversive planes known have m a prime power. In fact, for all the known inversive planes (and we will give some constructions below) the contraction at a point is actually the desarguesian affine plane $AG_2(\mathbf{F}_q)$, where $q = m$.

Notice that we have specifically restricted our definition to the finite case, since we are primarily dealing with designs. However, inversive planes need not be finite if defined in the following way, which is entirely equivalent to Definition 8.5.1 when the sets are finite:

Definition 8.5.2 *An* **inversive plane** *(or Möbius plane)* $\mathcal{I}$ *is a set* $\mathcal{P}$ *of points with a set of subsets of* $\mathcal{P}$ *that are called* **circles** *having the properties:*

(1) any three distinct points of $\mathcal{I}$ *are together on a unique circle;*

(2) if P and Q are points of $\mathcal{I}$ and C is a circle containing P but not Q, then there is a unique circle of $\mathcal{I}$ that contains both P and Q and meets C only at P;

(3) there are four points not together on a circle.

In the finite case it is easy to see the equivalence of these definitions.

Example 8.5.1 Let $\mathcal{S}$ be a sphere in euclidean 3-space and let $\mathcal{P}$ be the set of points on the surface of $\mathcal{S}$. Let $\mathcal{I}$ be the structure with point set $\mathcal{P}$ and for circles the intersections of planes in the 3-space that meet $\mathcal{S}$ in more than one point (i.e. are neither exterior planes nor tangent planes). Then $\mathcal{I}$ is an inversive plane.

The construction of the infinite inversive plane of this example points the way to a similar construction for finite inversive planes. In fact, this construction gives the only currently known finite inversive planes, the so-called **egglike** planes. Starting with the projective space $PG_3(\mathbf{F}_q)$, for any q, recall (see Definition 3.7.4) that an *ovoid* $\mathcal{O}$ is a set of $q^2 + 1$ points with the property that for every point $P \in \mathcal{O}$, the union of all lines that meet $\mathcal{O}$ at P form a plane, the tangent plane at P. Then every plane is either a tangent plane or meets $\mathcal{O}$ in a $(q + 1)$-arc in the plane. A structure $\mathbf{I}(\mathcal{O})$ is defined to have point set $\mathcal{O}$ and for blocks the intersections of non-tangent planes with $\mathcal{O}$. It follows easily that $\mathbf{I}(\mathcal{O})$ is an inversive plane (an egglike plane), and that its contraction at any point is the desarguesian affine plane of order q.

Every elliptic quadric is an ovoid, and for q odd these are the only ovoids, by results of Barlotti and Panella: see Dembowski [85, 1.4.50]. For $q > 2$ and even, there are other ovoids, for example the Tits ovoids, as described in [85, 1.4.56]. Thus inversive planes on $q^2 + 1$ points certainly exist for all prime powers q. Another way of describing those obtained from a quadric fits into a more general pattern of the so-called *spherical* designs of Witt, which we will give later in this section.

We turn now to the codes of the finite inversive planes and find that, as we have defined them, they are of no interest.

Theorem 8.5.1 *Let $\mathcal{I}$ be an inversive plane with parameters 3-$(m^2+1, m+1, 1)$. If p is a prime then*

$$\mathrm{rank}_p(\mathcal{I}) = \begin{cases} m^2 & \text{if } p \text{ divides } (m+1) \\ m^2 + 1 & \text{if } p \text{ does not divide } (m+1) \end{cases}.$$

Proof: The order of $\mathcal{I}$, as a 2-design, is $\lambda_1 - \lambda_2 = m^2 - 1$. If p does not divide $(m^2 - 1)$, then also p does not divide $(m + 1)$, the block size and so $\mathrm{rank}_p(\mathcal{I}) = m^2 + 1$ by Theorem 2.4.1.

Now suppose p divides $(m^2 - 1)$. Consider a contraction at a point, i.e. an affine plane π of order m. Clearly $\mathrm{rank}_p(\mathcal{I}) \geq \mathrm{rank}_p(\pi)$. Now p does not divide m, which is the order of π, so $\mathrm{rank}_p(\pi) = m^2$. Thus $\mathrm{rank}_p(\mathcal{I}) \geq m^2$ if p divides $(m^2-1) = (m-1)(m+1)$. If p divides $(m+1)$ then $C_p(\mathcal{I}) \subseteq (\mathbf{F}_p \jmath)^\perp$ and $\mathrm{rank}_p(\mathcal{I}) = m^2$. This completes the proof. $\square$

Thus the codes generated by the incidence matrices of inversive planes are not interesting. However, there are other, subtler methods of using coding theory in this instance: see Delsarte [82], for example.

Inversive planes have been intensively studied; a treatment of the work that has been done is in Dembowski [85, Chapter 6] and a more recent account is given by Thas [273].

The alternative construction of inversive planes arising from elliptic quadrics, giving the *Miquelian planes*, is a special case of the *spherical designs* of Witt [300]: these are designs of parameters 3-$(q^d + 1, q + 1, 1)$ for $q > 2$, with point set the points of a projective line $PG_1(\mathbf{F}_{q^d})$ over $\mathbf{F}_{q^d}$ and for blocks the projective sub-lines over $\mathbf{F}_q$. (An equivalent definition can be found in Beth *et al.* [37, Section III.6].) Designs with these parameters are also found in Key and Wagner [163] and Hanani [132]. In case $d > 2$, the codes may provide some information, but have apparently not been studied. In the case of the spherical designs, and for those in [163], there is a contraction at a point which is the affine design of points and lines of the affine space $AG_d(\mathbf{F}_q)$ and thus its codes are known. The method of extension of designs discussed in [163] follows a method described earlier in

Assmus and Key [9] which is a generalization of the method of extension of the projective plane $PG_2(\mathbf{F}_4)$ to a 3-(22,6,1) design, which leads to the large Witt designs (see 8.6 below). A further generalization is discussed in Key [159].

8.6 Witt designs and Golay codes

For $t > 3$ there are only a handful of Steiner systems known. Of these the most celebrated and heavily studied are the **Witt designs** — or **Mathieu designs** — with $t = 4$ and 5 and parameters 4-(11,5,1), 4-(23,7,1), 5-(12,6,1) and 5-(24,8,1), the last two being extensions of the first two. The usual divisibility constraint shows that the latter two do not extend.

Classically these designs were obtained as follows: for the smaller designs start with the affine plane $AG_2(\mathbf{F}_3)$, a 2-(9,3,1) design. This extends to the inversive plane with parameters 3-(10,4,1). The inversive plane then extends twice to the 4-design and then the 5-design. For the large Witt designs start with the projective plane $PG_2(\mathbf{F}_4)$, a 2-(21,5,1) design. This extends once to a 3-(22,6,1) design, essentially by including an orbit of ovals under the group $PSL_3(\mathbf{F}_4)$. A further extension to a 4-(23,7,1) design is obtained by including some Fano planes inside the original projective plane; the extension to a 5-(24,8,1) design is achieved by including so-called "double lines", i.e. the symmetric difference of two lines in the original plane.

The precise construction of these unique designs dates back to Witt [300]; this classical approach has been clearly explained in many papers and books that are readily available: for example, Hughes and Piper [144] give the constructions in detail, as does Lüneburg [196], where great attention is paid to the groups involved. Thus we will not include this method of construction, but take a different approach that makes use of the relationship of coding theory to design theory that we have already established.

We have already met *all* of these designs and their existence is easily derived from algebraic coding theory — as we have already seen. Paige [229] was the first to note the connection of the large Witt designs with the binary Golay code, and Assmus and Mattson [16] described the connection of the small designs and the Mathieu groups, M_{11} and M_{12}, with the ternary Golay code. Both these codes are quadratic-residue codes and, from that theory, the existence of the designs is immediate. It is fair to say that the interest of design theorists in coding theory was aroused by these discoveries.

Even the classical approach to the large Witt design is elucidated by algebraic coding theory. We give a brief explanation, in coding-theoretical terms, of how to extend the projective plane of order 4.

Let C be the binary code of the plane of order 4, i.e. $C = C_2(PG_2(\mathbf{F}_4))$. The weight enumerator is easily calculated (see, for example, Assmus and van Lint [15]) and $\widehat{C}$, the extended $[22, 10, 6]$ self-orthogonal code, has weight enumerator

$$X^0 + 21(X^6 + X^{16}) + 210(X^8 + X^{14}) + 280(X^{10} + X^{12}) + X^{22}.$$

It follows from the MacWilliams relations that $\widehat{C}^\perp$, a $[22, 12, 6]$ binary code, has weight enumerator

$$X^0 + 189(X^6 + X^{16}) + 570(X^8 + X^{14}) + 1288(X^{10} + X^{12}) + X^{22}.$$

Now the 189 weight-6 vectors are the 21 extended lines and the 168 ovals. Since $\widehat{C}$ is of codimension 2 in its orthogonal there are three self-dual codes between $\widehat{C}$ and $\widehat{C}^\perp$. Each is obtained by adjoining one new vector of $\widehat{C}^\perp$ to $\widehat{C}$ and this new vector can always be taken to be the incidence vector of an oval because of the fact that there are ovals that meet oddly — in fact three, any two of which meet three times. Since the ovals are all projectively equivalent, the three codes are isomorphic and hence the 168 ovals are shared equally. Hence, each of the three self-dual codes has 77 weight-6 vectors. By the self-orthogonality the supports of any two meet evenly and, because the minimum weight is 6, either twice or not at all. It follows immediately that these supports yield a 3-$(22, 6, 1)$ design. We have, incidentally, shown that the 2-rank of this design is 11. Since the Pascal triangle of any design with these parameters shows that the binary code of such a design is self-orthogonal and since its contraction is the projective plane, which is unique, we have even, incidentally, established the uniqueness of the design.

We have already — in Chapter 7, Theorem 7.4.3 — discussed the uniqueness of the Hadamard matrix of order 3 (size 12) and in so doing constructed the small Witt designs and determined their automorphism groups. Notice that the codes involved here are ternary but that the ternary Golay codes $\mathcal{G}_{11}$ and $\mathcal{G}_{12}$ are *not* the codes over $\mathbf{F}_3$ of the 4-$(11, 5, 1)$ and 5-$(12, 6, 1)$ designs, the connection being more subtle. See Theorem 8.5.1 and Corollary 7.4.3.

For the large Witt designs, the Golay codes *are* the codes of the associated designs. We first prove the uniqueness of the binary $[24,12,8]$ Golay code, using a proof recorded in van Lint [190]. In fact, the result is more general and is as follows:

Theorem 8.6.1 *Let C be a binary, possibly non-linear, code of length 24 and minimum distance 8. If $|C| = 2^{12}$, then C is a translate of $\mathcal{G}_{24}$ and thus, essentially, linear.*

Proof: First notice that we do have instances of such codes: for example, the extended quadratic-residue code $\hat{\mathcal{Q}}(23, 2)$ is such a code, by Exercise 2.10.1 (3). However, we need not even assume that the given code is linear, for, as the result indicates, linearity will follow. Put another way, if C contains the zero vector, then it *is* the Golay code.

If C is punctured in any position, a $(23, 2^{12}, 7)$ code is obtained and such a code must be a 3-error-correcting perfect code. Using a translation we may assume that C contains the zero vector. The weight distribution is then determined and, in the notation of Definition 2.11.1,

$$A_0 = A_{23} = 1, A_7 = A_{16} = 253, A_8 = A_{15} = 506, A_{11} = A_{12} = 1288.$$

For example, the number of vectors of weight 7 must account, in their spheres of radius 3, for all the weight-4 vectors of the ambient space and hence there are $\binom{23}{4}/\binom{7}{4} = 253$ such vectors. From this it follows that the code C has words of weights $0, 8, 12, 16$ and 24 but of no other weights. The same is true for the code $C + c$ for any $c \in C$, since $C + c$ must also contain the zero vector. Thus the distance between any two codewords is also divisible by 4 and hence C is self-orthogonal and doubly-even. The linear span of C has dimension at most 12 and since C already has 2^{12} vectors, it is itself linear.

To write down a generator matrix for C, choose first any vector $c \in C$ of weight 12. The projection of C onto the coordinate positions where c has entries 0 will have kernel $\{0, c\}$ and thus the image is of dimension 11. Because of the self-orthogonality of C this image consists only of even-weight vectors of length 12. Thus there is a generator matrix for C of the form

$$G = \begin{pmatrix} I_{12} & P \end{pmatrix} \text{ where } P = \begin{pmatrix} 0 & \jmath \\ \jmath^t & A \end{pmatrix},$$

and where A is an 11×11 matrix and $\jmath$ is the all-one vector of length 11. Further, since C has minimum weight 8, each row of A must have at least six 1's and hence exactly six 1's. With this information and what we know about C, it is an easy matter to show that A is in fact the incidence matrix of a symmetric $(11,6,3)$ design. That this design is unique follows from the observation that it is a Hadamard 2-design whose complement extends uniquely to a Hadamard 3-design, a 3-$(12,6,3)$ design. Then Theorem 7.4.3 tells us that there is only one equivalence class of Hadamard matrices of size 12 (order 3) and hence also of symmetric $(11,6,3)$ designs, since there is a five-fold transitive group acting. It now follows that C is the extended Golay code, $\mathcal{G}_{24}$. $\square$

Corollary 8.6.1 *The supports of the words of weight 8 in $\mathcal{G}_{24}$ form the blocks of a Steiner system* 5-$(24, 8, 1)$, *called a Witt or Mathieu design.*

                                    759
                                506 253
                            330 176   77
                        210 120   56   21
                    130   80   40   16   5
                 78   52   28   12   4   1
              46   32   20   8   4   0   1
           30   16   16   4   4   0   0   1
           30   0   16   0   4   0   0   0   1

Figure 8.2: Block intersection numbers for the 5-$(24, 8, 1)$ design

The Pascal triangle[2] for a design with these parameters is given in Figure 8.2. Now we continue to give the proof from van Lint [190] that there is only one design with these parameters.

Theorem 8.6.2 *The Steiner system* 5-$(24, 8, 1)$ *is unique (up to isomorphism).*

Proof: Since we have a code with the properties required for Theorem 8.6.1, we have, by the corollary, a design $\mathcal{D}$ with these parameters. Looking at Figure 8.2, we see that the binary code C of $\mathcal{D}$ is self-orthogonal and doubly-even, since blocks of $\mathcal{D}$ meet in zero, two or four points. If $d^{\perp}$ is the minimum weight of $C^{\perp}$ then Lemma 2.4.2 tells us that $d^{\perp} \geq 1 + \lambda_1/\lambda_2 = 1 + 253/77 > 4$, and hence that the minimum weight of C is 8.

We need now show only that C has dimension 12, for then Theorem 8.6.1 will yield the uniqueness of the design. Clearly C contains the all-one vector $\jmath$ and so the weight-8 vectors and their complements already force the dimension of C to be at least 11. In fact we know that the projective plane $PG_2(\mathbf{F}_4)$, obtained by deriving three times, has code of dimension precisely 10: see Theorem 6.4.2. Using the information displayed in Figure 8.2 it is easy to see that C has dimension 12. $\square$

We have clearly only touched on this important topic and neglected entirely the important connection with the Leech lattice. But there are numerous good sources for this material; the interested reader should certainly consult them to find out more. MacWilliams and Sloane [203, Chapter 20] deal with the codes, the groups and the designs in some detail and give a large bibliography up to that time. For an *ab initio* approach from the point

---

[2]We give the usual version of the Pascal triangle, which is the mirror image of that described in Section 1.2.

of view of the Mathieu group, but with a coding-theoretical perspective, see
Conway [73]; see also the chapter in Conway and Sloane [75, Chapter 11].
More work has been done since then and we mention a construction involv-
ing the edge-graph of a icosahedron in the book by Brouwer, Cohen and
Neumaier [51] (see also Curtis [76]) and a generalization of this method by
Curtis [77] leading to the construction of self-dual codes.

Bibliography

[1] R. W. Ahrens and G. Szekeres. On a combinatorial generalization of 27 lines associated with a cubic surface. *J. Austral. Math. Soc.*, 10:485–492, 1969.

[2] W. O. Alltop. An infinite class of 5-designs. *J. Combin. Theory*, 12:390–395, 1972.

[3] J. André. Über nicht-Desarguessche Ebenen mit transitiver Translationsgruppe. *Math. Z.*, 60:156–186, 1954.

[4] B. R. Andriamanalimanana. *Ovals, Unitals and Codes*. PhD thesis, Lehigh University, 1979.

[5] E. Artin. *Geometric Algebra*. New York: Wiley Interscience, 1957.

[6] E. F. Assmus, Jr. The binary code arising from a 2-design with a nice collection of ovals. *IEEE Trans. Inform. Theory*, 29:367–369, 1983.

[7] E. F. Assmus, Jr. On the theory of designs. In J. Siemons, editor, *Surveys in Combinatorics, 1989*, pages 1–21. Cambridge: Cambridge University Press, 1989. London Mathematical Society Lecture Note Series 141.

[8] E. F. Assmus, Jr. On the Reed-Muller codes. *Discrete Math.*, 106/107:25–33, 1992.

[9] E. F. Assmus, Jr. and J. D. Key. On an infinite class of Steiner systems with $t = 3$ and $k = 6$. *J. Combin. Theory, Ser. A*, 42:55–60, 1986.

[10] E. F. Assmus, Jr. and J. D. Key. Arcs and ovals in the hermitian and Ree unitals. *European J. Combin.*, 10:297–308, 1989.

[11] E. F. Assmus, Jr. and J. D. Key. Affine and projective planes. *Discrete Math.*, 83:161–187, 1990.

[12] E. F. Assmus, Jr. and J. D. Key. Baer subplanes, ovals and unitals. In Dijen Ray-Chaudhuri, editor, *Coding Theory and Design Theory, Part I*, pages 1–8. New York: Springer-Verlag, 1990. IMA Volumes in Mathematics and its Applications, 20.

[13] E. F. Assmus, Jr. and J. D. Key. Translation planes and derivation sets. *J. Geom.*, 37:3–16, 1990.

[14] E. F. Assmus, Jr. and J. D. Key. Hadamard matrices and their designs: a coding-theoretic approach. *Trans. Amer. Math. Soc.*, 330:269–293, 1992.

[15] E. F. Assmus, Jr. and J. H. van Lint. Ovals in projective designs. *J. Combin. Theory, Ser. A*, 27:307–324, 1979.

[16] E. F. Assmus, Jr. and H. F. Mattson, Jr. Perfect codes and the Mathieu groups. *Arch. Math.*, 17:122–135, 1966.

[17] E. F. Assmus, Jr. and H. F. Mattson, Jr. On tactical configurations and error-correcting codes. *J. Combin. Theory*, 2:243–257, 1967.

[18] E. F. Assmus, Jr. and H. F. Mattson, Jr. New 5-designs. *J. Combin. Theory*, 6:122–151, 1969.

[19] E. F. Assmus, Jr. and H. F. Mattson, Jr. Coding and combinatorics. *SIAM Review*, 16:349–388, 1974.

[20] E. F. Assmus, Jr., H. F. Mattson, Jr., and R. J. Turyn. *Research to Develop the Algebraic Theory of Codes*. Applied Research Laboratory, Sylvania Electronic Systems, June 1967. No. AFCRL-67-0365. Contract No. AF19(628)-5998.

[21] E. F. Assmus, Jr. and A. R. Prince. Biplanes and near biplanes. *J. Geom.*, 40:1–14, 1991.

[22] E. F. Assmus, Jr. and H. E. Sachar. Ovals from the point of view of coding theory. In M. Aigner, editor, *Higher Combinatorics*, pages 213–216. Dordrecht: D. Reidel, 1977. Proceedings of the NATO Conference, Berlin 1976.

[23] E. F. Assmus, Jr. and C. J. Salwach. The (16,6,2) designs. *Internat. J. Math. & Math. Sci.*, 2:261–281, 1979.

[24] R. Baer. Homogeneity of projective planes. *Amer. J. Math*, 64:137–152, 1942.

[25] R. Baer. Projectivities of finite projective planes. *Amer. J. Math*, 69:653–684, 1947.

[26] R. Baer. *Linear Algebra and Projective Geometry*. New York: Academic Press. 1952.

[27] B. Bagchi and N. S. N. Sastry. Even order inversive planes, generalized quadrangles and codes. *Geom. Dedicata*, 22:137–147, 1987.

[28] S. Bagchi and B. Bagchi. Designs from pairs of finite fields: I. A cyclic unital $U(6)$ and other regular Steiner 2-designs. *J. Combin. Theory, Ser. A*, 52:51–61, 1989.

[29] R. D. Baker and G. L. Ebert. Intersection of unitals in the desarguesian plane. In *Proceedings of the S.E. Conference on Combinatorics, Graph Theory and Computing*, 1989.

[30] A. Barlotti. Un'estensione del teorema di Segre-Kustaanheimo. *Boll. Un. Mat. Ital., Gruppo IV, Serie III*, 10:498–506, 1955.

[31] L. M. Batten. *Combinatorics of Finite Geometries*. Cambridge: Cambridge University Press, 1986.

[32] L. Bénéteau. Topics about 3-Moufang loops and Hall triple systems. *Simon Stevin*, 54(2):107–128, 1980.

[33] T. Berger and P. Charpin. The automorphism group of the generalized Reed-Muller codes. Paris: INRIA Rapports de Recherche No. 1363, 1991.

[34] E. R. Berlekamp. Factoring polynomials over finite fields. *Bell System Tech. J.*, 46:1853–1859, 1967.

[35] E. R. Berlekamp and L. R. Welch. Weight distributions of the cosets of the $(32, 6)$ Reed-Muller code. *IEEE Trans. Inform. Theory*, 18:203–207, 1972.

[36] S. D. Berman. On the theory of group codes. *Kibernetika*, 3(1):31–39, 1967.

[37] Th. Beth, D. Jungnickel, and H. Lenz. *Design Theory*. Mannheim, Wien, Zürich: Bibliographisches Institut Wissenschaftsverlag, 1985.

[38] V. N. Bhat and S. S. Shrikhande. Non-isomorphic solutions of some balanced incomplete block designs. I. *J. Combin. Theory*, 9:174–191, 1970.

[39] N. L. Biggs and A. T. White. *Permutation Groups and Combinatorial Structures*. Cambridge: Cambridge University Press, 1979. London Mathematical Society Lecture Notes Series 33.

[40] R. E. Blahut. Transform techniques for error control codes. *IBM J. Res. Develop.*, 23:299–315, 1979.

[41] R. E. Blahut. *Theory and Practice of Error Control Codes*. New York: Addison-Wesley, 1983.

[42] I. F. Blake and R. C. Mullin. *The Mathematical Theory of Coding.* New York: Academic Press, 1975.

[43] R. E. Block. On the orbits of collineation groups. *Math. Z.*, 96:33–49, 1967.

[44] A. Blokhuis, A. Brouwer, and H. Wilbrink. Hermitian unitals are codewords. *Discrete Math.*, 97:63–68, 1991.

[45] R. C. Bose and D. K. Ray-Chaudhuri. On a class of error correcting binary group codes. *Inform. and Control*, 3:68–79, 1960.

[46] R. C. Bose and S. S. Shrikhande. A note on a result in the theory of code construction. *Inform. and Control*, 2:183–194, 1959.

[47] R. C. Bose and S. S. Shrikhande. On the construction of sets of mutually orthogonal latin squares and the falsity of a conjecture of Euler. *Trans. Amer. Math. Soc.*, 95:191–209, 1960.

[48] R. Brauer. On the connection between the ordinary and the modular characters of groups of finite order. *Ann. Math.*, 42:926–935, 1941.

[49] W. G. Bridges. Algebraic duality theorems with combinatorial applications. *Linear Algebra Appl.*, 22:157–162, 1978.

[50] A. E. Brouwer. Some unitals on 28 points and their embeddings in projective planes of order 9. In M. Aigner and D. Jungnickel, editors, *Geometries and Groups*, pages 183–188. Berlin: Springer-Verlag, 1981. Lecture Notes in Mathematics, 893.

[51] A. E. Brouwer, A. M. Cohen, and A. Neumaier. *Distance-Regular Graphs.* Ergebnisse der Mathematik und ihrer Grenzgebiete, Folge 3, Band 18. Berlin, New York: Springer-Verlag, 1989.

[52] R. H. Bruck. Construction problems in finite projective spaces. In A. Barlotti, editor, *Finite Geometric Structures and their Applications*, pages 105–188. C.I.M.E., Edizioni Cremonese, Roma, 1973. Corso tenuto a Bressannone dal 18 al 27 Giugno 1972.

[53] R. H. Bruck and R. C. Bose. The construction of translation planes from projective spaces. *J. Algebra*, 1:85–102, 1964.

[54] A. A. Bruen. Blocking sets in finite projective planes. *SIAM J. Appl. Math.*, 21:380–392, 1971.

[55] A. A. Bruen and J. W. P. Hirschfeld. Intersections in projective space I: Combinatorics. *Math. Z.*, 193:215–225, 1986.

[56] A. A. Bruen and U. Ott. On the p-rank of incidence matrices and a question of E. S. Lander. *Contemp. Math.*, 111:39–45, 1990.

[57] F. Buekenhout. Existence of unitals in finite translation planes of order q^2 with a kernel of order q. *Geom. Dedicata*, 5:189–194, 1976.

[58] F. Buekenhout, A. Delandtsheer, and J. Doyen. Finite linear spaces with flag-transitive groups. *J. Combin. Theory, Ser. A*, 49:268–293, 1988.

[59] F. Buekenhout, A. Delandtsheer, J. Doyen, P. B. Kleidman, M. W. Liebeck, and J. Saxl. Linear spaces with flag-transitive automorphism groups. Preprint.

[60] W. Burau. Über die zur Kummerkonfiguration analogen Schemata von 16 Punkten und 16 Blöcken und ihre Gruppen. *Abh. Math. Sem. Univ. Hamburg*, 26:129–144, 1963.

[61] A. R. Calderbank, P. Delsarte, and N. J. A. Sloane. A strengthening of the Assmus-Mattson theorem. *IEEE Trans. Inform. Theory*, 37:1261–1268, 1991.

[62] P. J. Cameron. Biplanes. *Math. Z*, 131:85–101, 1973.

[63] P. J. Cameron. *Parallelisms of Complete Designs*. Cambridge: Cambridge University Press, 1976. London Mathematical Society Lecture Notes Series 23.

[64] P. J. Cameron and J. H. van Lint. *Graphs, Codes and Designs*. Cambridge: Cambridge University Press, 1980. London Mathematical Society Lecture Notes Series 43.

[65] P. Camion. *Difference Sets in Elementary Abelian Groups*. Les Presses de l'Université de Montréal, 1979. Séminaire de Mathématiques Supérieures, Département de Mathématiques et de Statistique, Université de Montréal.

[66] J. Cannon. *Cayley: A Language for Group Theory*. Department of Mathematics, University of Sydney, July 1982.

[67] P. Charpin. *Codes cycliques étendus invariants sous le groupe affine*. Thèse de Doctorat d'État, Université Paris VII, 1987.

[68] P. Charpin. A new description of some polynomial codes: the primitive generalized Reed-Muller code. Technical report, Université Paris VII, 1985.

[69] P. Charpin. Une généralisation de la construction de Berman des codes de Reed et Muller p-aires. *Communications in Algebra*, 16:2231–2246. 1988.

[70] P. Charpin. Codes cycliques étendus affines-invariants et antichaines d'un ensemble partiellement ordonné. *Discrete Math.*, 80:229–247, 1990.

[71] W. E. Cherowitzo. Hyperovals in desarguesian planes of even order. *Ann. of Discrete Math.*, 37:87–94, 1988.

[72] W. E. Cherowitzo. Hyperovals in the translation planes of order 16. *J. Combin. Math. & Combin. Comput.*, 9:39–55, 1991.

[73] J. H. Conway. Three lectures on exceptional groups. In M. B. Powell and G. Higman, editors, *Finite Simple Groups*, pages 215–247. New York, London: Academic Press, 1971. Proceedings of an Instructional Conference Organized by the London Mathematical Society — a NATO Advanced Study Institute.

[74] J. H. Conway and V. Pless. On the enumeration of self-dual codes. *J. Combin. Theory, Ser. A*, 28:26–53, 1980.

[75] J. H. Conway and N. J. A. Sloane. *Sphere Packings, Lattices and Groups*. Grundlehren der mathematischen Wissenschaften 290. New York: Springer-Verlag, 1988.

[76] R. T. Curtis. The regular dodecahedron and the binary Golay code. *Ars Combin.*, 29B:55–64, 1990.

[77] R. T. Curtis. On graphs and codes. *Geom. Dedicata*, 41:127–134, 1992.

[78] T. Czerwinski and D. J. Oakden. The translation planes of order twenty-five. *J. Combin. Theory, Ser. A*, 59:193–217, 1992.

[79] M. Dehon. Ranks of incidence matrices of t-designs $S_\lambda(t, t + 1, \lambda)$. *European J. Combin.*, 1:97–100, 1980.

[80] P. Delsarte. A geometric approach to a class of cyclic codes. *J. Combin. Theory*, 6:340–358, 1969.

[81] P. Delsarte. On cyclic codes that are invariant under the general linear group. *IEEE Trans. Inform. Theory*, 16:760–769, 1970.

[82] P. Delsarte. Majority logic decodable codes derived from finite inversive planes. *Inform. and Control*, 18:319–325, 1971.

[83] P. Delsarte, J. M. Goethals, and F. J. MacWilliams. On generalized Reed-Muller codes and their relatives. *Inform. and Control*, 16:403–442, 1970.

[84] P. Delsarte, J. M. Goethals, and J. J. Seidel. Spherical codes and designs. *Geom. Dedicata*, 6:363–388, 1977.

[85] P. Dembowski. *Finite Geometries*. Ergebnisse der Mathematik und ihrer Grenzbegiete, Band 44. Berlin, Heidelberg, New York: Springer-Verlag, 1968.

[86] P. Dembowski and A. Wagner. Some characterizations of finite projective spaces. *Arch. Math.*, 11:465–469, 1960.

[87] U. Dempwolff and A. Reifart. The classification of the translation planes of order 16, I. *Geom. Dedicata*, 15:137–153, 1983.

[88] R. H. F. Denniston. Some new 5-designs. *Bull. London Math. Soc.*, 8:263–267, 1976.

[89] L. E. Dickson. *Linear Groups with an Exposition of the Galois Field Theory*. New York: Dover Publications, 1958. (With an introduction by Wilhelm Magnus).

[90] J. Dieudonné. *La Géométrie des Groupes Classiques*. Ergebnisse der Mathematik und ihrer Grenzgebiete, Neue Folge, Band 5. Berlin, Göttingen, Heidelberg: Springer-Verlag, second edition, 1963.

[91] J. Dieudonné. *Sur les Groupes Classiques*. Actualités scientifiques et industrielles 1040. Paris: Hermann, third edition, 1973.

[92] J. F. Dillon. Private communication.

[93] J. F. Dillon. A survey of bent functions. NSAL-S-203,092.

[94] J. F. Dillon. *Elementary Hadamard Difference Sets*. PhD thesis, University of Maryland, 1974.

[95] J. F. Dillon. Elementary Hadamard difference sets. *Congressus Numerantium*, 14:237–249, 1975.

[96] J. F. Dillon and J. R. Schatz. Block designs with the symmetric difference property. In Robert L. Ward, editor, *Proceedings of the NSA Mathematical Sciences Meetings*, pages 159–164. The United States Government, 1987.

[97] S. T. Dougherty. *Nets and their codes*. PhD thesis, Lehigh University, 1992.

[98] J. Doyen. Linear spaces and Steiner systems. In M. Aigner and D. Jungnickel, editors, *Geometries and Groups*, pages 30–42. Berlin: Springer-Verlag, 1981. Lecture Notes in Mathematics, 893.

[99] J. Doyen, X. Hubaut, and M. Vandensavel. Ranks of incidence matrices of Steiner triple systems. *Math. Z.*, 163:251–259, 1978.

[100] G. L. Ebert. Translation planes of order q^2: asymptotic estimates. *Trans. Amer. Math. Soc.*, 238:301–308, 1978.

[101] J. C. Fisher, J. W. P. Hirschfeld, and J. A. Thas. Complete arcs in planes of square order. *Ann. Discrete Math.*, 30:243–250, 1986.

[102] M. J. Ganley. A class of unitary block designs. *Math. Z.*, 128:34–42, 1972.

[103] D. Ghinelli-Smit. Functions on symmetric designs. *Ars Combin.*, 24B:217–230, 1987.

[104] D. Gluck. Affine planes and permutation polynomials. In Dijen Ray-Chaudhuri, editor, *Coding Theory and Design Theory, Part II*, pages 99–100. Springer-Verlag, 1990. IMA Volumes in Mathematics and its Applications, 21.

[105] J. M. Goethals and P. Delsarte. On a class of majority-logic decodable cyclic codes. *IEEE Trans. Inform. Theory*, 14:182–188, 1968.

[106] J. M. Goethals and J. J. Seidel. Strongly regular graphs derived from combinatorial designs. *Canad. J. Math.*, 22:597–614, 1970.

[107] M. J. E. Golay. Notes on digital coding. *Proc. IRE*, 37:657, 1949.

[108] M. J. E. Golay. Anent codes, priorities, patents, etc. *Proc. IEEE*, 64:572, 1976.

[109] R. L. Graham and F. J. MacWilliams. On the number of information symbols in difference-set cyclic codes. *Bell System Tech. J.*, 45:1057–1070, 1966.

[110] K. Grey. Further results on designs carried by a code. *Ars Combin.*, 26B:133–152, 1988.

[111] B. H. Gross. Intersection triangles and block intersection numbers of Steiner systems. *Math. Z.*, 139:87–104, 1974.

[112] K. W. Gruenberg and A. J. Weir. *Linear Geometry.* Graduate Texts in Mathematics: 49. New York: Springer Verlag, second edition, 1977.

[113] J. Hadamard. Résolution d'une question relative aux déterminants. *Bull. Sci. Math.*, 2:240–246, 1893.

[114] A. J. Hahn and O. T. O'Meara. *The Classical Groups and K-Theory.* Grundlehren der mathematischen Wissenschaften 291. New York: Springer-Verlag, 1989.

[115] M. Hall, Jr. Projective planes. *Trans. Amer. Math. Soc.*, 54:229–277, 1943.

[116] M. Hall, Jr. Cyclic projective planes. *Duke Math. J.*, pages 1079–1090, 1947.

[117] M. Hall, Jr. A survey of difference sets. *Proc. Amer. Math. Soc*, 7:975–986, 1956.

[118] M. Hall, Jr. Automorphisms of Steiner triple systems. *Proc. Sympos. Pure Math.*, 6:47–66, 1962.

[119] M. Hall, Jr. Note on the Mathieu group M_{12}. *Arch. Math.*, 13:334–340, 1962.

[120] M. Hall, Jr. Ovals in the desarguesian plane of order 16. *Ann. Mat. Pura Appl. (4)*, 102:159–176, 1975. CLXXIV della Raccolta sotto gli auspici del Consiglio Nazionale delle Ricerche.

[121] M. Hall, Jr. Semi-automorphisms of Hadamard matrices. *Math. Proc. Camb. Phil. Soc.*, 77:459–473, 1975.

[122] M. Hall, Jr. *Combinatorial Theory*. New York: Wiley, second edition, 1986.

[123] M. Hall, Jr. and H. J. Ryser. Cyclic incidence matrices. *Canad. Math. J.*, 3:495–502, 1951.

[124] N. Hamada. The rank of the incidence matrix of points and d-flats in finite geometries. *J. Sci. Hiroshima Univ. Ser. A-I*, 32:381–396, 1968.

[125] N. Hamada. On the p-rank of the incidence matrix of a balanced or partially balanced incomplete block design and its applications to error correcting codes. *Hiroshima Math. J.*, 3:153–226, 1973.

[126] N. Hamada. The geometric structure and the p-rank of an affine triple system derived from a nonassociative Moufang loop with the maximum associative center. *J. Combin. Theory, Ser. A*, 30:285–297, 1981.

[127] N. Hamada and H. Ohmori. On the BIB design having the minimum p-rank. *J. Combin. Theory, Ser. A*, 18:131–140, 1975.

[128] R. W. Hamming. Error detecting and error correcting codes. *Bell System Tech. J.*, 29:147–160, 1950.

[129] R. W. Hamming. *Coding and Information Theory*. Englewood Cliffs, N.J.: Prentice Hall, 1980.

[130] H. Hanani. On quadruple systems. *Canad. J. Math*, 12:145–157, 1960.

[131] H. Hanani. The existence and construction of balanced incomplete block designs. *Ann. Math. Statist.*, 32:361–386, 1961.

[132] H. Hanani. A class of three-designs. *J. Combin. Theory, Ser. A*, 26:1–19, 1979.

[133] C. Hering. On codes and projective designs. Technical Report 344, Kyoto University Mathematics Research Institute Seminar Notes, 1979.

[134] R. Hill. *A First Course in Coding Theory*. Oxford Applied Mathematics and Computing Science Series. Oxford: Oxford University Press, 1986.

[135] G. Hillebrandt. The p-rank of $(0, 1)$-matrices. *J. Combin. Theory, Ser. A*, 60:131–139, 1992.

[136] J. W. P. Hirschfeld. *Projective Geometries over Finite Fields*. Oxford: Oxford University Press, 1979.

[137] J. W. P. Hirschfeld. *Finite Projective Spaces of Three Dimensions*. Oxford: Oxford University Press, 1985.

[138] G. Hiss. On the incidence matrix of the Ree unital. Preprint.

[139] A. Hocquenghem. Codes correcteurs d'erreurs. *Chiffres*, 2:147–158, 1959.

[140] G. Hölz. Construction of designs which contain a unital. *Arch. Math.*, 37:179–183, 1981.

[141] D. A. Huffman. The synthesis of linear sequential coding networks. In Colin Cherry, editor, *Information Theory*. London: Butterworths Scientific Publishers, 1956. Papers read at a Symposium on 'Information Theory' held in London in 1955.

[142] D. R. Hughes. On t-designs and groups. *Amer. J. Math.*, 87:761–778, 1965.

[143] D. R. Hughes and F. C. Piper. *Projective Planes*. Graduate Texts in Mathematics 6. New York: Springer-Verlag, 1973.

[144] D. R. Hughes and F. C. Piper. *Design Theory*. Cambridge: Cambridge University Press, 1985.

[145] B. Huppert. *Endliche Gruppen I*. Berlin, Heidelberg: Springer Verlag, 1967.

[146] Q. M. Hussain. On the totality of the solutions for the symmetrical incomplete block designs $\lambda = 2, k = 5$ or 6. *Sankhyā*, 7:204–208, 1945.

[147] N. Ito, J. S. Leon, and J. Q. Longyear. Classification of 3-(24,12,5) designs and 24-dimensional Hadamard matrices. *J. Combin. Theory, Ser. A*, 31:66–93, 1981.

[148] D. Jungnickel and V. D. Tonchev. On symmetric and quasi-symmetric designs with the symmetric difference property and their codes. *J. Combin. Theory, Ser. A*, 59:40–50, 1992.

[149] W. M. Kantor. Plane geometries associated with certain 2-transitive groups. *J. Algebra*, 37:489–521, 1975.

[150] W. M. Kantor. Symplectic groups, symmetric designs and line ovals. *J. Algebra*, 33:43–58, 1975.

[151] W. M. Kantor. Homogeneous designs and geometric lattices. *J. Combin. Theory, Ser. A*, 38:66–74, 1985.

[152] I. Kaplansky. *Linear Algebra and Geometry—A Second Course.* Boston: Allyn and Bacon, 1969.

[153] T. Kasami, S. Lin, and W. W. Peterson. Some results on cyclic codes which are invariant under the affine group and their applications. *Inform. and Control*, 11:475–496, 1967.

[154] T. Kasami, S. Lin, and W. W. Peterson. New generalizations of the Reed-Muller codes. Part I: Primitive codes. *IEEE Trans. Inform. Theory*, 14:189–199, 1968.

[155] T. Kasami, S. Lin, and W. W. Peterson. Polynomial codes. *IEEE Trans. Inform. Theory*, 14:807–814, 1968.

[156] B. C. Kestenband. Unital intersections in finite projective planes. *Geom. Dedicata*, 11:107–117, 1981.

[157] J. D. Key. A class of 1-designs. *European J. Combin.*, 14:37–41, 1993.

[158] J. D. Key. Hermitian varieties as codewords. *Des. Codes Cryptogr.*, 1:255–259, 1991.

[159] J. D. Key. Extendable Steiner systems. *Geom. Dedicata*, 41:201–205, 1992.

[160] J. D. Key and K. Mackenzie. An upper bound for the p-rank of a translation plane. *J. Combin. Theory, Ser. A*, 56:297–302, 1991.

[161] J. D. Key and K. Mackenzie. Ovals in the designs $W(2^m)$. *Ars Combin.*, 33:113–117, 1992.

[162] J. D. Key and E. E. Shult. Steiner triple systems with doubly transitive automorphism groups: a corollary to the classification theorem for finite simple groups. *J. Combin. Theory, Ser. A*, 36:105–110, 1984.

[163] J. D. Key and A. Wagner. On an infinite class of Steiner systems constructed from affine spaces. *Arch. Math.*, 47:376–378, 1986.

[164] R. E. Kibler. A summary of noncyclic difference sets, $k < 20$. *J. Combin. Theory, Ser. A*, 25:62–67, 1978.

[165] H. Kimura. Classification of Hadamard matrices of order 28 with Hall sets. Preprint.

[166] H. Kimura. On equivalence of Hadamard matrices. *Hokkaido Math. J.*, 17:139–146, 1988.

[167] H. Kimura. New Hadamard matrix of order 24. *Graphs and Combin.*, 5:235–242, 1989.

[168] H. Kimura and H. Ohmori. Construction of Hadamard matrices of order 28. *Graphs Combin.*, 2:247–257, 1986.

[169] T. P. Kirkman. On a problem in combinations. *Cambridge and Dublin Math. J.*, 2:191–204, 1847.

[170] E. Kleinfeld. Techniques for enumerating Veblen-Wedderburn systems. *J. Assoc. Comput. Mach.*, 7:330–337, 1960.

[171] M. Klemm. Über die Reduktion von Permutationsmoduln. *Math. Z.*, 143:113–117, 1975.

[172] M. Klemm. Über den p-Rang von Inzidenzmatrizen. *J. Combin. Theory, Ser. A*, 43:138–139, 1986.

[173] H. Koch. On self-dual, doubly even codes of length 32. *J. Combin. Theory, Ser. A*, 51:63–76, 1989.

[174] H. Koch. On self-dual doubly-even extremal codes. *Discrete Math.*, 83:291–300, 1990.

[175] G. Korchmáros. Old and new results on ovals in finite projective planes. In A. D. Keedwell, editor, *Surveys in Combinatorics, 1991*, pages 41–72. Cambridge: Cambridge University Press, 1991. London Mathematical Society Lecture Note Series 166.

[176] C. W. H. Lam. The search for a finite projective plane of order 10. *Amer. Math. Monthly*, 98:305–318, 1991.

[177] C. W. H. Lam, G. Kolesova, and L. Thiel. A computer search for finite projective planes of order 9. *Discrete Math.*, 92:187–195, 1991.

[178] C. W. H. Lam, L. Thiel, and A. Pautasso. On self-dual ternary codes generated by the inequivalent Hadamard matrices of order 24. Preprint.

[179] C. W. H. Lam, L. Thiel, and S. Swiercz. The non-existence of finite projective planes of order 10. *Canad. J. Math.*, 41:1117–1123, 1989.

[180] C. W. H. Lam, L. Thiel, S. Swiercz, and J. McKay. The non-existence of ovals in a projective plane of order 10. *Discrete Math.*, 45:319–321, 1983.

[181] E. S. Lander. *Symmetric Designs: an Algebraic Approach.* Cambridge: Cambridge University Press, 1983. London Mathematical Society Lecture Notes Series 74.

[182] P. Landrock and O. Manz. Classical codes as ideals in group algebras. *Des. Codes Cryptogr.*, 2:273–285, 1992.

[183] J. S. Leon, V. Pless, and N. J. A. Sloane. Self-dual codes over $GF(5)$. *J. Combin. Theory, Ser. A*, 32:178–194, 1982.

[184] J. S. Leon, V. Pless, and N. J. A. Sloane. On ternary self-dual codes of length 24. *IEEE Trans. Inform. Theory*, 27:176–180, 1981.

[185] R. Lidl and H. Niederreiter. *Introduction to Finite Fields and their Applications.* Cambridge: Cambridge University Press, 1986.

[186] J. H. van Lint. *Coding Theory.* Lecture Notes in Mathematics, 201. Berlin: Springer-Verlag, 1970.

[187] J. H. van Lint. A survey of perfect codes. *Rocky Mountain J. Math.*, 5:199–224, 1975.

[188] J. H. van Lint. *Introduction to Coding Theory.* Graduate Texts in Mathematics 86. New York: Springer-Verlag, 1982.

[189] J. H. van Lint. Algebraic geometric codes. In Dijen Ray-Chauddhuri, editor, *Coding Theory and Design Theory, Part I*, pages 137–162. New York: Springer-Verlag, 1990. IMA Volumes in Mathematics and its Applications, 20.

[190] J. H. van Lint. Codes and combinatorial designs. In D. Jungnickel et al., editors, *Design Theory, Coding Theory and Group Theory.* New York: Wiley Interscience.

[191] J. H. van Lint and G. van der Geer. *Introduction to Coding Theory and Algebraic Geometry.* DMV Seminar Band 12. Basel: Birkhäuser Verlag, 1988.

[192] P. Lorimer. A projective plane of order 16. *J. Combin. Theory*, 16:334–347, 1974.

[193] H. Lüneburg. Charakterisierungen der endlichen desarguesschen projektiven Ebenen. *Math. Z.*, 85:419–450, 1964.

[194] H. Lüneburg. Some remarks concerning the Ree group of type (G_2). *J. Algebra*, 3:256–259, 1966.

[195] H. Lüneburg. Lectures on projective planes. Technical report, University of Illinois at Chicago Circle, 1968/69.

[196] H. Lüneburg. *Transitive Erweiterungen endlicher Permutationsgruppen.* Lecture Notes in Mathematics, 84. Berlin: Springer-Verlag, 1969.

[197] H. Lüneburg. *Translation Planes.* New York: Springer-Verlag, 1980.

[198] L. Lunelli and M. Sce. *k*-Archi completi nei piani proietivi desarguesiani di rango 8 e 16. Technical report, Centro Calcoli Numerici, Politecnico di Milano, 1958.

[199] R. J. McEliece. The reliability of computer memories. *Scientific American*, 252:2–7, 1985.

[200] K. Mackenzie. *Codes of Designs.* PhD thesis, University of Birmingham, 1989.

[201] S. MacLane and G. Birkoff. *Algebra: Second Edition.* New York: Collier Macmillan, 1979.

[202] F. J. MacWilliams and H. B. Mann. On the p-rank of the design matrix of a difference set. *Inform. and Control*, 12:474–489, 1968.

[203] F. J. MacWilliams and N. J. A. Sloane. *The Theory of Error-Correcting Codes.* Amsterdam: North-Holland, 1983.

[204] F. J. MacWilliams, N. J. A. Sloane, and J. G. Thompson. Good self-dual codes exist. *Discrete Math.*, 3:153–162, 1972.

[205] S. S. Magliveras and D. M. Leavitt. Simple 6-(33,8,36) designs from $P\Gamma L_2(32)$. In *Computational Group Theory*, pages 337–352. New York: Academic Press, 1984.

[206] J. A. Maiorana. A classification of the cosets of the Reed-Muller code $\mathcal{R}(1,6)$. *Math. Comp.*, 57:403–414, 1991.

[207] H. B. Mann. *Addition Theorems: The Addition Theorems of Group Theory and Number theory.* Interscience Tracts in Pure and Applied Mathematics: 18. New York: Interscience Publishers, 1965.

[208] A. Maschietti. Hyperovals and Hadamard designs. *J. Geom.*, 44:107–116, 1992.

[209] J. L. Massey. Book Review: Theory and Practice of Error Control Codes, by R. E. Blahut. *IEEE Trans. Inform. Theory*, 31:553–554, 1985.

[210] R. Mathon. Constructions of cyclic Steiner 2-designs. *Ann. Discrete Math.*, 34:353–362, 1987.

[211] H. F. Mattson, Jr. Book Review: The Theory of Error-Correcting Codes, by F. J. MacWilliams and N. J. A. Sloane. *SIAM Review*, 22:513–519, 1980.

[212] H. F. Mattson, Jr. and G. Solomon. A new treatment of Bose-Chaudhuri codes. *J. Soc. Indust. Appl. Math.*, 9:654–669, 1961.

[213] N. S. Mendelsohn and B. Wolk. A search for a non-desarguesian plane of prime order. In C. A. Baker and L. M. Batten, editors, *Finite Geometries*, pages 199–208. New York: Marcel Dekker, 1985. Lecture Notes in Pure and Applied Mathematics, 103.

[214] P. K. Menon. Difference sets in abelian groups. *Proc. Amer. Math. Soc.*, 11:368–376, 1960.

[215] R. Metz. On a class of unitals. *Geom. Dedicata*, 8:125–126, 1979.

[216] E. H. Moore. Concerning triple systems. *Math. Ann.*, 43:271–285, 1893.

[217] E. H. Moore. Tactical memoranda I-III. *Amer. J. Math.*, 18:264–303, 1896.

[218] G. E. Moorhouse. Bruck nets, codes, and characters of loops. *Des. Codes Cryptogr.*, 1:7–29, 1991.

[219] B. Mortimer. The modular permutation representations of the known doubly transitive groups. *Proc. London Math. Soc. (3)*, 41:1–20, 1980.

[220] D. W. Newhart. On minimum weight codewords in QR codes. *J. Combin. Theory, Ser. A*, 48:104–119, 1988.

[221] C. W. Norman. Nonisomorphic Hadamard designs. *J. Combin. Theory, Ser. A*, 21:336–344, 1976.

[222] M. E. O'Nan. Automorphisms of unitary block designs. *J. Algebra*, 20:495–511, 1972.

[223] T. G. Ostrom. Semi-translation planes. *Trans. Amer. Math. Soc.*, 111:1–18, 1964.

[224] T. G. Ostrom. *Finite Translation Planes*. Lecture Notes in Mathematics, 158. Berlin: Springer-Verlag, 1970.

[225] T. G. Ostrom and A. Wagner. On projective and affine planes with transitive collineation groups. *Math. Z.*, 71:186–199, 1959.

[226] U. Ott. Endliche zyklische Ebenen. *Math. Z.*, 144:195–215, 1975.

[227] U. Ott. Some remarks on representation theory in finite geometry. In M. Aigner and D. Jungnickel, editors, *Geometries and Groups*, pages

68–110. Berlin: Springer-Verlag, 1981. Lecture Notes in Mathematics, 893.

[228] U. Ott. An elementary introduction to algebraic methods for finite projective planes. Technical Report 50, Università di Roma *La Sapienza*, Marzo 1984. Seminario di Geometrie Combinatorie, diretto da G. Tallini.

[229] L. J. Paige. A note on the Mathieu groups. *Canad. J. Math.*, 9:15–18, 1957.

[230] R. E. A. C. Paley. On orthogonal matrices. *J. Math. Phys.*, 12:311–320, 1933.

[231] G. Panella. Caratterizzazione delle quadriche di uno spazio (tridimensionale) lineare sopra un corpo finito. *Boll. Uni. Mat. Ital., Gruppo IV, Serie III*, 10:507–513, 1955.

[232] D. S. Passman. *Permutation Groups*. New York: W.A. Benjamin Inc., 1968.

[233] S. E. Payne and J. A. Thas. *Finite Generalized Quadrangles*. Research Notes in Mathematics 110. Boston: Pitman, 1984.

[234] T. Penttila and I. Pinneri. Private communication.

[235] W. W. Peterson. Error-correcting codes. *Scientific American*, 206:96–108, 1962.

[236] F. Piper. Unitary block designs. In R.M. Wilson, editor, *Graph Theory and Combinatorics*, pages 98–105. Pitman, 1979. Research Notes in Math., 34.

[237] V. Pless. Symmetry codes over GF(3) and new 5-designs. *J. Combin. Theory*, 12:119–142, 1972.

[238] V. Pless. *The Theory of Error-Correcting Codes*. New York: John Wiley and Sons, 1989. Second Edition.

[239] V. Pless and N. J. A. Sloane. Binary self-dual codes of length 24. *Bull. Amer. Math. Soc.*, 80:1173–1178, 1974.

[240] A. Pott. Applications of the DFT to abelian difference sets. *Arch. Math.*, 51:283–288, 1988.

[241] E. Prange. *Cyclic error-correcting codes in two symbols*. Electronics Research Directorate, Air Force Cambridge Research Center, September 1957. AFCRC-TN-57-103. ASTIA Document AD133749.

[242] E. Prange. *An algorism for factoring $x^n - 1$ over a finite field.* Electronics Research Directorate, Air Force Cambridge Research Center, October 1959. AFCRC-TN-59-775.

[243] E. Prange. *The use of coset equivalence in the analysis and decoding of group codes.* Electronics Research Directorate, Air Force Cambridge Research Center, June 1959. AFCRC-TN-59-164.

[244] B. Qvist. Some remarks concerning curves of second degree in a finite plane. *Ann. Acad. Sci. Fenn. Ser. AI,* (134), 1952.

[245] D. K. Ray-Chaudhuri and Richard M. Wilson. On t-designs. *Osaka J. Math.,* 12:737–744, 1975.

[246] R. Ree. A family of simple groups associated with the simple Lie algebra of type (G_2). *Amer. J. Math.,* 83:432–462, 1961.

[247] K. J. Rose. *Generalized Reed-Muller codes and finite geometries.* PhD thesis, Lehigh University, 1993.

[248] O. S. Rothaus. On "bent" functions. *J. Combin. Theory, Ser. A,* 20:300–305, 1976.

[249] L. D. Rudolph. A class of majority logic decodable codes. *IEEE Trans. Inform. Theory,* 13:305–307, 1967.

[250] H. J. Ryser. *Combinatorial Mathematics.* Mathematical Association of America, Wiley, 1963.

[251] H. Sachar. The F_p span of the incidence matrix of a finite projective plane. *Geom. Dedicata,* 8:407–415, 1979.

[252] R. Safavi-Naini and I. F. Blake. Generalized t-designs and weighted majority decoding. *Inform. and Control,* 42:261–282, 1979.

[253] R. Safavi-Naini and I. F. Blake. On designs from codes. *Utilitas Math.,* 14:49–63, 1979.

[254] U. Scarpis. Sui determinanti di valore massimo. *Rendiconti Reale Istituto Lombardo di Scienze e Lettere (Milan Rendiconti),* 31:1441–1446, 1898.

[255] T. Schaub. *A linear complexity approach to cyclic codes.* PhD thesis, Swiss Federal Institute of Technology, Zürich, 1988. Diss. ETH No. 8730.

[256] W. M. Schmidt. *Equations over Finite Fields: An elementary approach.* Berlin: Springer-Verlag, 1976.

[257] R. Schoof and M. van der Vlugt. Hecke operators and the weight distributions of certain codes. *J. Combin. Theory, Ser. A*, 57:163–186, 1991.

[258] M. P. Schützenberger. A non-existence theorem for an infinite family of symmetrical block designs. *Ann. Eugenics*, pages 286–287, 1949.

[259] B. Segre. Ovals in a finite projective plane. *Canad. J. Math.*, 7:414–416, 1955.

[260] C. E. Shannon. A mathematical theory of communication. *Bell System Tech. J.*, 27:379–423,623–656, 1948.

[261] E. P. Shaughnessy. *Associated t-designs and automorpism groups of certain linear codes*. PhD thesis, Lehigh University, 1969.

[262] M. S. Shrikhande and S. S. Sane. *Quasi-Symmetric Designs*. Cambridge: Cambridge University Press, 1991. London Mathematical Society Lecture Notes Series, 164.

[263] S. S. Shrikhande and N. K. Singh. On a method of constructing incomplete block designs. *Sankhyā, A*, 24:25–32, 1962.

[264] J. Siemons. Orbits in finite incidence structures. *Geom. Dedicata*, 14:87–94, 1983.

[265] J. Singer. A theorem in finite projective geometry and some applications to number theory. *Trans. Amer. Math. Soc.*, 43:377–385, 1938.

[266] D. Slepian. Some further theory of group codes. *Bell System Tech. J.*, 39:1219–1252, 1960.

[267] K. J. C. Smith. On the p-rank of the incidence matrix of points and hyperplanes in a finite projective geometry. *J. Combin. Theory*, 7:122–129, 1969.

[268] J. Steiner. Combinatorische Aufgabe. *Crelle's Journal*, XLV:181–182, 1853.

[269] J. J. Sylvester. Thoughts on Inverse Orthogonal Matrices, simultaneous Sign-succession, and Tessellated Pavements in two or more colours, with applications to Newton's Rule, Ornamental Tile-work, and the Theory of Numbers. *The London, Edinburgh, and Dublin Philosophical Magazine and Journal of Science*, 34:461–475, December 1867.

[270] L. Teirlinck. On projective and affine hyperplanes. *J. Combin. Theory, Ser. A*, 28:290–306, 1980.

[271] L. Teirlinck. Non-trivial t-designs without repeated blocks exist for all t. *Discrete Math.*, 65:301–311, 1987.

[272] J. A. Thas. Extensions of finite generalized quadrangles. *Symposia Mathematica*, 28:127–143, 1986. Published by Istituto Nazionale di Alta Matematica Francesco Severi and distributed by Academic Press.

[273] J. A. Thas. Solution of a classical problem on finite inversive planes. In W.M. Kantor, R.A. Liebler, S.E. Payne, and E.E. Shult, editors, *Finite Geometries, Buildings, and Related Topics*, pages 145–159. Oxford: Oxford University Press, 1990.

[274] J. G. Thompson. Fixed point free involutions and finite projective planes. In Michael J. Collins, editor, *Finite Simple Groups II*, pages 321–337. New York: Academic Press, 1980.

[275] A. Tietäväinen. On the nonexistence of perfect codes over finite fields. *SIAM J. Appl. Math.*, 24:88–96, 1973.

[276] J. Tits. Les groupes simples de Suzuki et de Ree. *Séminaire Bourbaki*, *13*, 210:1–18, 1960/61.

[277] J. A. Todd. A combinatorial problem. *J. Math. Phys.*, 12:321–333, 1933.

[278] J. A. Todd. *Projective and Analytic Geometry*. Pitman, 1947.

[279] V. D. Tonchev. Hadamard matrices of order 28 with automorphisms of order 7. *J. Combin. Theory, Ser. A*, 40:62–81, 1985.

[280] V. D. Tonchev. Quasi-symmetric 2-(31,7,7) designs and a revision of Hamada's conjecture. *J. Combin. Theory, Ser. A*, 42:104–110, 1986.

[281] V. D. Tonchev. *Combinatorial Configurations*. Pitman Monographs and Surveys in Pure and Applied Mathematics, 40. New York: Longman, 1988.

[282] V. D. Tonchev. Unitals in the Hölz design on 28 points. *Geom. Dedicata*, 38:357–363, 1991.

[283] R. J. Turyn. Character sums and difference sets. *Pacific J. Math.*, 15:319–346, 1965.

[284] Ju. L. Vasil'ev. On nongroup close-packed codes. *Probl. Kibernet.*, 8:337–339, 1962. Translated from the Russian in *Probleme der Kybernetik*, **8**(1965), 375-378.

[285] O. Veblen and J. H. Maclaglan-Wedderburn. Non-desarguesian and non-pascalian geometries. *Trans. Amer. Math. Soc.*, 8:379–388, 1907.

[286] O. Veblen and J. W. Young. *Projective Geometry: Volumes I and II*. Boston: Ginn and Co., 1918.

[287] H. L. de Vries. Some Steiner Quadruple Systems $S(3,4,16)$ such that all 16 derived Steiner Triple Systems $S(2,3,15)$ are isomorphic. *Ars Combin.*, 24A:107–129, 1987.

[288] A. Wagner. Orbits on finite incidence structures. *Symposia Mathematica*, 28:219–229, 1986.

[289] H. N. Ward. On Ree's series of simple groups. *Trans. Amer. Math. Soc.*, 121:62–89, 1966.

[290] H. N. Ward. Quadratic residue codes and symplectic groups. *J. Algebra*, 29:150–171, 1974.

[291] H. N. Ward. Quadratic residue codes in their prime. *J. Algebra*, 150:87–100, 1992.

[292] E. J. Weldon, Jr. New generalizations of the Reed-Muller codes. Part II: Nonprimitive codes. *IEEE Trans. Inform. Theory*, 14:199–205, 1968.

[293] M. A. Wertheimer. *Designs in Quadrics*. PhD thesis, University of Pennsylvania, 1986.

[294] M. A. Wertheimer. Oval designs in quadrics. *Contemp. Math.*, 111:287–297, 1990.

[295] H. Weyl. *The Classical Groups*. Princeton: Princeton University Press, 1946.

[296] H. Wielandt. *Finite Permutation Groups*. New York: Academic Press, 1964.

[297] H. Wilbrink. A characterization of the classical unitals. In N. L. Johnson, M. J. Kallaher, and C. T. Long, editors, *Finite Geometries*, pages 445–454. Marcel Dekker,Inc., 1983. Lecture Notes in Pure and Applied Mathematics, 82.

[298] H. S. Wilf. The 'Snake Oil' method for proving combinatorial identities. In J. Siemons, editor, *Surveys in Combinatorics, 1989*, pages 208–217. Cambridge: Cambridge University Press, 1989. London Mathematical Society Lecture Note Series 141.

[299] R. M. Wilson. Inequalities in $S_\lambda(t,k,v)$. Lecture notes, IMA, Minnesota, 1988.

[300] E. Witt. Über Steinersche Systeme. *Abh. Math. Sem. Univ. Hamburg*, 12:265–275, 1938.

[301] H. P. Young. Affine triple systems and matroid designs. *Math. Z.*, 132:343–359. 1973.

Glossary

337

Index of Names

Index of Terms

absolute,
    point or block, 11
    points and hyperplanes, 95
    line, 109
adjacent, 3
affine part, 201
algebraic geometry, 32
alphabet, 27
annihilator map, 95
anti-flag, 2
arc, 19
    complete, 20
    maximal size, 76
    n-, 75
Assmus-Mattson theorem, 81, 85, 87
automorphism, 11
    anti-, 11
    group, 11

Baer
    involution, 216
    segment, *240*-242
    subplane, *202*, 208-211, 215, 220-223, 229, 239-243, 245-246
    subset, *202*, 215
bent function, 275-278, *277*, 282, 284, 292
biplane, 5, 6, 9-11, 15, 19, 39, 55, *120*, 126, 283
Blahut's theorem, 69, 70
block, 1

block design, 2
block intersection numbers, 9
block-tactical, 21
blocking set, *210*, 245-246
Block's theorem, 20-23, 122, 202
Boolean function, 141
bound,
    BCH, 73
    Singleton, 29, 30, 33, 37, 74, 75
    sphere packing, 30, 33, 73, 81, 259
    square-root, 78, 80-81, 87
boundary map, 45
Brauer's theorem, 125, 128, 215
Bruck-Ryser-Chowla theorem, 121, 249
Bruck's theorem, 202
Burnside's lemma, 123

CAYLEY language, 235, 285, 292, 301, 305
central collineation, 214-218, *216*
    axis of, 216
    centre of, 216
channel, 25-26
    noisy communications, 26
    symmetric, 28
characteristic function, 13
characteristic number, 267
check symbols, 37
circle, 310
circuit, 50

344

Printed in the United States
By Bookmasters